CAVITY EXPANSION METHODS IN GEOMECHANICS

Cavity Expansion Methods in Geomechanics

by

Hai-Sui Yu

School of Civil Engineering,
University of Nottingham, U. K.

KLUWER ACADEMIC PUBLISHERS

DORDRECHT / BOSTON / LONDON

Library of Congress Cataloging-in-Publication Data

ISBN 0-412-79990-1

Published by Kluwer Academic Publishers,
P.O. Box 17, 3300 AH Dordrecht, The Netherlands.

Sold and distributed in North, Central and South America
by Kluwer Academic Publishers,
101 Philip Drive, Norwell, MA 02061, U.S.A.

In all other countries, sold and distributed
by Kluwer Academic Publishers,
P.O. Box 322, 3300 AH Dordrecht, The Netherlands.

Cover illustration: Circular opening with an internal support pressure

Printed on acid-free paper

Printed in the Netherlands.

This book is dedicated to
my late parents
for many sacrifices they made
in supporting my education

FOREWORD

by James K. Mitchell
University Distinguished Professor Emeritus, Virginia Tech

Analysis of the expansion of cylindrical and spherical cavities in soil and rock provides a surprisingly versatile and accurate geomechanics approach for study of important problems in geotechnical engineering. Among them are the axial and lateral capacity of deep foundations, interpretation of pressuremeter and cone penetration tests for determination of soil state and properties *in-situ*, and analysis of stability and deformations associated with excavation and tunnelling. A realisation in the latter part of the Twentieth Century that improved understanding and quantification of many problems in soil and rock mechanics could be obtained by application of cavity expansion theory marks a milestone in the development of our field.

In this very thorough and comprehensive treatment of the subject, Professor Yu has provided a valuable resource for students, researchers, and engineering practitioners alike. A full range of assumptions and solution procedures has been examined. In the first part of the book analytical solutions for stresses and deformations are developed using elastic theory, theory of elastic-perfectly plastic behaviour, critical state theory, and for the cases of strain hardening and strain softening materials. Numerical analysis using the finite element method is also described. The assumptions are clearly stated, and the details of the derivations are presented for all of the cases analysed. Limitations in the results are identified. Like all theories in geomechanics, idealisations of geometry, soil properties and behaviour are necessary; nonetheless, the results using cavity expansion theory reflect field behaviour as well or better than many other soil and rock mechanics theories in common use in geotechnical practice.

For those readers interested primarily in the results of the derivations and their applications, Professor Yu has included a summary at the end of each chapter in which what has been done is listed, the key equations are identified, and where the solutions are used in the geotechnical applications presented in the second part of the book is stated. A very comprehensive list of references makes it possible to trace all the information to its source.

The second part of the book, Geotechnical Applications, contains numerous illustrations of how the theoretical solutions can be applied to real problems in *in-situ* testing, foundation engineering, and underground construction, as well as many comparisons between predicted and measured behaviour. The ability to treat the results of cone penetration and pressuremeter tests in sand and clay on a realistic theoretical basis enhances their value for site characterisation and determination of relevant soil mechanical properties.

I am unaware of any other treatment of cavity expansion analysis that is so complete and definitive as in this book by Professor Yu.

J.K.M.
April 28, 2000

PREFACE

Cavity expansion theory is concerned with the stress and displacement fields around cavities embedded in either linear or nonlinear media. The procedures that make use of cavity expansion theory in solving practical problems are termed cavity expansion methods. No mathematical theories can completely describe the complex world around us, and therefore each theory is only aimed at a certain type of problem, describes its essential features, and ignores what is of minor importance. As a result, the theory will be invalid or inaccurate when a neglected influence becomes important. Over the last few decades, cavity expansion theory has found many applications in the analysis and design of a variety of geotechnical problems. This book arose from my belief that there is a need for the geotechnical community to have a unified presentation of cavity expansion theory and its applications in geomechanics. Accordingly, the book attempts to summarise and present the major developments in the field of cavity expansion theory and its geotechnical applications. Much of this research monograph is based on work carried out over the last two decades.

The book is intended primarily as a reference book for civil, mining and petroleum engineers who are interested in cavity expansion methods and their applications. As cavity expansion problems have long been used as a classic example in the teaching of elasticity and plasticity theories, the solutions presented in the book will also be of interest to students and researchers in the fields of applied mechanics and mechanical engineering.

As indicated by its table of contents, the book is divided into two parts. The first part, Chapters 2 to 7, presents fundamental solutions for the expansion and contraction of cavities in soil and rock. Whilst Chapters 2 to 6 cover some of the key analytical solutions for cavity expansion in geomaterials modelled by elastic, elasto-plastic and viscoelastic/viscoplastic theories, Chapter 7 also provides a brief summary of the basic numerical formulations for finite element analysis of cavity expansion problems. To facilitate the use and application of cavity expansion theory, a finite element program, CAVEXP, will be made available for readers of this book.

The second part of the book, containing Chapters 8 to 11, summarises the major applications of cavity expansion methods in soil and rock mechanics. While new areas of application may continue to emerge, the well-established areas of application are in the fields of pile foundations and earth anchors, tunnels and underground excavations, *in-situ* soil testing, and wellbore instability. All these applications have been covered in some detail. Due to space limitations, it is not possible to include every single application that has been published in the literature. Instead the aim of the application section is to present typical examples to show how cavity expansion solutions can be used to provide simple and useful frameworks for the analysis and design of complex geotechnical problems.

The preparation of the book reflects many years of study and research. In this process, I have benefited much from discussions and collaborations with many colleagues. Those concerned will know that I appreciate their help and assistance with considerable gratitude.

In particular, I wish to thank Professor Jim Mitchell for his constant support over the years and also for his many constructive comments on the original manuscript. His contribution in the form of a Foreword to the book is deeply appreciated.

I am grateful to the late Professor Peter Wroth, Professors Ted Brown and Guy Houlsby for introducing me to the challenging field of soil and rock mechanics. Professor Ted Brown also read some chapters of the original manuscript and I thank him for his detailed comments.

I am indebted to Professor Scott Sloan and Professor Ian Collins for being my mentors in the early stages of my academic career. Their advice and encouragement in those early years has been most valuable.

I also want to thank Professor Kerry Rowe, Professor W.F. Chen, Professor John Carter, Professor Bruce Kutter, and Professor Mark Randolph for having been a constant source of inspiration and encouragement.

A few sections of the book were prepared in the summer of 1999 when I was on sabbatical leave at MIT. I would like to record my thanks to Professor Andrew Whittle for his hospitality during my visit.

I am very grateful to Miss Kylie Ebert for her excellent proofreading of the manuscript. Dr. Mark Allman and Dr. Bailin Wu also read some chapters of the book and I thank them for their comments. Special thanks are also extended to Ms. Petra van Steenbergen and Ms. Manja Fredriksz of Kluwer Academic Publishers for their assistance during the final stage of this project.

Finally I like to say 'thankyou' to my wife, Xiu-li, and children, Christina and Thomas, for their love and support without which this book could not have been written.

Hai-Sui Yu
Newcastle, Australia
April 2000

TABLE OF CONTENTS

Part II: Geotechnical Applications

CHAPTER 1

INTRODUCTION

1.1 SCOPE AND AIMS

Cavity expansion theory is concerned with the theoretical study of changes in stresses, porewater pressures and displacements caused by the expansion and contraction of cylindrical or spherical cavities. Cavity expansion in soil or rock is a fundamental problem in theoretical geomechanics primarily because it provides a useful and simple tool for modelling many complex geotechnical problems.

Although steady progress has been made since the 1950's and 1960's, much of the important work in relation to cavity expansion theory has been undertaken in the past 20 years. Progress has been made on two fronts: (a) the development of fundamental solutions for cavity expansion in geomaterials, and (b) the application of cavity expansion theory to a wide range of geotechnical problems. In relation to the first aspect, numerous analytical solutions have been developed for cavity expansion in soil and rock based on constitutive models of various complexities. Furthermore, these analytical solutions have been widely applied to solve a large number of practical problems in geotechnical engineering. Examples of these problems include the investigation of the capacity of pile foundations and earth anchors, the interpretation of *in-situ* soil testing, analysis of the behaviour of tunnels and underground excavations, and the prediction of borehole instability.

Despite the volume of research in relation to cavity expansion theory, it is often difficult for the practising engineer to understand and apply these theoretical solutions to geotechnical problems. This is because major findings are usually reported in technical papers and conference proceedings which may be difficult to access. At the same time, many solutions have been developed using different notations which restricts the comparison of various methods.

This book aims to summarise and present the major developments and achievements in cavity expansion theory in a single volume, and also discusses its application to geotechnical engineering. Throughout this book cavity expansion solutions are presented and developed using a consistent set of notations which enables them to be easily compared and understood. It is hoped that this publication will facilitate the further development and application of cavity expansion solutions in geomechanics.

1.2 CAVITY EXPANSION THEORY

The expansion or contraction of cavities in soils or rocks is a one-dimensional boundary value problem. To solve it using the principles of continuum mechanics, a mathematical constitutive model is needed to describe the stress-strain behaviour of soil or rock. Geomaterials such as soil and rock are some of the oldest and most complex construction materials. Unlike other materials in civil engineering such as steel or concrete, the mechanical properties of geomaterials are more difficult to measure. A description of soil behaviour can be achieved by developing a constitutive model and a satisfactory technique for measuring the required soil properties. As far as soil modelling is concerned, a tremendous amount of research has been carried out. Although further refinement may be needed in some cases, overall it is believed that satisfactory progress has been made in soil modelling. In contrast, much more work is still needed in relation to soil testing. Although rapid progress has been made in the past 20 years, the current state-of-the art of soil testing is still far from satisfactory.

The most widely used theories for developing soil and rock models are the assumptions of elasticity and plasticity. Many existing soil and rock models can be divided into three groups:

(a) elastic models (linear or nonlinear);

(b) viscoelastic or viscoelastic-plastic models; and

(c) elastic-plastic models (perfectly plastic or strain hardening/softening).

All these models may be used to adequately describe the stress–strain behaviour of soils or rocks, however the most appropriate one will vary according to the problem under consideration and the required level of accuracy.

1.3 APPLICATION TO GEOMECHANICS

Cavity expansion theory has been widely applied in the areas of *in-situ* soil testing, deep foundations, tunnels and underground excavations in soil and rock, and well-bore instability in the oil industry. Accordingly, Chapters 8–11 of this book are devoted to a detailed description of these main applications.

1.3.1 *In-situ* soil testing

Pressuremeters and cone penetrometers are the two most widely used instruments for measuring soil properties *in-situ*. The first method accurately measures soil stiffness and strength while the second can be used to obtain approximate soil profiles quickly. Since the pioneering work of Menard (1957) and Gibson and Anderson (1961), the theory of cylindrical cavity expansion has been adopted by geotechnical researchers and engineers as the most important interpretation method for self-boring pressuremeter tests in soil and rock (Clarke, 1995). Spherical cavity expansion is also widely used as a simple method for predicting the cone tip resistance in the cone penetration test (CPT) (Yu and Mitchell, 1998).

The cone pressuremeter is a relatively new *in-situ* testing device that combines the standard cone penetrometer with a pressuremeter module incorporated behind the cone tip. The idea of mounting a pressuremeter module behind the tip of the cone penetrometer was first introduced in the early 1980's. This development combines the merits of both the standard cone and the pressuremeter into a single instrument. The cone pressuremeter can be installed by standard CPT jacking equipment and enables pressuremeter tests to be carried out as part of routine CPT operations. Although the analysis of cone pressuremeter tests is more difficult than that of self-boring pressuremeter tests, significant progress has been made in recent years. By using large strain cavity expansion solutions, methods have been developed to derive soil properties from measured cone pressuremeter results in clay (Houlsby and Withers, 1988) and sand (Yu *et al.*, 1996).

1.3.2 Pile foundations and earth anchors

The prediction of end-bearing and shaft capacities of a driven pile in soils remains a difficult problem in geotechnical engineering. This is partly because the analysis of pile installation in soil is a large strain problem which involves strong material and geometric nonlinearities. Most research work has focused upon developing a rigorous method for modelling the behaviour of driven piles in clay, for example, the strain path method of Baligh (1985) and the steady state finite element method developed by Yu *et al.* (2000). In contrast, less success has been achieved in the rigorous prediction of pile behaviour in sand. Due to the lack of rigorous analysis methods, many semi-analytical or empirically based methods are still widely in use in the design and construction of piles in soil. In particular, following the early suggestion of Bishop *et al* (1945), Hill (1950) and Gibson (1950), solutions of the limit pressures of spherical and cylindrical cavities are used to predict the end bearing and shaft capacities of piles in soils and rock (see, for example, Vesic, 1972; Randolph *et al.*, 1979; Yu and Houlsby, 1991; Collins *et al.*, 1992; Carter and Kulhawy, 1992; Randolph *et al*, 1994).

In addition to the modelling of pile behaviour, cavity expansion theory can also be used to estimate the pull-out capacity of earth anchors (Vesic, 1971). In this book, a new procedure is developed for applying cavity expansion solutions to predict the capacity of plate anchors buried in clay and sand. This simplified solution procedure is found to compare well with other more sophisticated numerical analyses (Merifield *et al.*, 2000).

1.3.3 Underground excavations and tunnelling

Cavity expansion theory has also been applied to the design and construction of tunnels and underground excavations in order to address the two most important design considerations – stability and serviceability. Stability requires that the tunnel is built without causing failure. In the past, elastic or elastic-plastic stress solutions around cavities have been widely used in the analysis and design of rock support in underground excavations (e.g. Terzaghi and Richart, 1952; Hoek and Brown, 1980; Brown *et al.*, 1983; Brady and Brown, 1993). This book will demonstrate that a simple cavity expansion approach is also a valid means of assessing the stability of tunnels in soil.

With regard to serviceability, displacement due to tunnelling must be sufficiently small so as to ensure that serious damage is not made to adjacent or overlying buildings and utilities. This is particularly important for tunnel construction in soft ground as the deformation caused by soft ground tunnelling may be very large. The action of tunnel excavations effectively reduces the *in-situ* stresses along the excavated circumference and can therefore be simulated by unloading cavities from an *in-situ* stress state. Increasing evidence suggests that soil behaviour in the plane perpendicular to the tunnel axis can be accurately modelled by a contracting cylindrical cavity, while movements ahead of an advancing tunnel heading can be better predicted by spherical cavity contraction theory. Examples of using cavity expansion solutions to predict tunnelling-induced ground movements are given by Rowe (1986), Mair and Taylor (1993), Verruijt and Booker (1996), Sagaseta (1998), Loganathan and Poulos (1998), Yu and Rowe (1999) and many others.

1.3.4 Wellbore instability

Wellbore instability during drilling is a major problem in the application of rock mechanics in petroleum engineering. It is well recognised that the value of lost equipment and time arising directly from wellbore instability is significant with estimates of losses over $500 million per year worldwide (Dusseault, 1994). It is expected that improved techniques for the analysis and prediction of wellbore instability could reduce these costs considerably.

As discussed by Bradley (1979) and Santarelli *et al.* (1986), stress-induced wellbore instabilities consist of three general types:

(a) hole size reduction due to ductile yield of the rock;

(b) hole enlargement due to brittle rock fracture or rupture; and

(c) unintentional hydraulic fracturing induced by excessive mud pressure.

It is common practice to adjust the internal wellbore pressure (i.e. mud pressure) to avoid borehole instabilities caused by rock fracturing and rupture.

The cavity expansion solutions obtained with elastic, poroelastic and plastic models have been used to study the problem of wellbore instabilities in soft rock (Charlez, 1997). For example, the basic steps in the elastic analysis of borehole instability are:

(a) determine the elastic stress field in rock around the wellbore;

(b) choose a suitable failure criterion for the rock concerned;

(c) compare the elastic stresses and the selected rock failure criterion.

If the failure criterion is satisfied anywhere in the rock the wellbore is regarded as unstable.

While all three major types of wellbore instability mentioned above can be considered using cavity expansion theory, this book will focus upon the instability associated with hole enlargement due to brittle rock fracture or rupture around the wellbore. Instability due to borehole size reduction caused by ductile yield of the rock will also be considered.

1.4 SIGN CONVENTIONS

In general, this book adopts the conventional geomechanics sign notation wherein the compressive stresses and pressures are considered to be positive. However, there is an exception to this convention in Chapter 3, which deals with elastic-perfectly plastic solutions. As many key papers in this area have been written using a tension positive notation, this book will also treat tension as positive. Although it is possible to convert the solutions of a tension positive notation into those of a compressive positive notation, the process is time consuming and increases the possibility of error.

1.5 SUMMARY

1. Cavity expansion theory is concerned with the theoretical study of changes in stresses, porewater pressures and displacements caused by the expansion and contraction of cylindrical or spherical cavities.

2. Numerous analytical and numerical solutions have been developed for cavity expansion in soils and rocks using constitutive models of various complexities.

Most of these constitutive models are based on the assumptions of elasticity, plasticity or viscoelasticity.

3. Cavity expansion theory provides a useful and simple tool for modelling many geotechnical problems including the investigation of the capacity of pile foundations and earth anchors, the interpretation of *in-situ* soil testing, analysis of the behaviour of tunnels and underground excavations, and the prediction of borehole instability.

4. Except in Chapter 3, this book adopts the conventional geomechanics sign convention wherein the compressive stresses and pressures are considered to be positive.

REFERENCES

Baligh, M.M. (1985). Strain path method. *Journal of Geotechnical Engineering,* ASCE, 111(9), 1108–1136 .

Bishop, R.F., Hill, R and Mott, N.F. (1945). The theory of indentation and hardness tests. *Proceedings of Physics Society,* 57, 147–159.

Bradley, W.B. (1979). Failure of inclined boreholes. *Journal of Energy Resource and Technology, Transaction of ASME,* 101, 232–239.

Brady, B.H.G. and Brown, E.T. (1993). *Rock Mechanics for Underground Mining.* 2nd Edition, Chapman and Hall, London.

Brown, E.T., Bray, J.W., Ladanyi, B. and Hoek, E. (1983). Ground response curves for rock tunnels. *Journal of Geotechnical Engineering*, ASCE 109(1), 15–39.

Carter, J.P. and Kulhawy, F.H. (1992). Analysis of laterally loaded shafts in rock. *Journal of Geotechnical Engineering*, ASCE, 118(6), 839–855.

Charlez, Ph. (1997). *Rock Mechanics. Vol 2: Petroleum Applications.* Editions Technip.

Clarke, B.G. (1995). *Pressuremeter in Geotechnical Design.* Chapman and Hall, London.

Collins, I.F., Pender, M.J. and Wang, Y. (1992). Cavity expansion in sands under drained loading conditions. *International Journal for Numerical and Analytical Methods in Geomechanics*, 16(1), 3–23.

Dusseault, M.B. (1994). Analysis of borehole stability. *Computer Methods and Advances in Geomechanics, Siriwardane and Zaman (ed), Balkema,* 125–137.

Gibson, R.E. (1950). Correspondence. *Journal of Institution of Civil Engineers,* 34, 382–383.

Gibson, R.E. and Anderson, W.F. (1961). *In-situ* measurement of soil properties with the pressuremeter. *Civil Engineering Public Works Reviews,* 56, 615–618.

Hill, R. (1950), *The Mathematical Theory of Plasticity.* Oxford University Press.

Hoek, E. and Brown, E.T. (1980). *Underground Excavations in Rock.* The Institution of Mining and Metallurgy, London, England.

Houlsby, G.T. and Withers, N.J. (1988). Analysis of the cone pressuremeter test in clay. *Geotechnique*, 38, 575–587.

Loganathan, N. and Poulos, H.G. (1998). Analytical prediction for tunnelling-induced ground movements in clays. *Journal of Geotechnical and Geoenvironmental Engineering,* ASCE, 124(9), 846–856.

Mair, R.J. and Taylor, R.N. (1993). Prediction of clay behaviour around tunnels using plasticity solutions. *Predictive Soil Mechanics* (Editors: G.T. Houlsby and A.N. Schofield), Thomas Telford, London, 449–463.

Menard, L. (1957). *An Apparatus for Measuring the Strength of Soils in Place.* MSc Thesis, University of Illinois.

Merifield R. S., Lyamin, A.V., Sloan S. W. and Yu H. S. (2000). Three dimensional lower bound solutions for stability of plate anchors in clay. Submitted to *Journal of Geotechnical and Geoenvironmental Engineering*, ASCE.

Randolph, M.F., Carter, J.P. and Wroth, C.P. (1979). Driven piles in clay – the effects of installation and subsequent consolidation. *Geotechnique*, 29, 361–393.

Randolph, M.F., Dolwin, J. and Beck, R. (1994). Design of driven piles in sand. *Geotechnique,* 44(3), 427–448.

Rowe, R.K. (1986). The prediction of deformations caused by soft ground tunnelling – Recent trends. *Canadian Tunnelling*, D. Eisenstein (ed.), 91–108.

Sagaseta, C. (1998). On the role of analytical solutions for the evaluation of soil deformation around tunnels. *Internal Report*, University of Cantabria, Santander, Spain.

Santarelli, F.J., Brown, E.T. and Maury, V. (1986). Analysis of borehole stresses using pressure-dependent, linear elasticity. *International Journal for Rock Mechanics and Mining Sciences,* 23(6), 445–449.

Terzaghi, K. and Richart F. E., Jr. (1952). Stresses in rock about cavities. *Geotechnique*, 2, 57–90.

Verruijt, A. and Booker, J.R. (1996). Surface settlements due to deformation of a tunnel in an elastic half plane. *Geotechnique,* 46(4), 753–756.

Vesic, A.S. (1971). Breakout resistance of objects embedded in ocean bottom. *Journal of the Soil Mechanics and Foundations Division,* ASCE, 97 (9), 1183–1205.

Vesic, A.S. (1972). Expansion of cavities in infinite soil mass. *Journal of the Soil Mechanics and Foundations Division,* ASCE, 98, 265–290.

Yu, H.S. and Houlsby, G.T. (1991). Finite cavity expansion in dilatant soil: loading analysis. *Geotechnique*, 41, 173–183.

Yu, H.S. and Mitchell, J.K. (1998). Analysis of cone resistance: review of methods. *Journal of Geotechnical and Geoenvironmental Engineering,* ASCE, 124(2), 140–149.

Yu, H.S. and Rowe, R.K. (1999). Plasticity solutions for soil behaviour around contracting cavities and tunnels. *International Journal for Numerical and Analytical Methods in Geomechanics*, 23, 1245–1279.

Yu, H.S., Herrmann, L.R. and Boulanger, R.W. (2000). Analysis of steady cone penetration in clay. *Journal of Geotechnical and Geoenvironmental Engineering,* ASCE, 126(7).

Yu, H.S., Schnaid, F. and Collins, I.F. (1996). Analysis of cone pressuremeter tests in sand. *Journal of Geotechnical Engineering,* ASCE, 122 (8), 623–632.

CHAPTER 2

ELASTIC SOLUTIONS

2.1 INTRODUCTION

This chapter presents some fundamental solutions for cavity expansion in elastic materials. These elastic solutions also serve as a necessary introduction to the development of nonlinear plastic solutions which are presented in later chapters of the book.

2.2 ELASTIC SOLUTIONS IN ISOTROPIC MEDIA

2.2.1 Expansion of a hollow sphere

The problem considered here is a classic problem in the theory of elasticity. Consider a hollow sphere under internal and external pressures p and p_0 respectively. The inner and outer radii of the sphere are denoted by a and b (see Figure 2.1).

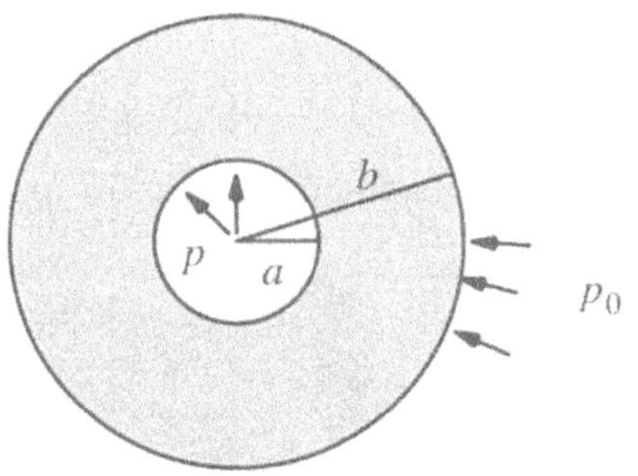

Figure 2.1: Cavity under uniform internal and external pressures

It is assumed that both the internal and external pressures are applied from zero initial values. We are interested in the stress and displacement fields in the sphere once the pressures have been applied. The solution for this problem is well known and can be found in many textbooks such as that by Timoshenko and Goodier (1970).

(a) Equilibrium equation and boundary conditions

The equation of equilibrium for a spherical cavity problem can be expressed in terms of radial and hoop stresses:

$$r\frac{d\sigma_r}{dr} + 2(\sigma_r - \sigma_\theta) = 0 \tag{2.1}$$

where σ_r and σ_θ are normal stresses acting in the radial and tangential directions respectively.

The radial stress is known on both internal and external boundaries. These boundary conditions can be written as:

$$\sigma_r|_{r=a} = p \tag{2.2}$$

$$\sigma_r|_{r=b} = p_0 \tag{2.3}$$

(b) Compatibility condition and stress-strain relations

Corresponding to the radial and tangential stresses, there are two normal strain components in these two directions which can be expressed as functions of the radial displacement u:

$$\varepsilon_r = -\frac{du}{dr} \quad \text{and} \quad \varepsilon_\theta = -\frac{u}{r} \tag{2.4}$$

Equation (2.4) can be used to eliminate the displacement u to give the following compatibility condition:

$$\varepsilon_r = \frac{d}{dr}(r\varepsilon_\theta) \tag{2.5}$$

For elastic materials, the stress-strain relations for spherical cavity problems are:

$$\varepsilon_r = \frac{1}{E}[\sigma_r - 2\nu\sigma_\theta] \tag{2.6}$$

$$\varepsilon_\theta = \frac{1}{E}[-\nu\sigma_r + (1-\nu)\sigma_\theta] \tag{2.7}$$

where E is Young's modulus and ν is Poisson's ratio.

(c) Solution procedure

All the governing equations that are needed for solving the expansion of a hollow sphere have been presented. Using these equations, several alternative procedures may be used to arrive at the final solution. In this case we follow the procedure used by Timoshenko and Goodier (1970).

Combining equations (2.1),(2.5) and (2.6)-(2.7) gives a differential equation in terms of the radial stress which has the following general solution:

$$\sigma_r = A + \frac{B}{r^3} \tag{2.8}$$

where A and B are integration constants. Substituting equation (2.8) into the equilibrium equation (2.1) leads to the following expression for the tangential stress:

$$\sigma_\theta = A - \frac{B}{2r^3} \tag{2.9}$$

Now the boundary conditions (2.2) and (2.3) can be used to determine the integration constants A and B. Applying the boundary conditions leads to:

$$A + \frac{B}{a^3} = p \tag{2.10}$$

$$A + \frac{B}{b^3} = p_0 \tag{2.11}$$

The above two equations can be used to solve for A and B as

$$A = \frac{-p_0 b^3 + pa^3}{a^3 - b^3} \tag{2.12}$$

$$B = \frac{(-p + p_0)a^3 b^3}{a^3 - b^3} \tag{2.13}$$

Substituting the above solutions of A and B into equations (2.8) and (2.9) results in the following solution for the stresses:

$$\sigma_r = -\frac{p_0 b^3 (r^3 - a^3)}{r^3 (a^3 - b^3)} - \frac{pa^3 (b^3 - r^3)}{r^3 (a^3 - b^3)} \tag{2.14}$$

$$\sigma_\theta = -\frac{p_0 b^3 (2r^3 + a^3)}{2r^3 (a^3 - b^3)} + \frac{pa^3 (b^3 + 2r^3)}{2r^3 (a^3 - b^3)} \tag{2.15}$$

The radial displacement u (positive when it is directed out of the cavity) can be determined from the tangential strain in equation (2.7):

$$u = -r\varepsilon_\theta = \frac{p - p_0}{2G(\frac{1}{a^3} - \frac{1}{b^3})}[\frac{1-2\nu}{(1+\nu)b^3} r + \frac{1}{2r^2}] \tag{2.16}$$

where $G = E/2(1 + \nu)$ is the shear modulus of the material.

(d) Special case: infinite medium

It is of particular interest to consider a special case when $b \rightarrow \infty$, as this case has many applications in geotechnical engineering. By putting $b \rightarrow \infty$ into equations (2.14), (2.15) and (2.16), we obtain the following solutions for stresses and displacement:

$$\sigma_r = p_0 + (p - p_0)(\frac{a}{r})^3 \tag{2.17}$$

$$\sigma_\theta = p_0 - \frac{1}{2}(p-p_0)(\frac{a}{r})^3 \tag{2.18}$$

$$u = \frac{p-p_0}{4G}(\frac{a}{r})^3 r \tag{2.19}$$

2.2.2 Expansion of a thick-walled cylinder

The analysis of cylinder expansion follows a same procedure as used for the expansion of a hollow sphere. The main difference lies on the equilibrium equation and stress-strain relations. As we only consider an infinitely long cylinder, the plane strain condition is assumed.

(a) Equilibrium equation and boundary conditions

The equilibrium equation for a cylindrical cavity problem can be expressed in terms of radial and hoop stresses as follows:

$$r\frac{d\sigma_r}{dr} + (\sigma_r - \sigma_\theta) = 0 \tag{2.20}$$

where σ_r and σ_θ are the normal stresses acting in the radial and tangential directions respectively. The stress boundary conditions are the same as for the spherical cavity expansion which are defined by (2.2) and (2.3).

(b) Compatibility condition and stress-strain relations

The radial and tangential strains for the cylindrical cavity are the same as for the spherical cavity, namely:

$$\varepsilon_r = -\frac{du}{dr} \quad \text{and} \quad \varepsilon_\theta = -\frac{u}{r}$$

which can be used to eliminate the displacement u to give the following compatibility condition:

$$\varepsilon_r = \frac{d}{dr}(r\varepsilon_\theta)$$

For elastic materials, the stress-strain relations for plane strain cylindrical cavity expansion are:

$$\varepsilon_r = \frac{1-\nu^2}{E}[\sigma_r - \frac{\nu}{1-\nu}\sigma_\theta] \tag{2.21}$$

$$\varepsilon_\theta = \frac{1-\nu^2}{E}[-\frac{\nu}{1-\nu}\sigma_r + \sigma_\theta] \tag{2.22}$$

where E is Young's modulus and ν is Poisson's ratio.

(c) Solution procedure

Combining the equations of equilibrium, compatibility and stress-strain relations gives a differential equation in terms of the radial stress which has the following general solution:

$$\sigma_r = C + \frac{D}{r^2} \tag{2.23}$$

where C and D are constants. By substituting equation (2.23) into the equilibrium equation (2.20), the following expression for the tangential stress is obtained:

$$\sigma_\theta = C - \frac{D}{r^2} \tag{2.24}$$

Now the boundary conditions (2.2) and (2.3) can be used to determine the integration constants C and D. Applying these two boundary conditions gives:

$$C + \frac{D}{a^2} = p \tag{2.25}$$

$$C + \frac{D}{b^2} = p_0 \tag{2.26}$$

which can be used to solve for C and D as follows:

$$C = \frac{-pa^2 + p_0 b^2}{b^2 - a^2} \tag{2.27}$$

$$D = \frac{a^2 b^2(-p_0 + p)}{b^2 - a^2} \tag{2.28}$$

Hence the following solution for the stresses is obtained:

$$\sigma_r = -\frac{p_0 b^2(r^2 - a^2)}{r^2(a^2 - b^2)} - \frac{pa^2(b^2 - r^2)}{r^2(a^2 - b^2)} \tag{2.29}$$

$$\sigma_\theta = -\frac{p_0 b^2(r^2 + a^2)}{r^2(a^2 - b^2)} + \frac{pa^2(b^2 + r^2)}{r^2(a^2 - b^2)} \tag{2.30}$$

The radial displacement u can be determined from the tangential strain in equation (2.7) as:

$$u = -r\varepsilon_\theta = \frac{p - p_0}{2G(\frac{1}{a^2} - \frac{1}{b^2})}\left[\frac{1-2\nu}{b^2} r + \frac{1}{r}\right] \tag{2.31}$$

where $G = E/2(1 + \nu)$ is the shear modulus of the material.

(d) Special case: infinite medium

By substituting $b \to \infty$ into equations (2.29), (2.30) and (2.31), we obtain the following solutions for the expansion of a cylindrical cavity in an infinite medium:

$$\sigma_r = p_0 + (p - p_0)(\tfrac{a}{r})^2 \tag{2.32}$$

$$\sigma_\theta = p_0 - (p - p_0)(\tfrac{a}{r})^2 \tag{2.33}$$

$$u = \frac{p - p_0}{2G}(\tfrac{a}{r})^2 r \tag{2.34}$$

2.2.3 Cylindrical cavity subject to biaxial *in-situ* stresses

In this section, we consider a cylindrical cavity located in an infinite medium subject to a biaxial stress condition. This problem will be analysed in two steps.

(a) A stress-free hole subject to a horizontal pressure at infinity

As we are interested in a cylindrical cavity, this case is the same as that considered by Timoshenko and Goodier (1970) on the effect of circular holes on stress distributions in a plate of unit-thickness (see Figure 2.2).

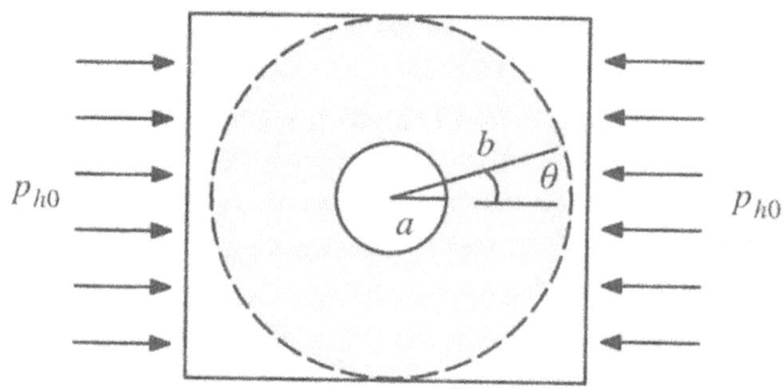

Figure 2.2: A stress free hole subject to a horizontal pressure

Figure 2.2 represents an unit-thickness plate which is subject to a uniform compression in the horizontal direction. Although having a small cylindrical cavity will change the stress distribution around the cavity, its influence on stresses will be very small at distances that are large compared with the cavity radius a. In other words, the stresses at radius b ($b \gg a$) are effectively the same as those in the plate without the cavity. In this way, we can derive the stresses at radius b as:

$$\sigma_r|_{r=b} = p_{h0}\cos^2\theta = \frac{1}{2}p_{h0} + \frac{1}{2}p_{h0}\cos 2\theta \tag{2.35}$$

$$\tau_{r\theta}|_{r=b} = -\frac{1}{2}p_{h0}\sin 2\theta \tag{2.36}$$

These stresses act around the outside of the ring having inner and outer radii a and b. We need to derive the stress distribution within the ring under these applied stresses at $r = b$.

For simplicity, the stresses applied at the outer boundary, as defined by (2.35) and (2.36) are divided into two parts. The first is due to the constant radial pressure as defined by:

$$\sigma_r\big|_{r=b} = \frac{1}{2}p_{h0} \quad \text{and} \quad \tau_{r\theta}\big|_{r=b} = 0 \tag{2.37}$$

and the stresses produced by the stresses (2.37) can be obtained from the obtained solutions (2.29) and (2.30) by setting $p = 0$ and $p_0 = \frac{1}{2}p_{h0}$.

The second part is due to the application of the following stresses:

$$\sigma_r\big|_{r=b} = \frac{1}{2}p_{h0}\cos 2\theta \quad \text{and} \quad \tau_{r\theta}\big|_{r=b} = -\frac{1}{2}p_{h0}\sin 2\theta \tag{2.38}$$

The stress distribution created by the stresses (2.38) can be obtained from the following stress function:

$$\Phi = f(r)\cos 2\theta \tag{2.39}$$

By putting the above stress function into the following governing equation for plane problems:

$$\left(\frac{\partial^2}{\partial r^2} + \frac{1}{r}\frac{\partial}{\partial r} + \frac{1}{r^2}\frac{\partial^2}{\partial \theta^2}\right)\left(\frac{\partial^2\Phi}{\partial r^2} + \frac{1}{r}\frac{\partial\Phi}{\partial r} + \frac{1}{r^2}\frac{\partial^2\Phi}{\partial \theta^2}\right) = 0 \tag{2.40}$$

we can obtain an ordinary differential equation in terms of the function f(r), which can be solved to be:

$$f(r) = Ar^2 + Br^4 + C\frac{1}{r^2} + D \tag{2.41}$$

Substituting (2.41) into (2.39) leads to the correct stress function:

$$\Phi = (Ar^2 + Br^4 + C\frac{1}{r^2} + D)\cos 2\theta \tag{2.42}$$

where A, B, C and D are integration constants.

Using the relationship between stresses and stress function, we derive the following expressions for stresses:

$$\sigma_r = \frac{1}{r}\frac{\partial\Phi}{\partial r} + \frac{1}{r^2}\frac{\partial^2\Phi}{\partial\theta^2} = -(2A + \frac{6C}{r^4} + \frac{4D}{r^2})\cos 2\theta \tag{2.43}$$

$$\sigma_\theta = \frac{\partial^2\Phi}{\partial r^2} = (2A + 12Br^2 + \frac{6C}{r^4})\cos 2\theta \tag{2.44}$$

$$\tau_{r\theta} = -\frac{\partial}{\partial r}\left(\frac{1}{r}\frac{\partial\Phi}{\partial\theta}\right) = (2A + 6Br^2 - \frac{6C}{r^4} - \frac{2D}{r^2})\sin 2\theta \tag{2.45}$$

With the radial and shear stresses at $r=a$ and $r=b$, the following boundary conditions are obtained:

$$2A + \frac{6C}{b^4} + \frac{4D}{b^2} = -\frac{1}{2}p_{h0} \tag{2.46}$$

$$2A + \frac{6C}{a^4} + \frac{4D}{a^2} = 0 \tag{2.47}$$

$$2A + 6Bb^2 - \frac{6C}{b^4} - \frac{2D}{b^2} = -\frac{1}{2}p_{h0} \tag{2.48}$$

$$2A + 6Ba^2 - \frac{6C}{a^4} - \frac{2D}{a^2} = 0 \tag{2.49}$$

If we consider an infinite medium, $b \rightarrow \infty$, then the above equations can be solved to give the following solutions for A, B, C, and D:

$$A = -\frac{p_{h0}}{4} \tag{2.50}$$

$$B = 0 \tag{2.51}$$

$$C = -\frac{a^4}{4}p_{h0} \tag{2.52}$$

$$D = \frac{a^2}{2}p_{h0} \tag{2.53}$$

Finally substituting the above solutions for A, B, C and D into the stress expressions (2.43)-(2.45) and then adding the stresses produced by the uniform pressure (2.37) on the outer boundary, we obtain the final solution for the stresses:

$$\sigma_r = \frac{p_{h0}}{2}(1 - \frac{a^2}{r^2}) + \frac{p_{h0}}{2}(1 + \frac{3a^4}{r^4} - \frac{4a^2}{r^2})\cos 2\theta \tag{2.54}$$

$$\sigma_\theta = \frac{p_{h0}}{2}(1 + \frac{a^2}{r^2}) - \frac{p_{h0}}{2}(1 + \frac{3a^4}{r^4})\cos 2\theta \tag{2.55}$$

$$\tau_{r\theta} = -\frac{p_{h0}}{2}(1 - \frac{3a^4}{r^4} + \frac{2a^2}{r^2})\sin 2\theta \tag{2.56}$$

(b) A cylindrical cavity subject to biaxial pressures at infinity

Figure 2.3 shows the case in which, in addition to the horizontal pressure p_{h0}, there is a second stress p_{v0} acting in the vertical direction at infinity. This case is of particular interest for tunnel problems.

The stresses produced by this second pressure in the vertical direction can be obtained in the same way as that just presented for the horizontal pressure. The final stresses produced by both p_{h0} and p_{v0} can be obtained by superposition as follows:

$$\sigma_r = \frac{p_{h0} + p_{v0}}{2}(1 - \frac{a^2}{r^2}) + \frac{p_{h0} - p_{v0}}{2}(1 + \frac{3a^4}{r^4} - \frac{4a^2}{r^2})\cos 2\theta \tag{2.57}$$

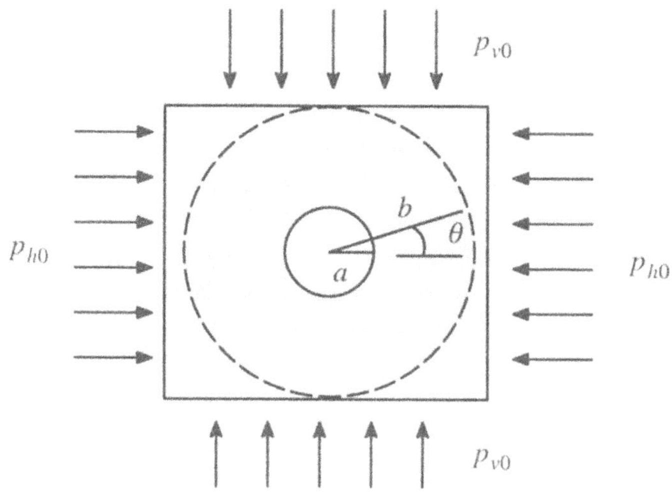

Figure 2.3: A stress free hole subject to biaxial pressures

$$\sigma_\theta = \frac{p_{h0}+p_{v0}}{2}(1+\frac{a^2}{r^2}) - \frac{p_{h0}-p_{v0}}{2}(1+\frac{3a^4}{r^4})\cos 2\theta \tag{2.58}$$

$$\tau_{r\theta} = -\frac{p_{h0}-p_{v0}}{2}(1-\frac{3a^4}{r^4}+\frac{2a^2}{r^2})\sin 2\theta \tag{2.59}$$

To determine the displacements, we need to use the stress-strain relations and compatibility condition. For the case considered here, the strains are expressed in terms of radial and tangential displacements u and v:

$$\varepsilon_r = -\frac{\partial u}{\partial r} \tag{2.60}$$

$$\varepsilon_\theta = -\frac{u}{r} - \frac{1}{r}\frac{\partial v}{\partial \theta} \tag{2.61}$$

Using the stress-strain relations for a plane strain cylindrical cavity results in:

$$-\frac{\partial u}{\partial r} = \frac{1-\nu^2}{E}[\sigma_r - \frac{\nu}{1-\nu}\sigma_\theta] \tag{2.62}$$

$$-\frac{u}{r} - \frac{1}{r}\frac{\partial v}{\partial \theta} = \frac{1-\nu^2}{E}[-\frac{\nu}{1-\nu}\sigma_r + \sigma_\theta] \tag{2.63}$$

Note that the radial displacement u is positive when it is directed out of the cavity and the tangential displacement v is positive in the clockwise direction.

With the displacement boundary conditions at infinity, equations (2.62) and (2.63) can be used to give the following solutions for the displacement fields:

$$u = -\frac{1-\nu^2}{E}\left\{p_m(r + \frac{a^2}{r}) + p_d(r-\frac{a^4}{r^3} + \frac{4a^2}{r})\cos 2\theta\right\}$$
$$+ \frac{\nu(1+\nu)}{E}\left\{p_m(r-\frac{a^2}{r})-p_d(r-\frac{a^4}{r^3})\cos 2\theta\right\} \quad (2.64)$$

$$v = \frac{p_d}{E}\left\{(1-\nu^2)(r + \frac{2a^2}{r} + \frac{a^4}{r^3}) + \nu(1+\nu)(r-\frac{2a^2}{r} + \frac{a^4}{r^3})\right\}\sin 2\theta \quad (2.65)$$

where

$$p_m = \frac{p_{h0} + p_{v0}}{2} \quad \text{and} \quad p_d = \frac{p_{h0}-p_{v0}}{2} \quad (2.66)$$

In particular, at the cavity wall $r=a$, the displacements are:

$$\frac{u}{a} = -\frac{1-\nu^2}{E}[(p_{h0} + p_{v0}) + 2(p_{h0}-p_{v0})\cos 2\theta] \quad (2.67)$$

$$\frac{v}{a} = \frac{2(1-\nu^2)}{E}(p_{h0}-p_{v0})\sin 2\theta \quad (2.68)$$

Constant pressure applied at the cavity wall

If the boundary condition at $r=a$ was a non-zero constant pressure p, the stresses can be obtained by superimposing the stress solution due to an internal pressure p (from the solution presented in the previous section) on to the previous results (2.57) and (2.58). The final stresses obtained are:

$$\sigma_r = \frac{p_{h0} + p_{v0}}{2}(1-\frac{a^2}{r^2}) + \frac{pa^2}{r^2} + \frac{p_{h0} - p_{v0}}{2}(1 + \frac{3a^4}{r^4}-\frac{4a^2}{r^2})\cos 2\theta \quad (2.69)$$

$$\sigma_\theta = \frac{p_{h0} + p_{v0}}{2}(1 + \frac{a^2}{r^2})-\frac{pa^2}{r^2} - \frac{p_{h0} - p_{v0}}{2}(1 + \frac{3a^4}{r^4})\cos 2\theta \quad (2.70)$$

2.3 ELASTIC SOLUTIONS IN ANISOTROPIC MEDIA

In addition to isotropic materials, the behaviour of a transversely isotropic material (often referred to as cross-anisotropy in soil mechanics) is also relevant to modelling soils and rocks (see, for example, Graham and Houlsby, 1983; Wu *et al.*, 1991). For this reason, analytical solutions are presented here for the expansion of cavities in a cross-anisotropic elastic material. Application of these solutions to the study of wellbore stability will be discussed in detail in Chapter 11.

2.3.1 Expansion of a hollow sphere

The solution for the expansion of a hollow sphere in a cross-anisotropic elastic material has been presented by Lekhnitskii (1963).

It is assumed that the geometry and loading conditions of the problem are the same as shown in Figure 2.1. The only difference is that now we are considering the case of a material possessing transverse isotropy in the direction of the radius. In other words, the surface (θ, ϕ) normal to the radius is regarded as the isotropic surface.

(a) Stress-strain relations

The stress-strain relationship for this type of cross-anisotropic material can be written as (Lekhnitskii, 1963; Van Cauwelaert, 1977):

$$\varepsilon_r = \frac{1}{E'}\sigma_r - \frac{\nu'}{E'}(\sigma_\phi + \sigma_\theta) \tag{2.71}$$

$$\varepsilon_\phi = -\frac{\nu'}{E'}\sigma_r - \frac{1}{E}(\sigma_\phi - \nu\sigma_\theta) \tag{2.72}$$

$$\varepsilon_\theta = -\frac{\nu'}{E'}\sigma_r + \frac{1}{E}(\sigma_\theta - \nu\sigma_\phi) \tag{2.73}$$

where E' = Young's modulus in the r direction; E = Young's modulus in the isotropic (θ, ϕ) plane; ν'= Poisson's ratio defining strain induced in the isotopic plane by stress applied along the radial direction; and ν= Poisson's ratio defining strain induced in the isotopic plane by stress applied along the plane.

Alternatively, the stress-strain relations (2.71)-(2.73) may also be written as follows:

$$\sigma_r = A_{11}\varepsilon_r + A_{12}\varepsilon_\phi + A_{12}\varepsilon_\theta \tag{2.74}$$

$$\sigma_\phi = A_{12}\varepsilon_r + A_{22}\varepsilon_\phi + A_{23}\varepsilon_\theta \tag{2.75}$$

$$\sigma_\theta = A_{12}\varepsilon_r + A_{23}\varepsilon_\phi + A_{22}\varepsilon_\theta \tag{2.76}$$

The coefficients of the stiffness matrix are:

$$A_{11} = \frac{E'(1-\nu)}{1-\nu-2\nu'^2\ E/E'} \tag{2.77}$$

$$A_{12} = \frac{E\nu'}{1-\nu-\nu'^2\ E/E'} \tag{2.78}$$

$$A_{22} = \frac{E}{1+\nu} \times \frac{1-\nu'^2\ E/E'}{1-\nu-2\nu'^2\ E/E'} \tag{2.79}$$

$$A_{23} = \frac{E}{1+\nu} \times \frac{\nu+\nu'^2\ E/E'}{1-\nu-2\nu'^2\ E/E'} \tag{2.80}$$

(b) Solution procedure

For a spherical cavity, the strains can be expressed as functions of the radial displacement u:

$$\varepsilon_r = -\frac{du}{dr} \quad , \quad \varepsilon_\phi = \varepsilon_\theta = -\frac{u}{r} \tag{2.81}$$

Using the condition $\sigma_\phi = \sigma_\theta$, the stress-strain relations and the equilibrium equation:

$$r\frac{d\sigma_r}{dr} + 2(\sigma_r - \sigma_\theta) = 0$$

we obtain the following differential equation in terms of the radial displacement u:

$$\frac{d^2u}{dr^2} + \frac{2}{r}\frac{du}{dr} - \frac{2(A_{22} + A_{23} - A_{12})}{A_{11}} \times \frac{u}{r^2} = 0 \tag{2.82}$$

which has the following general solution

$$u = Ar^{n-\frac{1}{2}} + \frac{B}{r^{n+\frac{1}{2}}} \tag{2.83}$$

where n is defined by:

$$n = \sqrt{\frac{1}{4} + \frac{2(A_{22} + A_{23} - A_{12})}{A_{11}}} \tag{2.84}$$

The stress expressions can then be obtained by substituting equations (2.83) and (2.81) into the stress-strain relations (2.74)-(2.76). The integration constants A and B can be derived by using the boundary conditions:

$$\sigma_r|_{r=a} = p$$

$$\sigma_r|_{r=b} = p_0$$

The derived values for A and B are:

$$A = \frac{1}{A_{11}(n-\frac{1}{2}) + 2A_{12}} \times \frac{-pa^{n+3/2} + p_0 b^{n+3/2}}{b^{2n} - a^{2n}} \tag{2.85}$$

$$B = \frac{1}{-A_{11}(n+\frac{1}{2}) + 2A_{12}} \times \frac{(-p_0 b^{3/2-n} + pa^{3/2-n})(ab)^{2n}}{b^{2n} - a^{2n}} \tag{2.86}$$

The final stresses are derived as follows:

$$\sigma_r = \frac{-pa^{n+3/2} + p_0 b^{n+3/2}}{b^{2n} - a^{2n}} r^{n-\frac{3}{2}}$$

$$+ \frac{(-p_0 b^{3/2-n} + pa^{3/2-n})(ab)^{2n}}{b^{2n}-a^{2n}} r^{-n-\frac{3}{2}} \tag{2.87}$$

$$\sigma_\theta = \sigma_\phi = \lambda_1 \frac{-pa^{n+3/2} + p_0 b^{n+3/2}}{b^{2n}-a^{2n}} r^{n-\frac{3}{2}}$$

$$+ \lambda_2 \frac{(-p_0 b^{3/2-n} + pa^{3/2-n})(ab)^{2n}}{b^{2n}-a^{2n}} r^{-n-\frac{3}{2}} \tag{2.88}$$

where

$$\lambda_1 = \frac{A_{22} + A_{23} + A_{12}(n-\frac{1}{2})}{A_{11}(n-\frac{1}{2}) + 2A_{12}} \tag{2.89}$$

$$\lambda_2 = \frac{A_{22} + A_{23} - A_{12}(n+\frac{1}{2})}{-A_{11}(n+\frac{1}{2}) + 2A_{12}} \tag{2.90}$$

It can be shown that for an isotropic material $n = 3/2$, the above solutions reduce to the previous solutions for isotropic materials.

2.3.2 Expansion of a thick-walled cylinder

The solution presented here for a cylinder with both internal and external pressures follows closely the procedure used by Wu *et al.* (1991). This is a special case of the general solutions derived by Lekhnitskii (1963) for a variety of loading conditions.

Unlike the solution that we have just presented for a hollow sphere that is only for the case of a material possessing transverse isotropy with the direction of the radius as the axis of symmetry, this section describes an analytical solution for cylinder expansion that can be used to deal with the case in which either axis (r, z *or* θ) could be treated as the axis of symmetry. In other words, the following three possibilities exist:

(a) the r (radial) direction is the axis of symmetry and the (z, θ) plane is the isotropic plane;

(b) the z (axial) direction is the axis of symmetry and the (r, θ) plane is the isotropic plane;

(c) the θ (tangential) direction is the axis of symmetry and the (r, z) plane is the isotropic plane.

It must be noted that while the first two cases find applications to geotechnical engineering, the third case is not very relevant to geomechanics problems.

(a) Stress-strain relations

The general stress-strain relationship for a cross-anisotropic material can be written as (Lekhnitskii, 1963; Van Cauwelaert, 1977):

$$\varepsilon_r = a_{11}\sigma_r + a_{12}\sigma_\theta + a_{13}\sigma_z \tag{2.91}$$

$$\varepsilon_\theta = a_{12}\sigma_r + a_{22}\sigma_\theta + a_{23}\sigma_z \tag{2.92}$$

$$\varepsilon_z = a_{13}\sigma_r + a_{23}\sigma_\theta + a_{33}\sigma_z \tag{2.93}$$

where the coefficients a_{ij} are expressed as simple functions of Young's moduli and Poisson's ratios (Lekhnitskii, 1963).

For plane strain, the strain in the z direction should be zero, which gives:

$$\sigma_z = -\frac{1}{a_{33}}(a_{13}\sigma_r + a_{23}\sigma_\theta) \tag{2.94}$$

Substituting equation (2.94) into equations (2.91) and (2.92) leads to:

$$\varepsilon_r = \beta_{11}\sigma_r + \beta_{12}\sigma_\theta \tag{2.95}$$

$$\varepsilon_\theta = \beta_{12}\sigma_r + \beta_{22}\sigma_\theta \tag{2.96}$$

where the coefficients β_{ij} are:

$$\beta_{ij} = a_{ij} - \frac{a_{i3}a_{j3}}{a_{33}}, \quad (i,j = 1,2) \tag{2.97}$$

(b) Solution procedure

For the cylindrical cavity, the strains can be expressed as functions of the radial displacement u:

$$\varepsilon_r = -\frac{du}{dr} \quad , \quad \varepsilon_\theta = -\frac{u}{r} \tag{2.98}$$

which are used to eliminate the displacement u to give the following compatibility condition:

$$\varepsilon_r = \frac{d}{dr}(r\varepsilon_\theta)$$

The equilibrium equation is:

$$r\frac{d\sigma_r}{dr} + (\sigma_r - \sigma_\theta) = 0$$

If we use the radial stress as the fundamental variable, we can obtain the following differential equation by combining the stress-strain relations, compatibility condition and the equilibrium equation:

$$\beta_{22} r^2 \frac{d^2\sigma_r}{dr^2} + 3\beta_{22} r \frac{d\sigma_r}{dr} - (\beta_{11} - \beta_{22})\sigma_r = 0 \tag{2.99}$$

Equation (2.99) can be solved to give the following general solution for the radial stress:

$$\sigma_r = A r^{n-1} + \frac{B}{r^{n+1}} \tag{2.100}$$

where n is defined by:

$$n = \sqrt{\frac{\beta_{11}}{\beta_{22}}} \tag{2.101}$$

Using the boundary conditions:

$$\sigma_r|_{r=a} = p$$

$$\sigma_r|_{r=b} = p_0$$

the integration constants A and B can be derived. The final solutions for the stresses are:

$$\sigma_r = \frac{p_0 - p(\frac{a}{b})^{n+1}}{1-(\frac{a}{b})^{2n}} (\frac{r}{b})^{n-1} + \frac{p - p_0(\frac{a}{b})^{n-1}}{1-(\frac{a}{b})^{2n}} (\frac{a}{r})^{n+1} \tag{2.102}$$

$$\sigma_\theta = n \frac{p_0 - p(\frac{a}{b})^{n+1}}{1-(\frac{a}{b})^{2n}} (\frac{r}{b})^{n-1} - n \frac{p - p_0(\frac{a}{b})^{n-1}}{1-(\frac{a}{b})^{2n}} (\frac{a}{r})^{n+1} \tag{2.103}$$

The radial displacement can be shown to be:

$$u = -r\varepsilon_\theta = -r(\beta_{12}\sigma_r + \beta_{22}\sigma_\theta) \tag{2.104}$$

For an isotropic material $n = 1$, the above solutions reduce to the previous cylinder expansion solutions for isotropic materials.

2.4 ELASTIC SOLUTIONS IN A SEMI-INFINITE HALF-SPACE

Although this book is predominantly concerned with cavities in an infinite medium, elastic solutions also exist for the expansion or contraction of cavities in a semi-infinite half-space. The problem of a cavity in a half-space is particularly relevant to the prediction of ground settlements caused by underground tunnelling. For this reason there is continuing interest in the geotechnical community in seeking analytical solutions for cavity unloading problems in an elastic half-space. Unlike cavity expansion in an infinite medium, the cavity problem in a half-space is a two dimensional problem so that analytical solutions are currently only possible in elastic materials.

2.4.1 Cylindrical cavity in a half-space

Verruijt and Booker (1996) presented a simple analytical solution considering two basic deformation loadings: (1) a uniform radial displacement (representing the ground loss that may occur during tunnel construction), and (2) an ovalization of the tunnel (which may be caused by an anisotropic initial stress state). These two types of displacement loading are shown in Figure 2.4.

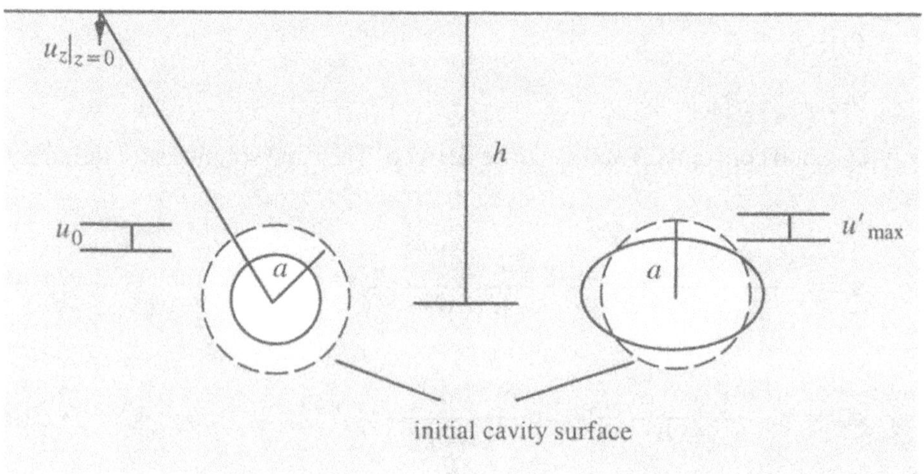

Figure 2.4: Cavity unloading in an elastic half-space

The method used by Verruijt and Booker (1996) is an extension of the virtual image approach proposed by Sagaseta (1987). While Sagaseta (1987) deals with incompressible materials (either elastic or plastic soils) with a uniform radial displacement, Verruijt and Booker's elastic analysis accounts for the effect of ovalization as well as the elastic compressibility defined by an arbitrary value of Poisson's ratio.

Before presenting the results, it is useful to define the two displacement input parameters. Assuming that the uniform inward radial displacement of the cavity wall is u_0, the cavity strain at the cavity wall should be $\varepsilon = u_0/a$, where a is the cavity radius. The cavity ovalization (or distortion) may be defined by the maximum inward displacement u'_{max} and this can be expressed by a normalised quantity $\delta = u'_{max}/a$. The solutions derived by Verruijt and Booker (1996) are in terms of the cavity strain ε and the normalised cavity distortion δ.

The solution for this problem can be derived in three steps. The first two steps are singular solutions from the theory of elasticity (Timoshenko and Goodier, 1970) for the points defined by $(0, h)$ and $(0, -h)$ in the (x, z) space, Figure 2.5. The expressions for the displacements in x and z directions, caused by the two types of displacement loading in the single point (i.e. tunnel) and its image can be shown to be:

$$u_x = -\varepsilon a^2 x(\frac{1}{r_1^2} + \frac{1}{r_2^2}) + \delta a^2 x[\frac{x^2 - k_0 z_1^2}{r_1^4} + \frac{x^2 - k_0 z_2^2}{r_2^4}] \quad (2.105)$$

$$u_z = -\varepsilon a^2(\frac{z_1}{r_1^2} + \frac{z_2}{r_2^2}) + \delta a^2[\frac{z_1(k_0 x^2 - z_1^2)}{r_1^4} + \frac{z_2(k_0 x^2 - z_2^2)}{r_2^4}] \quad (2.106)$$

where $k_0 = \nu/(1 - \nu)$, $z_1 = z - h$ and $z_2 = z + h$.

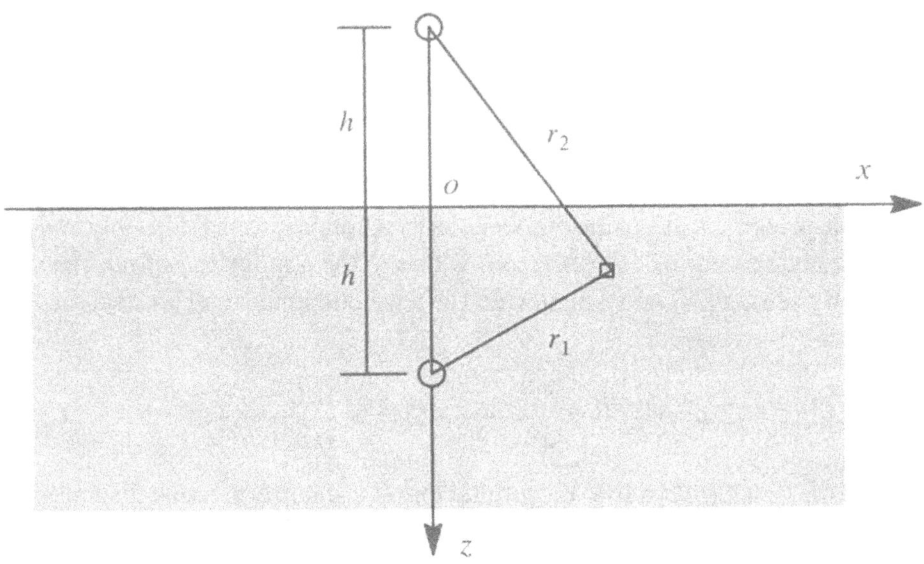

Figure 2.5: A singularity and its image

Because of the symmetry of the solutions for the singularity and its image, they produce zero shear stress $\sigma_{xz} = 0$ and zero vertical displacement $u_z = 0$ on the free surface z=0. However, they do produce a non-zero normal stress on the surface z=0, which can be shown to be:

$$\sigma_z|_{z=0} = -4G\varepsilon a^2 \frac{x^2 - h^2}{(x^2 + h^2)^2} - \frac{8Gm\delta a^2}{1 + m} \times \frac{h^2(3x^2 - h^2)}{(x^2 + h^2)^3} \tag{2.107}$$

where G is shear modulus and m is defined by $m = 1/(1 - 2\nu)$.

In reality, however, on the surface z=0 the normal stress is zero. As a result, a third step of solution is to apply a normal stress on the z=0 surface which has the same magnitude as equation (2.107), but in the opposite direction. As noted by Verruijt and Booker (1996), problems of this type are most conveniently solved using Fourier transforms (Sneddon, 1951). The resulting displacement solutions are:

$$u_x = -\frac{2\varepsilon a^2 x}{m}\left(\frac{1}{r_2^2} - \frac{2mzz_2}{r_2^4}\right) - \frac{4\delta a^2 xh}{1 + m}\left(\frac{z_2}{r_2^4} + \frac{mz(x^2 - 3z_2^2)}{r_2^6}\right) \tag{2.108}$$

$$u_z = \frac{2\varepsilon a^2}{m}\left(\frac{(m + 1)z_2}{r_2^2} - \frac{mz(x^2 - z_2^2)}{r_2^4}\right)$$

$$-2\delta a^2 h\left(\frac{x^2 - z_2^2}{r_2^4} + \frac{m}{1 + m}\frac{2zz_2(3x^2 - z_2^2)}{r_2^6}\right) \tag{2.109}$$

The final solution for the displacements due to ground loss and tunnel distortion ε and δ is the sum of equations (2.108), (2.109) and (2.105), (2.106). Once the displacements are known, it is a simple matter to derive all the strain and stress components in elasticity.

On the free surface z=0, $z_1 = -z_2$, so from equation (2.106) we know that the vertical displacement due to the two singularity solutions is zero. The only non-zero vertical displacement on the plane z=0 is due to the application of normal stress (2.107). By setting z=0 into equation (2.109), we obtain the total vertical displacement on the free surface:

$$u_z|_{z=0} = 2\varepsilon a^2 \frac{1 + m}{m} \times \frac{h}{x^2 + h^2} - 2\delta a^2 \times \frac{h(x^2 - h^2)}{(x^2 + h^2)^2} \tag{2.110}$$

in which the first term is due to the ground loss (i.e. a uniform radial displacement) and the second term is due to the cavity distortion.

For incompressible materials, Poisson's ratio is equal to 0.5 and therefore the factor $(1+m)/m$ is 1. In this case, the first term of equation (2.110) reduces to the solution derived by Sagaseta (1987) for ground loss.

It is also interesting to note that using the solution (2.110) for ground loss only, the cavity wall displacement may be linked to the surface settlement in a simple way:

$$\frac{u_z|_{z=0}}{u_0} = \frac{2(1+m)}{m} \times \frac{\frac{h}{a}}{(\frac{x}{a})^2 + (\frac{h}{a})^2} \tag{2.111}$$

It will be shown in Chapter 10 that the above equation can be used in conjunction with cavity expansion solutions in an infinite medium to estimate the surface settlement profile due to the construction of a shallow tunnel.

2.4.2 Spherical cavity in a half-space

Using the image source method of Mindlin and Cheng (1950) and the concept of cavity expansion source of Hopkins (1960), Keer *et al.* (1998) derived a solution for the expansion of a spherical cavity in a half-space.

The concept of cavity expansion source was first used by Hopkins (1960). Consider a cavity under a uniform pressure p on the internal spherical surface with radius a. The radial stress at radius R is known to be $\sigma_R = p(a/R)^3$, which results in the integral for any spherical surface when the same centre is a constant. Therefore:

$$S = \int\int \sigma_R R ds \tag{2.112}$$

This constant is defined as a cavity expansion source at the centre of the spherical cavity.

For a spherical cavity subject to an internal pressure p, a cavity expansion source at the centre of the cavity is therefore $S = 4\pi a^3 p$ (Keer *et al.*, 1998). For an infinite medium, the stress and displacement fields induced by this source for a radius of R from the centre of the cavity are:

$$\sigma_R = \frac{S}{4\pi R^3} \tag{2.113}$$

$$\sigma_{R\phi} = 0 \tag{2.114}$$

$$\sigma_\theta = \sigma_\phi = -\frac{S}{8\pi R^3} \tag{2.115}$$

$$u = \frac{S}{16\pi G R^2} \tag{2.116}$$

To derive the stress and displacement solutions for a cavity in a half-space we can put another source S at the image point as shown in Figure 2.6. On the free surface (z=0), the only non-zero stress component due to the cavity source and its image source is the normal stress, which is shown to be:

$$\sigma_z = -\frac{S}{8\pi}\left(\frac{1}{R_0^3} - \frac{3h^2}{R_0^5}\right) \tag{2.117}$$

where $R_0 = \sqrt{r^2 + h^2} = \sqrt{x^2 + y^2 + h^2}$.

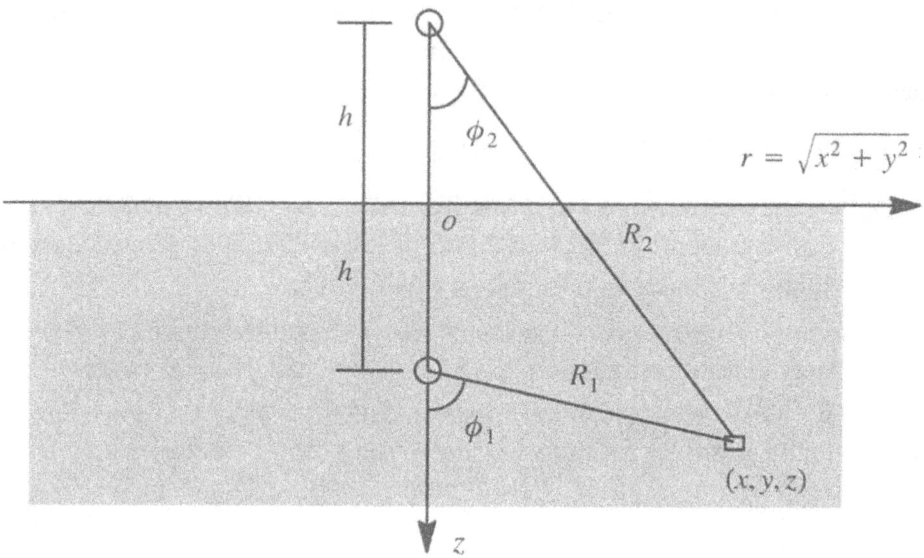

Figure 2.6: Image sources for traction free on the boundary

Using the axially symmetric stress function *f(r,z)* of Kassir and Sih (1975), the final stress and displacement solutions are written as:

$$\sigma_r = -2G[(1\text{–}2\nu)\frac{\partial^2 f}{\partial r^2} - 2\nu\frac{\partial^2 f}{\partial z^2} + z\frac{\partial^3 f}{\partial r^2 \partial z}] \tag{2.118}$$

$$\sigma_{rz} = 2Gz\frac{\partial^3 f}{\partial r \partial z^2} \tag{2.119}$$

$$\sigma_z = -2G[-\frac{\partial^2 f}{\partial z^2} + z\frac{\partial^3 f}{\partial z^3}] \tag{2.120}$$

$$\sigma_\theta = -2G[\frac{1}{r}\frac{\partial f}{\partial r} + 2\nu\frac{\partial^2 f}{\partial r^2} + \frac{z}{r}\frac{\partial^2 f}{\partial r \partial z}] \tag{2.121}$$

and

$$u_r = (1\text{–}2\nu)\frac{\partial f}{\partial r} + z\frac{\partial^2 f}{\partial r \partial z} \tag{2.122}$$

$$u_z = -2(1\text{–}\nu)\frac{\partial f}{\partial r} + z\frac{\partial^2 f}{\partial z^2} \tag{2.123}$$

in which the stress function f is a harmonic function. The shear stress $\sigma_{rz} = 0$ on the free surface z=0. By comparing equations (2.120) and (2.117) and requiring that the normal stress is zero at the free surface, the following stress function is derived:

$$f = -\frac{S}{8\pi G R_2} \tag{2.124}$$

By substituting the above stress function back into equations (2.118) to (2.123) and using the principle of superposition, we obtain stress and displacement solutions for a spherical cavity in an infinite half-space. For example, the stress results are (Keer *et al.*, 1998):

$$\sigma_r = -\frac{S}{8\pi R_1^3}[3\cos^2\phi_1 - 2] - \frac{S}{8\pi R_2^3}[3\cos^2\phi_2 - 2]$$

$$-[(1-2\nu)\frac{\partial^2 F}{\partial r^2} - 2\nu\frac{\partial^2 F}{\partial z^2} + z\frac{\partial^3 F}{\partial r^2 \partial z}] \tag{2.125}$$

$$\sigma_{rz} = -\frac{3S}{8\pi R_1^3}\sin\phi_1\cos\phi_1 - \frac{3S}{8\pi R_2^3}\sin\phi_2\cos\phi_2 + z\frac{\partial^3 F}{\partial r \partial z^2} \tag{2.126}$$

$$\sigma_z = -\frac{S}{8\pi R_1^3}[1 - 3\cos^2\phi_1] - \frac{S}{8\pi R_2^3}[1 - 3\cos^2\phi_2]$$

$$+\frac{\partial^2 F}{\partial z^2} - z\frac{\partial^3 F}{\partial z^3} \tag{2.127}$$

$$\sigma_\theta = -\frac{S}{8\pi R_1^3} - \frac{S}{8\pi R_2^3} - \frac{1}{r}\frac{\partial F}{\partial r} - 2\nu\frac{\partial^2 F}{\partial r^2} - \frac{z}{r}\frac{\partial^2 F}{\partial r \partial z} \tag{2.128}$$

in which

$$F = \frac{S}{4\pi R_2} \tag{2.129}$$

$$R_1 = \sqrt{x^2 + y^2 + (z-h)^2} = \sqrt{r^2 + (z-h)^2} \tag{2.130}$$

$$R_2 = \sqrt{x^2 + y^2 + (z+h)^2} = \sqrt{r^2 + (z+h)^2} \tag{2.131}$$

$$\cos\phi_1 = \frac{z-h}{R_1} \tag{2.132}$$

$$\cos\phi_2 = \frac{z+h}{R_2} \tag{2.133}$$

Interested readers can refer to the paper by Keer *et al.* (1998) for further details.

2.5 SUMMARY

1. Cavity expansion problems can be solved by considering the equations of equilibrium, compatibility, stress-strain relationship, and stress boundary conditions. Assuming a linear elastic stress-strain relationship, cavity expansion solutions can generally be obtained in closed form. This chapter presented a number of important elastic cavity expansion solutions, many of which will be used in later chapters as a necessary starting point for deriving elastic-plastic and viscoelastic solutions.

2. The final solutions for stresses and displacement around a spherical cavity in an isotropic medium are given by equations (2.14), (2.15) and (2.16).

3. The final solutions for stresses and displacement around a cylindrical cavity in an isotropic medium are given by equations (2.29), (2.30) and (2.31).

4. The stresses and displacement around a cylindrical cavity in an infinite medium, when subject to biaxial stresses, are given by equations (2.57)-(2.59) and (2.64)-(2.65) respectively.

5. It is possible to derive closed form solutions for stresses and displacement around cavities in an anisotropic medium. The final solutions for stresses and displacements around a cylindrical cavity in such a medium are given by equations (2.102), (2.103) and (2.104). These elastic solutions will be used in Chapter 11 to study the wellbore instability in rock.

6. This book is predominantly concerned with cavities in an infinite medium, though elastic solutions also exist for cavity expansion in a semi-infinite half-space. Such two-dimensional elastic solutions are given by equations (2.105)-(2.109) for cylindrical cavities, and equations (2.125)-(2.128) for spherical cavities. The problem of a cavity in a half-space is relevant to geotechnical engineering. In particular, it will be shown in Chapter 10 that the displacement solutions (2.105)-(2.109) can be used successfully to predict ground settlements caused by underground tunnelling.

REFERENCES

Graham, J. and Houlsby, G.T. (1983). Anisotropic elasticity of a natural clay. *Geotechnique,* 33(2), 165–180.

Hopkins, H.G. (1960). Dynamic expansion of spherical cavities in metals. in: *Progress in Solid Mechanics,* Vol 1, (Editors: I.N. Sneddon and R.Hill), North-Holland, Amsterdam.

Keer, L.M., Xu, Y. and Luk, V.K. (1998). Boundary effects in penetration or perforation. *Journal of Applied Mechanics,* ASME, 65, 489–496.

Kassir, M.K. and Sih, G.C. (1975). Three-dimensional crack problems. in: *Mechanics of Fracture,* Vol 2, Noordhoff International Publishing, Dordrecht, The Netherlands.

Lekhnitskii, S.G. (1963). *Theory of Elasticity of an Anisotropic Elastic Body.* Holden-Day, Inc.

Mindlin, R.D. and Cheng, D.H. (1950). Nuclei of strain in semi-infinite solid. *Journal of Applied Physics,* 21, 926–930.

Sagaseta, C. (1987). Analysis of undrained soil deformation due to ground loss. *Geotechnique,* 37(3), 301–320.

Sneddon, I.N. (1951). *Fourier Transforms.* New York, McGraw-Hill.

Timoshenko, S.P. and Goodier, J.N. (1970). *Theory of Elasticity.* 3rd edition, McGraw-Hill.

Van Cauwelaert, F. (1977). Coefficients of deformations of an anisotropic body. *Journal of the Engineering Mechanics Division*, ASCE, 103(EM5), 823–835.

Verruijt, A. and Booker, J.R. (1996). Surface settlements due to deformation of a tunnel in an elastic half plane. *Geotechnique,* 46(4), 753–756.

Wu, B.L., King, M.S. and Hudson, J.A. (1991). Stress-induced ultrasonic wave velocity anisotropy in a sandstone. *International Journal for Rock Mechanics and Mining Sciences*, 28(1), 101–107.

CHAPTER 3

ELASTIC-PERFECTLY PLASTIC SOLUTIONS

3.1 INTRODUCTION

This chapter describes some basic cavity expansion solutions in elastic-perfectly plastic soils. Both cylindrical and spherical cavities in either a finite or an infinite soil mass are considered. Note that a tension positive notation is used in this chapter.

When modelled by perfect plasticity, the soil behaviour is generally assumed to be under either drained or undrained conditions. Under undrained loading, cohesive soil is often modelled by a total stress analysis using the Tresca yield criterion. On the other hand, the drained behaviour of sand can be accurately modelled by an effective stress analysis using the Mohr-Coulomb yield criterion. As a result, this chapter focuses on the elastic-plastic solutions for cavity expansion problems using both Tresca and Mohr-Coulomb plasticity models. Shown in Figure 3.1*a* are the Tresca, von Mises and Mohr-Coulomb yield surfaces plotted in the π plane. As discussed by Hill (1950), once the solutions for Tresca materials are obtained, the corresponding solutions for von Mises materials can also be derived. This is because for spherical cavities the von Mises yield criterion is identical to the Tresca criterion. For the expansion of an infinitely long, plane strain cylindrical cavity, the von Mises criterion can be approximated by the Tresca criterion provided the shear strength is increased by 15%.

3.2 SOLUTIONS FOR TRESCA CRITERION

A comprehensive large strain solution of stresses and displacements for the expansion of spherical and cylindrical cavities in a Tresca material has been presented by Hill (1950). While we will generally follow the solution procedure of Hill (1950), this section includes a non-zero external pressure, and therefore is a generalization of Hill's solution which only considers a zero-initial stress condition. In addition, Hill's solution for the expansion of a cylindrical cavity from zero radius involves some simplifying assumptions. The solution presented here removes these unnecessary assumptions and is therefore more rigorous.

It is also noted that the large strain solutions described in this section are valid for any value of Poisson's ratio, and therefore, for Tresca materials the present solution is more general than that of Gibson and Anderson (1961).

3.2.1 Expansion of a spherical cavity in a finite medium

(a) Stress analysis

Assume the current internal and external radii of a shell are denoted by a and b (see Figure 3.1b). Initially the radii of the internal and external boundaries are a_0 and b_0 and a hydrostatic pressure p_0 acts throughout the soil. The internal pressure is increased monotonically from its initial value p_0 on the internal surface. An essential task of the analysis is to determine the variation of internal and external radii with the internal pressure p. As the internal pressure increases from p_0 the material will initially behave in an elastic manner, and the elastic solutions for both stresses and displacements are shown to be:

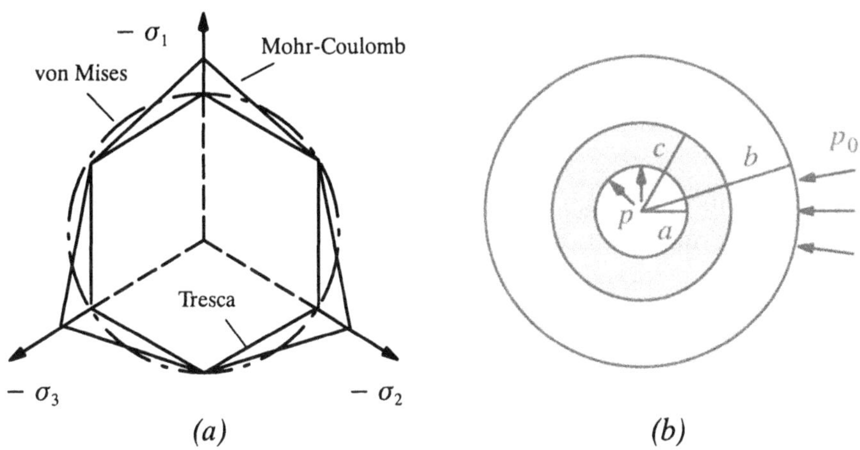

Figure 3.1: Yield surfaces and expansion of a spherical cavity

$$\sigma_r = -p_0 - (p - p_0)\frac{(\frac{b_0}{r})^3 - 1}{(\frac{b_0}{a_0})^3 - 1} \tag{3.1}$$

$$\sigma_\theta = \sigma_\phi = -p_0 + (p - p_0)\frac{\frac{1}{2}(\frac{b_0}{r})^3 - 1}{(\frac{b_0}{a_0})^3 - 1} \tag{3.2}$$

The radial displacement is

$$u = r - r_0 = \frac{(p - p_0)}{E} \frac{(1\text{–}2\nu)r + \frac{(1+\nu)b_0^3}{2r^2}}{(\frac{b_0}{a_0})^3\text{–}1} \tag{3.3}$$

where E is Young's modulus and ν is Poisson's ratio.

The Tresca yield criterion is expressed by the maximum and minimum principal stresses σ_1 and σ_3 as follows:

$$\sigma_1\text{–}\sigma_3 = Y \tag{3.4}$$

where $Y = 2s_u$ and s_u is undrained shear strength.

It is easily seen that the yield condition (3.4) is first satisfied at the internal boundary. Substituting the elastic stress solutions (3.1) and (3.2) into equation (3.4) and knowing that $\sigma_1 = \sigma_\theta$ and $\sigma_3 = \sigma_r$, the internal pressure required to cause yielding at the internal boundary is:

$$p = p_{1y} = p_0 + \frac{2Y}{3}[1\text{-}(\frac{a_0}{b_0})^3] \tag{3.5}$$

The displacements at both internal and external boundaries when the internal boundary is deformed plastically are:

$$u|_{r=a_0} = \frac{Ya_0}{E}[\frac{2(1\text{–}2\nu)a_0^3}{3b_0^3} + \frac{1+\nu}{3}] \tag{3.6}$$

$$u|_{r=b_0} = \frac{Y(1\text{–}\nu)a_0}{Eb_0^2} \tag{3.7}$$

If the internal pressure is further increased, a plastic region will spread into the sphere; the radius of the plastic region at any instant is denoted by c. In the outer elastic region, the stresses are still of the form:

$$\sigma_r = \text{–}A[(\frac{b_0}{r})^3\text{–}1] - p_0 \tag{3.8}$$

$$\sigma_\theta = \sigma_\phi = A[\frac{1}{2}(\frac{b_0}{r})^3 + 1] - p_0 \tag{3.9}$$

where A is a constant that can be determined by the condition that the material just on the elastic side of the plastic boundary must be on the point of yielding. Substituting equations (3.8) and (3.9) into equation (3.4), A is found to be:

$$A = \frac{2Yc^3}{3b_0^3} \tag{3.10}$$

The elastic stress distribution is therefore obtained by substituting (3.10) into (3.8) and (3.9):

$$\sigma_r = -\frac{2Yc^3}{3b_0^3}[(\frac{b_0}{r})^3-1] - p_0 \tag{3.11}$$

$$\sigma_\theta = \sigma_\phi = \frac{2Yc^3}{3b_0^3}[\frac{1}{2}(\frac{b_0}{r})^3 + 1] - p_0 \tag{3.12}$$

The displacement in the elastic region is given by:

$$u = \frac{2Yc^3}{3Eb_0^3}[(1-2\nu)r + \frac{(1+\nu)b_0^3}{2r^2}] \tag{3.13}$$

From equations (3.11) to (3.13), it is seen that the solution in the elastic region is dependent only on the radius of elastic-plastic boundary c. In the plastic region, we have the equilibrium equation:

$$\frac{\partial \sigma_r}{\partial r} = \frac{2(\sigma_\theta - \sigma_r)}{r} \tag{3.14}$$

Substituting the yield condition (3.4) into the equilibrium equation (3.14) results in:

$$\sigma_r = 2Y \ln r + B \tag{3.15}$$

where B is another constant that can be determined using the condition that the radial stress must be continuous across the elastic-plastic boundary. Equating (3.11) and (3.15) at $r = c$ leads to:

$$B = -2Y \ln c - \frac{2Y}{3}[1-(\frac{c}{b_0})^3] - p_0 \tag{3.16}$$

By substituting (3.16) into (3.15) and (3.4), the following solution for stresses in the plastic region is obtained:

$$\sigma_r = -2Y \ln(\frac{c}{r}) - \frac{2Y}{3}[1-(\frac{c}{b_0})^3] - p_0 \tag{3.17}$$

$$\sigma_\theta = Y - 2Y\ln(\frac{c}{r}) - \frac{2Y}{3}[1-(\frac{c}{b_0})^3] - p_0 \tag{3.18}$$

By substituting $r = a$ into equation (3.17), the internal pressure needed to produce plastic flow to a radius c is found:

$$p = 2Y\ln(\frac{c}{a}) + \frac{2Y}{3}[1-(\frac{c}{b_0})^3] + p_0 \tag{3.19}$$

(b) Displacement analysis

When calculating the displacement of any individual particle it is convenient to take the movement of the plastic boundary as the scale of 'time' or progress of the expansion, since the parameter c appears in the formulae for the stresses. We consider the velocity V of a particle, meaning that the particle is displaced by an amount Vdc when the plastic boundary moves outwards a further distance dc. V can be expressed directly in terms of the total displacement u, which is a function of both the current radius r and plastic radius c so that:

$$du = \frac{\partial u}{\partial c}dc + \frac{\partial u}{\partial r}dr = (\frac{\partial u}{\partial c} + V\frac{\partial u}{\partial r})dc \tag{3.20}$$

where r and c are taken as two independent variables. Equating the above equation to Vdc we obtain the expression for the particle velocity:

$$V = \frac{\frac{\partial u}{\partial c}}{1-\frac{\partial u}{\partial r}} \tag{3.21}$$

Now the compressibility equation in the plastic region is:

$$d\varepsilon_r + d\varepsilon_\theta + d\varepsilon_\phi = \frac{1-2\nu}{E}(d\sigma_r + d\sigma_\theta + d\sigma_\phi) \tag{3.22}$$

To evaluate the increments of stress and strain we must follow a given element. Thus:

$$d\varepsilon_r = \frac{\partial(du)}{\partial r} = \frac{\partial V}{\partial r}dc \tag{3.23}$$

$$d\varepsilon_\theta = d\varepsilon_\phi = \frac{du}{r} = \frac{Vdc}{r} \tag{3.24}$$

$$d\sigma_r = (\frac{\partial \sigma_r}{\partial c} + V\frac{\partial \sigma_r}{\partial r})dc \tag{3.25}$$

$$d\sigma_\theta = d\sigma_\phi = (\frac{\partial \sigma_\theta}{\partial c} + V\frac{\partial \sigma_\theta}{\partial r})dc \tag{3.26}$$

Hence the compressibility condition is written as follows:

$$\frac{\partial V}{\partial r} + \frac{2V}{r} = \frac{1-2\nu}{E}(\frac{\partial}{\partial c} + V\frac{\partial}{\partial r})(\sigma_r + 2\sigma_\theta) \tag{3.27}$$

Substituting the expressions for stresses in the plastic region (3.17) and (3.18) into (3.27) leads to:

$$\frac{\partial V}{\partial r} + \frac{2V}{r} = 6(1-2\nu)\frac{Y}{E}[\frac{V}{r} - \frac{1}{c}(1-\frac{c^3}{b_0^3})] \tag{3.28}$$

It is noted that the velocity is known on the plastic boundary from the solution for the displacement in the elastic region. Thus from (3.13) and (3.21), we obtain:

$$V_{r=c} = \frac{Y}{E}[2(1-2\nu)\frac{c^3}{b_0^3} + (1 + \nu)] \tag{3.29}$$

With the above boundary condition, equation (3.28) can be integrated to give the following solution for velocity V:

$$V = \frac{3(1-\nu)Yc^2}{Er^2} - \frac{2(1-2\nu)Y}{E}(1-\frac{c^3}{b_0^3})\frac{r}{c} \tag{3.30}$$

For the cavity wall $r=a$, we have $V=da/dc$, so that:

$$\frac{da}{dc} = \frac{3(1-\nu)Yc^2}{Ea^2} - \frac{2(1-2\nu)Y}{E}(1-\frac{c^3}{b_0^3})\frac{a}{c} \tag{3.31}$$

After integration, we can express the cavity expansion in terms of the radius of plastic boundary:

$$(\frac{a}{a_0})^3 = 1 + \frac{3(1-\nu)Yc^3}{Ea_0^3} - \frac{2(1-2\nu)Y}{E} \times [3\ln(\frac{c}{a_0}) + 1-(\frac{c}{b_0})^3] \tag{3.32}$$

(c) Solutions for cavity expansion in an infinite medium

For the special case of a spherical cavity expanding from zero radius in an infinite medium, the stresses are functions of r/a only, and the ratio of plastic radius to the current cavity radius remains constant. As a result, equation (3.31) can be directly used to give the plastic radius:

$$\frac{c}{a} = [\frac{E}{3(1-\nu)Y}]^{1/3} \tag{3.33}$$

Substituting the above solution into equation (3.19) leads to the following solution for the constant internal cavity pressure:

$$p_{\lim} = \frac{2Y}{3}[1 + \ln(\frac{E}{3(1-\nu)Y})] + p_0 \tag{3.34}$$

3.2.2 Expansion of a cylindrical cavity in a finite medium

(a) Stress analysis

Consider a long, plane strain cylindrical tube expanded by an internal pressure. As for the expansion of a spherical cavity, let a and b be the current internal and external radii (initial values are a_0 and b_0), and let c be the radius of the plastic boundary. At present, no restrictions are placed on the amount of expansion. As the internal pressure increases from the initial value p_0, the material will first behave elastically and the general radial displacement, measured from the initial hydrostatic state, is in the form:

$$du = Ar + \frac{B}{r} \tag{3.35}$$

The elastic solutions for stress changes from the initial hydrostatic state in cylindrical coordinates (r, θ, z) are:

$$d\sigma_r = \frac{E}{(1+\nu)(1-2\nu)}[A-(1-2\nu)\frac{B}{r^2}] \tag{3.36}$$

$$d\sigma_\theta = \frac{E}{(1+\nu)(1-2\nu)}[A+(1-2\nu)\frac{B}{r^2}] \tag{3.37}$$

$$d\sigma_z = \nu(d\sigma_r + d\sigma_\theta) \tag{3.38}$$

where constants A and B are determined by the stress boundary conditions $\sigma_r = -p$ on $r = a_0$ and $\sigma_r = -p_0$ on $r = b_0$:

$$A = \frac{(1+\nu)(1-2\nu)(p-p_0)}{E(b_0^2/a_0^2-1)} \tag{3.39}$$

$$B = \frac{(1+\nu)b_0^2(p-p_0)}{E(b_0^2/a_0^2-1)} \tag{3.40}$$

The final expressions for the stresses are then :

$$\sigma_r = -p_0-(p-p_0)(\frac{b_0^2}{r^2}-1)/(\frac{b_0^2}{a_0^2}-1) \tag{3.41}$$

$$\sigma_\theta = -p_0+(p-p_0)(\frac{b_0^2}{r^2}+1)/(\frac{b_0^2}{a_0^2}-1) \tag{3.42}$$

The displacement is:

$$u = \frac{(1+\nu)(p-p_0)}{E(\frac{b_0^2}{a_0^2}-1)}[(1-2\nu)r+\frac{b_0^2}{r}] \tag{3.43}$$

Yielding begins to occur on the internal surface when the internal cavity pressure is:

$$p = p_{1y} = \frac{Y}{2}(1-\frac{a_0^2}{b_0^2})+p_0 \tag{3.44}$$

When the internal pressure exceeds the above value the cylinder will become partially plastic. The stresses in the elastic region are still of the form:

$$\sigma_r = -D(\frac{b_0^2}{r^2}-1)-p_0 \tag{3.45}$$

$$\sigma_\theta = D(\frac{b_0^2}{r^2}+1)-p_0 \tag{3.46}$$

Using the condition that the material immediately on the elastic side of the plastic boundary $r{=}c$ is on the point of yielding, the constant D is determined as:

$$D = \frac{Yc^2}{2b_0^2} \tag{3.47}$$

Therefore the stresses in the elastic region are:

$$\sigma_r = -\frac{Yc^2}{2b_0^2}(\frac{b_0^2}{r^2}-1) - p_0 \tag{3.48}$$

$$\sigma_\theta = \frac{Yc^2}{2b_0^2}(\frac{b_0^2}{r^2} + 1) - p_0 \tag{3.49}$$

The radial displacement in the elastic region is:

$$u = \frac{(1 + \nu)Yc^2}{2Eb_0^2}[(1-2\nu)r + \frac{b_0^2}{r}] \tag{3.50}$$

In the plastic region the equation of equilibrium combined with the yield condition leads to:

$$\frac{\partial\sigma_r}{\partial r} = \frac{(\sigma_\theta-\sigma_r)}{r} = \frac{Y}{r} \tag{3.51}$$

which can be used to give the following stresses in the plastic zone:

$$\sigma_r = - p_0-\frac{1}{2}Y-Y\ln(\frac{c}{r}) + \frac{c^2Y}{2b_0^2} \tag{3.52}$$

$$\sigma_\theta = - p_0 + \frac{1}{2}Y-Y\ln(\frac{c}{r}) + \frac{c^2Y}{2b_0^2} \tag{3.53}$$

The internal cavity pressure during the elastic-plastic stage of cavity expansion is given by applying equation (3.52) at the cavity wall:

$$\frac{p - p_0}{Y} = \frac{1}{2} + \ln(\frac{c}{a})-\frac{c^2}{2b_0^2} \tag{3.54}$$

(b) Displacement analysis

The compressibility equation in the plastic region for a plane strain cylindrical cavity is:

$$d\varepsilon_r + d\varepsilon_\theta = \frac{(1 + \nu)(1-2\nu)}{E}(d\sigma_r + d\sigma_\theta) \tag{3.55}$$

To evaluate the increments of stress and strain we must follow a given particle by treating r and c as two independent variables and therefore:

$$d\varepsilon_r = \frac{\partial(du)}{\partial r} = \frac{\partial V}{\partial r}dc$$

$$d\varepsilon_\theta = \frac{du}{r} = \frac{Vdc}{r}$$

$$d\sigma_r = (\frac{\partial\sigma_r}{\partial c} + V\frac{\partial\sigma_r}{\partial r})dc$$

$$d\sigma_\theta = (\frac{\partial\sigma_\theta}{\partial c} + V\frac{\partial\sigma_\theta}{\partial r})dc$$

Hence the compressibility condition can be written as follows:

$$\frac{\partial V}{\partial r} + \frac{V}{r} = \frac{(1+\nu)(1-2\nu)}{E}(\frac{\partial}{\partial c} + V\frac{\partial}{\partial r})(\sigma_r + \sigma_\theta) \tag{3.56}$$

Substituting the expressions for stresses in the plastic region (3.52) and (3.53) into (3.56) leads to:

$$\frac{\partial V}{\partial r} + \frac{V}{r} = \frac{(1+\nu)(1-2\nu)Y}{E}[\frac{2c}{b_0^2} - \frac{2}{c} + \frac{2V}{r}] \tag{3.57}$$

It is noted that the velocity is known on the plastic boundary from the solution for the displacement in the elastic region. Thus from (3.21) and (3.50), we obtain

$$V_{r=c} = \frac{Y}{E}[(1+\nu)(1-2\nu)\frac{c^2}{b_0^2} + (1+\nu)] \tag{3.58}$$

With the above boundary condition, equation (3.57) can be integrated to give the following solution for velocity V:

$$V = [(m - \frac{m}{1-m})\frac{c^2}{b_0^2} + \frac{m}{1-m} + \frac{(1+\nu)Y}{E}](\frac{r}{c})^{2m-1}$$
$$+ \frac{m}{1-m}(\frac{c^2}{b_0^2} - 1)\frac{r}{c} \tag{3.59}$$

where $m = (1+\nu)(1-2\nu)Y/E$.

At the cavity wall $r=a$, $V=da/dc$, so that:

$$\frac{da}{dc} = [(m - \frac{m}{1-m})\frac{c^2}{b_0^2} + \frac{m}{1-m} + \frac{(1+\nu)Y}{E}](\frac{a}{c})^{2m-1}$$

$$+\frac{m}{1-m}\left(\frac{c^2}{b_0^2}-1\right)\frac{a}{c} \tag{3.60}$$

Without further assumptions, the above equation cannot be integrated analytically. However, as will be shown in the next section for some special cases the displacement equation (3.60) can be integrated to give closed form solutions.

(c) Solutions for cavity expansion in an infinite medium

Cavity expansion from zero initial radius

For the special case of a cylindrical cavity expanding from zero radius in an infinite medium, the stresses are functions of *r/a* only and the ratio of plastic radius to the current cavity radius remains constant. Hence:

$$\frac{da}{dc} = \frac{a}{c} \tag{3.61}$$

As a result, equation (3.60) can be directly used to give the plastic radius:

$$\frac{c}{a} = \left[\frac{E}{Em + (1 - m)(1 + \nu)Y}\right]^{\frac{1}{2(1-m)}} \tag{3.62}$$

For the incompressible undrained case, Poisson's ratio is 0.5 and therefore m=0, equation (3.62) reduces to:

$$\frac{c}{a} = \left[\frac{E}{(1 + \nu)Y}\right]^{1/2} = \left(\frac{G}{s_u}\right)^{1/2} \tag{3.63}$$

Substituting the above solution into equation (3.54) leads to the following well known solution for the constant internal pressure for undrained cavity expansion:

$$p_{\lim} = s_u\left[1 + \ln\left(\frac{G}{s_u}\right)\right] + p_0 \tag{3.64}$$

Finite cavity expansion in an infinite soil mass

For the special case when a cylindrical cavity is expanded from a finite radius in an infinite medium, it is also possible to derive a closed form solution. By using $1/b_0 = 0$, equation (3.60) can be integrated to give:

$$\left(\frac{c}{a_0}\right)^{2(1+m)} = \frac{n-1-m}{n-(1 + m)\left(\frac{a}{c}\right)^2} \tag{3.65}$$

where $n = 2(1 - \nu^2)Y/E$.

By combining equations (3.65) and (3.54), the cavity expansion curve can be constructed.

For undrained clays, it is often assumed that Poisson's ratio is equal to 0.5. In this case, equation (3.65) reduces to:

$$(\frac{c}{a})^2 = (\frac{a_0}{a})^2 + \frac{1}{n}[1 - (\frac{a_0}{a})^2] \tag{3.66}$$

Substituting the above equation into equation (3.54) leads to the following cavity expansion relationship:

$$\frac{p - p_0}{Y} = \frac{1}{2} + \frac{1}{2}\ln[\frac{G}{s_u}[1 - (\frac{a_0}{a})^2] + (\frac{a_0}{a})^2] \tag{3.67}$$

where G is shear modulus and s_u is the undrained shear strength. It is noted that for the special case of a cylindrical cavity in an infinite, incompressible undrained clay, the present solution (3.67) is identical to the solution of Gibson and Anderson (1961).

3.2.3 Contraction of cavities in an infinite medium

The solutions presented so far consider the case in which the internal cavity pressure gradually increases. This section presents some fundamental solutions for the contraction of cavities in soils using the Tresca yield criterion. For simplicity, attention is focused on the unloading of cavities in an infinite medium.

(a) Contraction from an *in-situ* stress state

It is assumed that an unbound Tresca medium contains a single cylindrical or spherical cavity. Initially the radius of the cavity is a_0 and a hydrostatic pressure p_0 acts throughout the soil, which is assumed to be homogeneous. The pressure inside of cavity is then decreased sufficiently slowly for dynamic effects to be negligible. This section is concerned with the distribution of stress and displacement in the soil as the cavity pressure reduces from its initial value. For simplicity, the symbol k is used to denote cylindrical (*k=1*) and spherical cavities (*k*=2) so that the solutions for both cylindrical and spherical cavities can be presented together.

Elastic response and initial yielding

As the internal cavity pressure *p* decreases from its *in-situ* value p_0, the deformation of the soil at first is purely elastic. Under conditions of radial symmetry the elastic stress-strain relationship can be expressed as:

$$\dot{\varepsilon}_r = \frac{\partial \dot{u}}{\partial r} = \frac{1}{M}[\dot{\sigma}_r - \frac{k\nu}{1-\nu(2-k)}\dot{\sigma}_\theta] \tag{3.68}$$

$$\dot{\varepsilon}_\theta = \frac{\dot{u}}{r} = \frac{1}{M}[-\frac{\nu}{1-\nu(2-k)}\dot{\sigma}_r + [1-\nu(k-1)]\dot{\sigma}_\theta] \tag{3.69}$$

where $M = E/(1 - \nu^2(2 - k))$.

The elastic solutions for stresses and displacement are as follows:

$$\sigma_r = -p_0-(p-p_0)(\frac{a}{r})^{1+k} \tag{3.70}$$

$$\sigma_\theta = -p_0 + \frac{p-p_0}{k}(\frac{a}{r})^{1+k} \tag{3.71}$$

$$u = \frac{p-p_0}{2kG}(\frac{a}{r})^{1+k}r \tag{3.72}$$

Since a tension positive notation is used throughout this chapter, the yield equation takes the form:

$$\sigma_r-\sigma_\theta = Y = 2s_u \tag{3.73}$$

As the internal pressure decreases further, initial yielding occurs at the cavity wall when the condition

$$p = p_{1y} = p_0-\frac{kY}{1 + k} \tag{3.74}$$

is satisfied.

Elastic-plastic stress analysis

After initial yielding at the cavity wall a plastic zone within the region $a \leq r \leq c$ forms around the inner wall of the cavity with the decrease of the cavity pressure p.

The stresses in the elastic zone can be shown to be as follows:

$$\sigma_r = -p_0-Br^{-(1+k)} \tag{3.75}$$

$$\sigma_\theta = -p_0 + \frac{B}{k}r^{-(1+k)} \tag{3.76}$$

On the other hand, the stresses in the plastic zone must satisfy equilibrium and yield condition and can be shown as:

$$\sigma_r = A-kY\ln r \tag{3.77}$$

$$\sigma_\theta = A - kY \ln r - Y \tag{3.78}$$

The continuity of stress components at the elastic-plastic interface can be used to determine the constants A and B in terms of the plastic radius c:

$$A = -p_0 + kY \ln c + \frac{kY}{1+k} \tag{3.79}$$

$$B = -\frac{kY}{1+k} c^{1+k} \tag{3.80}$$

Substituting equation (3.79) into (3.77) and applying it at the cavity wall leads to a relationship between the cavity pressure p and the plastic radius c:

$$\frac{p_0 - p}{Y} = k \ln \frac{c}{a} + \frac{k}{1+k} \tag{3.81}$$

Elastic-plastic displacement analysis

The results obtained above can not be used to calculate the distribution of stresses until the displacement field is known. The displacement in the elastic zone is shown to be:

$$u = \frac{p_{1y} - p_0}{2kG} \left(\frac{c}{r}\right)^{1+k} r = -\frac{Y}{2(1+k)G} \left(\frac{c}{r}\right)^{1+k} r \tag{3.82}$$

Therefore, at the elastic-plastic interface the displacement is:

$$u|_{r=c} = c - c_0 = -\frac{Yc}{2(1+k)G} \tag{3.83}$$

For incompressible undrained clays, it is assumed that Poisson's ratio is 0.5, which leads to the following incompressibility condition:

$$r_0^{1+k} - r^{1+k} = c_0^{1+k} - c^{1+k} \tag{3.84}$$

where r_0 and r denote the radii of a soil particle before and after cavity contraction.

By using the elastic displacement solution at the plastic boundary (3.83) and the radius of the reverse plastic zone (3.81), the incompressibility condition (3.84) can be used to give the following displacement field in the plastic region:

$$\left(\frac{r_0}{r}\right)^{1+k} = 1 + \exp\left[\frac{(1+k)(p_0 - p)}{kY} - 1\right]$$

$$\times [(1 + \frac{Y}{2(1+k)G})^{1+k} - 1](\frac{a}{r})^{1+k} \tag{3.85}$$

For the cavity wall $r = a$ and $r_0 = a_0$, equation (3.85) reduces to the following relationship between cavity pressure and cavity displacement:

$$(\frac{a_0}{a})^{1+k} = 1 + \exp[\frac{(1+k)(p_0-p)}{kY} - 1]$$

$$\times [(1 + \frac{Y}{2(1+k)G})^{1+k} - 1] \tag{3.86}$$

The solutions (3.85) and (3.86) are derived with no restriction on the magnitude of deformation and are therefore rigorous large strain solutions. However, if strains are assumed to be small in the plastic zone, the large strain solutions (3.85) and (3.86) can be simplified to:

$$\frac{u}{a} = -\frac{Y}{2(1+k)G}(\frac{c}{a})^{1+k}(\frac{a}{r})^k$$

$$= -\frac{Y}{2(1+k)G}(\frac{a}{r})^k \exp[\frac{(1+k)(p_0-p)}{kY} - 1] \tag{3.87}$$

for displacement at any radius, and

$$\frac{u_a}{a} = -\frac{Y}{2(1+k)G}\exp[\frac{(1+k)(p_0-p)}{kY} - 1] \tag{3.88}$$

for displacement at the cavity wall. The small strain solutions (3.87) and (3.88) are the same as those presented by Mair and Taylor (1993), but no derivation was given in their paper.

(b) Contraction from a limiting plastic state

It is assumed that in an infinite Tresca medium a single cylindrical or spherical cavity has been expanded from zero radius to the current radius a'. It is known from the loading solutions presented in previous sections that cavity expansion from zero radius takes place at a constant pressure $p_{\lim}$. This section presents the solution for the relationship between cavity pressure and cavity radius (or strain) when cavity pressure decreases from the limit value $p_{\lim}$. The solution presented here was first derived by Houlsby and Withers (1988) and Jefferies (1988).

Initial conditions and elastic unloading

Referring to Figure 3.2, at the beginning of unloading the radial and hoop stresses at the cavity wall are:

$$\sigma_r = -p_{\lim} \tag{3.89}$$

$$\sigma_\theta = -p_{\lim} + Y \tag{3.90}$$

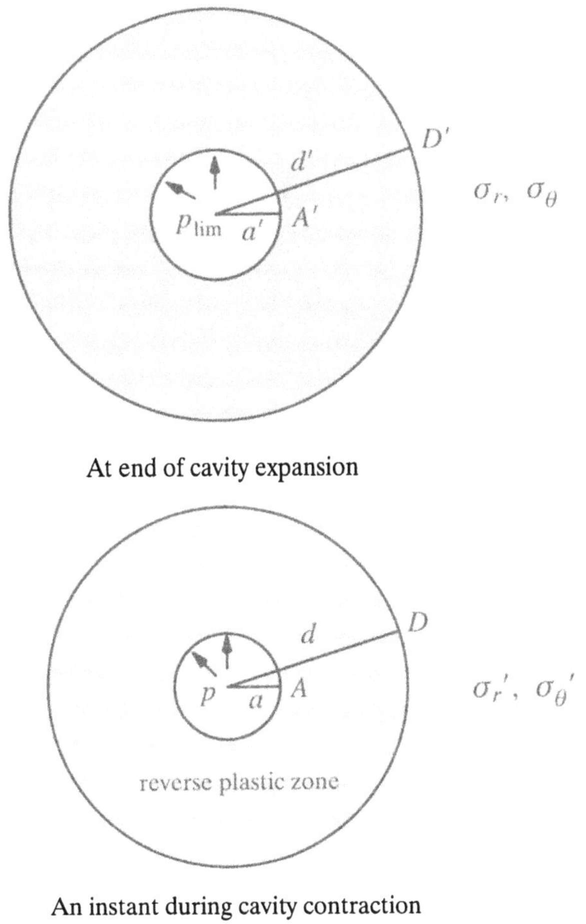

Figure 3.2: Contraction of cavities from a limiting plastic state

In the loading plastic zone, the following relations also hold (see Section 3.2):

$$\sigma_\theta - \sigma_r = Y \tag{3.91}$$

$$\sigma_{r1} - \sigma_{r2} = kY \ln \frac{r_1}{r_2} \tag{3.92}$$

By setting $\sigma_{r1} = -p_{\lim}$ for $r_1 = a'$, equation (3.92) reduces to:

$$\sigma_{r2} = -p_{\lim} - kY\ln\frac{a'}{r_2} \tag{3.93}$$

To distinguish the stresses between the stages of loading and unloading, the stresses during unloading are denoted by a prime. The initial phase of cavity contraction is elastic and the stresses after the cavity pressure is reduced by Δp are:

$$\sigma'_r = \sigma_r + \Delta p\left(\frac{a}{r}\right)^{1+k} \tag{3.94}$$

$$\sigma'_\theta = \sigma_\theta - \frac{\Delta p}{k}\left(\frac{a}{r}\right)^{1+k} \tag{3.95}$$

The elastic displacement caused by the application of Δp at the cavity wall is:

$$\Delta u = r - r' = -\frac{\Delta p}{2kG}\left(\frac{a}{r}\right)^{1+k} r \tag{3.96}$$

Initial yielding

The elastic stage ends when the stresses from (3.94) and (3.95) satisfy the reverse (unloading) yield condition which is:

$$\sigma'_r - \sigma'_\theta = Y \tag{3.97}$$

From equations (3.91), (3.94) and (3.95), it can be shown that the condition (3.97) is first satisfied at the cavity wall when the cavity pressure reduction Δp is:

$$\Delta p = p_{\lim} - p = \frac{2kY}{1+k} \tag{3.98}$$

At the initiation of reverse yielding, the hoop strain at the cavity wall is:

$$\Delta \varepsilon_\theta = \frac{\Delta u}{a} = 1 - \frac{a'}{a} = -\frac{Y}{(1+k)G} = -\frac{2}{(1+k)I_r} \tag{3.99}$$

where the quantity $I_r = G/s_u$ is known as soil rigidity index.

Elastic-plastic stress analysis

When the cavity pressure reduction exceeds the value defined in (3.98), a reverse (unloading) plastic zone will form around the cavity wall. It is assumed that the size of the reverse plastic region is denoted by an outer radius d which corresponds to the radius d' at the end of cavity expansion. At present, we assume that the outer radius of the reverse plastic zone is always smaller than that of the loading plastic zone. From this assumption, it follows that the initial stresses before cavity contraction (or end of cavity expansion) at every point in the reverse plastic zone can be determined by the elastic-plastic solutions in the loading plastic region. The validity of this assumption will be discussed in the next section when solutions for the Mohr-Coulomb criterion are considered.

Within the region of reverse plasticity, both equilibrium and yield conditions can be used to give the following relationship between radial stresses at two different radii:

$$\sigma'_{r1} - \sigma'_{r2} = -kY\ln\frac{r_1}{r_2} \tag{3.100}$$

By setting $\sigma'_{r1} = -p$ for $r_1 = a$, and $r_2 = d$, equation (3.100) reduces to:

$$\sigma'_d = -p + kY\ln\frac{a}{d} \tag{3.101}$$

At the end of cavity expansion (or beginning of cavity contraction), the radial stress at point D can be obtained from equation (3.93) by setting $r_2 = d'$:

$$\sigma_d = -p_{\lim} - kY\ln\frac{a'}{d'} \tag{3.102}$$

Also, at point D the soil is just becoming plastic due to unloading and therefore from equation (3.98), the following relationship between radial stresses before and after unloading at point D can be obtained:

$$\sigma'_d - \sigma_d = \frac{2kY}{1+k} \tag{3.103}$$

Combining equations (3.101), (3.102) and (3.103) gives the current internal cavity pressure:

$$p = p_{\lim} - \frac{2kY}{1+k} - kY(\ln\frac{d}{a} + \ln\frac{d'}{a'}) \tag{3.104}$$

Elastic-plastic displacement analysis

The incompressibility condition in the reverse plastic region gives:

$$d'^{1+k} - a'^{1+k} = d^{1+k} - a^{1+k} \tag{3.105}$$

which can also be written as:

$$(\frac{d'}{d})^{1+k} - 1 = (\frac{a}{d})^{1+k}[(\frac{a'}{a})^{1+k} - 1] \tag{3.106}$$

From equation (3.99), the change of hoop strain during unloading of point D from end of loading to plastic unloading results in:

$$\ln\frac{d'}{d} = \frac{Y}{(1+k)G} = \frac{2}{(1+k)I_r} \tag{3.107}$$

By definition, the cavity strain reduction should be:

$$\varepsilon_{\max} - \varepsilon = \ln\frac{a'}{a} \tag{3.108}$$

Substitution of equations (3.106), (3.107) and (3.108) into (3.104) provides the complete plastic cavity contraction curve:

$$p = p_{\lim} - \frac{2kY}{1+k}$$

$$\times \left[1 + \ln\left(\sinh\frac{1+k}{2}(\varepsilon_{\max} - \varepsilon)\right) - \ln\left(\sinh\frac{Y}{2G}\right)\right] \tag{3.109}$$

3.3 SOLUTIONS FOR MOHR-COULOMB CRITERION

In this section, analytical solutions for large strain expansion of cavities in Mohr-Coulomb materials are presented, following the same sequence as that adopted in the previous section for the Tresca criterion.

3.3.1 Expansion of a spherical cavity in a finite medium

Initially the radii of inner and outer boundaries of a hollow sphere are a_0 and b_0 respectively. A hydrostatic pressure p_0 acts throughout the soil which is assumed to be homogeneous. This subsection is concerned with stress and strain fields around the cavity when the internal pressure p increases slowly from its initial value p_0. The solution presented was derived by Yu (1993).

The soil is modelled as an isotropic dilatant elastic-perfectly plastic material. It behaves elastically and obeys Hooke's law until the onset of yielding which is determined by the Mohr-Coulomb yield criterion. If the principal stress components satisfy the inequalities $\sigma_i \le \sigma_j \le \sigma_k$, the Mohr-Coulomb yield function takes the form:

$$a\sigma_k - \sigma_i = Y \tag{3.110}$$

where $a = (1 + \sin\phi)/(1 - \sin\phi)$ and for the Mohr-Coulomb criterion the quantity Y takes the form $Y = 2C\cos\phi/(1 - \sin\phi)$ in which C and ϕ are soil cohesion and friction angle. It is noted that when the friction angle is zero, the Mohr-Coulomb yield function reduces to the Tresca criterion and the definition for Y also reduces to the definition used before in the Tresca solutions.

At any time in any position in the soil around the cavity with inner and outer radii of a and b, the stresses must satisfy the following equation of equilibrium:

$$2(\sigma_\theta - \sigma_r) = r\frac{\partial \sigma_r}{\partial r} \tag{3.111}$$

which is subject to two boundary conditions:

$$\sigma_r|_{r=a} = -p \tag{3.112}$$

$$\sigma_r|_{r=b} = -p_0 \tag{3.113}$$

(a) Elastic solution

As the cavity pressure increases from its initial value, the deformation of the soil is at first purely elastic. Under conditions of radial symmetry the elastic stress-strain relationship may be expressed as:

$$\dot{\varepsilon}_r = \frac{\partial \dot{u}}{\partial r} = \frac{1}{E}[\dot{\sigma}_r - 2\nu\dot{\sigma}_\theta] \tag{3.114}$$

$$\dot{\varepsilon}_\theta = \frac{\dot{u}}{r} = \frac{1}{E}[-\nu\dot{\sigma}_r + (1-\nu)\dot{\sigma}_\theta] \tag{3.115}$$

The solution of equations (3.111), (3.114) and (3.115), subject to the boundary conditions (3.112) and (3.113), can be shown to be:

$$\sigma_r = -p_0 + (p-p_0)[\frac{1}{(\frac{b}{a})^3-1} - \frac{1}{(\frac{r}{a})^3-(\frac{r}{b})^3}] \tag{3.116}$$

$$\sigma_\theta = -p_0 + (p-p_0)[\frac{1}{(\frac{b}{a})^3-1} + \frac{1/2}{(\frac{r}{a})^3-(\frac{r}{b})^3}] \tag{3.117}$$

$$u = r - r_0 = \frac{p-p_0}{2G(\frac{1}{a^3}-\frac{1}{b^3})}[\frac{1-2\nu}{(1+\nu)b^3}r + \frac{1}{2r^2}] \tag{3.118}$$

Using a tension positive notation, the Mohr-Coulomb yield condition during cavity expansion is:

$$\alpha\sigma_\theta - \sigma_r = Y \tag{3.119}$$

When the cavity pressure increases further, initial yielding occurs first at the cavity wall when the stresses satisfy the yield condition. This happens when the cavity pressure reaches the following value:

$$p = p_{1y} = p_0 + \frac{2(b^3 - a^3)[Y + (\alpha - 1)p_0]}{(2 + \alpha)b^3 + 2(\alpha - 1)a^3} \tag{3.120}$$

(b) Elastic-plastic stress analysis

After initial yielding occurs at the cavity wall, a plastic zone will form around the inner cavity wall with an increase in the applied pressure p. As is the case with the Tresca material, the outer radius of the plastic zone is denoted by c.

Stresses in the plastic region

The stress components, which satisfy the equilibrium equation (3.111) and the yield condition (3.119), are found to be:

$$\sigma_r = \frac{Y}{\alpha - 1} + Ar^{-\frac{2(\alpha-1)}{\alpha}} \tag{3.121}$$

$$\sigma_\theta = \frac{Y}{\alpha - 1} + \frac{A}{\alpha}r^{-\frac{2(\alpha-1)}{\alpha}} \tag{3.122}$$

where A is a constant of integration.

Stresses in the elastic region

The stress components in the elastic region can be obtained from equations (3.111)–(3.115) as follows:

$$\sigma_r = -p_0 + B\left(\frac{1}{b^3} - \frac{1}{r^3}\right) \tag{3.123}$$

$$\sigma_\theta = -p_0 + B\left(\frac{1}{b^3} + \frac{1}{2r^3}\right) \tag{3.124}$$

where B is the second constant of integration.

The continuity of stress components at the elastic-plastic interface can be used to determine the constants A and B:

$$A = -[Y + (\alpha - 1)p_0]$$

$$\times \left[\frac{1}{a-1} - \frac{(\frac{c}{b})^3 - 1}{(a-1)(\frac{c}{b})^3 + \frac{2+a}{2}}\right] c^{\frac{2(a-1)}{a}} \tag{3.125}$$

$$B = \frac{Y + (a-1)p_0}{\frac{a-1}{b^3} + \frac{2+a}{2c^3}} \tag{3.126}$$

Combining equations (3.112) and (3.121) gives the radius of elastic-plastic boundary:

$$\frac{c}{a} = \left\{ \frac{Y + (a-1)p}{(a-1)\left[\frac{1}{a-1} - \frac{(\frac{c}{b})^3 - 1}{(a-1)(\frac{c}{b})^3 + \frac{a+2}{2}}\right](Y + (a-1)p_0)} \right\}^{\frac{a}{2(a-1)}} \tag{3.127}$$

The entire sphere becomes plastic when the elastic-plastic interface reaches the outer boundary $c = b$. When this occurs, the internal cavity pressure can be obtained from equation (3.127) as follows:

$$p = \frac{Y + (a-1)p_0}{a-1}\left[\left(\frac{b}{a}\right)^{\frac{2(a-1)}{a}} - 1\right] + p_0 \tag{3.128}$$

(c) Elastic-plastic displacement analysis

The solution obtained so far for the stress field in the plastic zone cannot be used to calculate the distribution of stresses until the displacement field is known.

By substituting equations (3.123) and (3.124) into (3.115), the displacement in the elastic zone is shown to be:

$$u = r - r_0 = \frac{r}{\delta[(a-1) + \frac{2+a}{2}(\frac{b}{c})^3]}\left[1 - 2\nu + \frac{1+\nu}{2}\left(\frac{b}{r}\right)^3\right] \tag{3.129}$$

where $\delta = E/[Y + (a-1)p_0]$. In particular, for the outer boundary of the sphere we have:

$$u(b) = b - b_0 = \frac{3(1-\nu)b}{2\delta[a - 1 + \frac{2+a}{2}(\frac{b}{c})^3]} \tag{3.130}$$

The determination of displacement fields in the plastic zone requires the use of a plastic flow rule. It is assumed that while yielding occurs, the total strain is decomposed additively into elastic and plastic components. Indices e and p are used to distinguish the elastic and plastic components of the total strains. Following Davis (1969), the soil is assumed to dilate plastically at a constant rate.

For the expansion of a hollow sphere, the non-associated flow rule may be expressed as:

$$\frac{\dot{\varepsilon}_r^p}{\dot{\varepsilon}_\theta^p} = \frac{\dot{\varepsilon}_r - \dot{\varepsilon}_r^e}{\dot{\varepsilon}_\theta - \dot{\varepsilon}_\theta^e} = -\frac{2}{\beta} \tag{3.131}$$

where $\beta = (1 + \sin\psi)/(1 - \sin\psi)$ and ψ is soil dilation angle.

Substituting the elastic strain solutions (3.114) and (3.115) into the plastic flow rule (3.131) results in:

$$\beta\dot{\varepsilon}_r + 2\dot{\varepsilon}_\theta = \frac{1}{E}[(\beta-2\nu)\dot{\sigma}_r + (2-2\nu-2\beta\nu)\dot{\sigma}_\theta] \tag{3.132}$$

The distribution of stress and strain in the soil at the initiation of plastic yielding is obtained from equations (3.114) to (3.117) by putting $p = p_{1y}$ defined in equation (3.120). Subject to this initial condition, equation (3.132) can be integrated to give the following total stress-strain relation:

$$\beta\varepsilon_r + 2\varepsilon_\theta = \frac{1}{E}[(\beta-2\nu)\sigma_r + (2-2\nu-2\beta\nu)\sigma_\theta + (\beta + 2-4\nu-2\beta\nu)p_0] \tag{3.133}$$

Of the many definitions of large strain, the logarithmic (or Hencky) definition is particularly attractive because of its simple physical interpretation for problems with no rotation. To account for the effect of large strain in the plastic zone, we follow Chadwick (1959) in adopting the logarithmic strain, namely:

$$\varepsilon_r = \ln(\frac{dr}{dr_0}) \tag{3.134}$$

$$\varepsilon_\theta = \ln\frac{r}{r_0} \tag{3.135}$$

Substituting equations (3.134)–(3.135) and (3.121)–(3.122) into (3.133) leads to:

$$\ln[(\frac{r}{r_0})^{\frac{2}{\beta}}\frac{dr}{dr_0}] = \ln\eta - \omega(\frac{c}{r})^{\frac{2(\alpha-1)}{\alpha}} \tag{3.136}$$

where

$$\eta = \exp\left\{\frac{(\beta + 2)(1-2\nu)}{\delta(\alpha-1)\beta}\right\} \tag{3.137}$$

$$\omega = \frac{1}{\delta\beta}(\beta-2\nu + \frac{2-2\nu-2\beta\nu}{\alpha})$$

$$\times \left[\frac{1}{\alpha-1} - \frac{(\frac{c}{b})^3-1}{(\alpha-1)(\frac{c}{b})^3 + \frac{2+\alpha}{2}}\right] \tag{3.138}$$

By means of the following transformation:

$$\theta = \omega(\frac{c}{r})^{\frac{2(\alpha-1)}{\alpha}} \tag{3.139}$$

$$\xi = (\frac{r_0}{c})^{\frac{\beta+2}{\beta}} \tag{3.140}$$

and use of equation (3.129) for $r = c$, equation (3.136) can be integrated over the interval [r, c], leading to:

$$\frac{\eta}{\gamma}\left\{[1-\frac{g}{\delta}]^{\frac{\beta+2}{\beta}} - (\frac{r_0}{c})^{\frac{\beta+2}{\beta}}\right\} = -\omega^{\gamma}\int_{\theta}^{\omega} e^{\theta}\theta^{-\gamma-1}d\theta \tag{3.141}$$

where $\gamma = \dfrac{\alpha(2 + \beta)}{2(\alpha - 1)\beta}$ and g is a function of b/c defined by:

$$g = \frac{1-2\nu}{\alpha-1 + \frac{2+\alpha}{2}(\frac{b}{c})^3} + \frac{(1 + \nu)/2}{(\alpha-1)(\frac{c}{b})^3 + \frac{2+\alpha}{2}} \tag{3.142}$$

By introducing the infinite series

$$e^{\theta} = \sum_{n=0}^{\infty}\frac{\theta^n}{n!} \tag{3.143}$$

and putting $r = a$ and $r_0 = a_0$ in equation (3.141), we have:

$$\frac{\eta}{\gamma}\left\{[1-\frac{g}{\delta}]^{\frac{\beta+2}{\beta}} - (\frac{a_0}{c})^{\frac{\beta+2}{\beta}}\right\} = \sum_{n=0}^{\infty} A_n^1 \tag{3.144}$$

in which A_n^1 is defined by:

$$A_n^1 = \begin{cases} \dfrac{2(\alpha-1)\omega^n}{\alpha n!}\ln\dfrac{c}{a} & \text{if } n = \gamma \\ \dfrac{\omega^n}{n!(n-\gamma)}[(\dfrac{c}{a})^{\frac{2(\alpha-1)}{\alpha}(n-\gamma)}-1] & \textit{otherwise} \end{cases} \tag{3.145}$$

(d) Fully plastic spheres

The displacement equations obtained above for a partially plastic sphere will no longer be valid after the entire sphere becomes plastic. This condition is defined by $c = b$. In this case, equation (3.133) derived from the plastic flow rule still holds. However it must be integrated over the interval [r,b], leading to:

$$\frac{\eta}{\gamma}\left\{(\frac{b_0}{b})^{\frac{\beta+2}{\beta}}-(\frac{r_0}{b})^{\frac{\beta+2}{\beta}}\right\} = -\omega^\gamma\int_\theta^\omega e^\theta\theta^{-\gamma-1}d\theta \tag{3.146}$$

where ω can be obtained from equation (3.138) by setting $c = b$ which gives:

$$\omega = \frac{1}{\delta\beta(\alpha-1)}(\beta-2\nu + \frac{2-2\nu-2\beta\nu}{\alpha}) \tag{3.147}$$

By putting $r = a$ and $r_0 = a_0$, we obtain the following displacement equation in the fully plastic sphere:

$$\frac{\eta}{\gamma}\left\{(\frac{b_0}{b})^{\frac{\beta+2}{\beta}}-(\frac{a_0}{b})^{\frac{\beta+2}{\beta}}\right\} = \sum_{n=0}^{\infty}A_n^2 \tag{3.148}$$

in which

$$A_n^2 = \begin{cases} \dfrac{2(\alpha-1)\omega^n}{\alpha n!}\ln\dfrac{b}{a} & \text{if } n = \gamma \\ \dfrac{\omega^n}{n!(n-\gamma)}[(\dfrac{b}{a})^{\frac{2(\alpha-1)}{\alpha}(n-\gamma)}-1] & \textit{otherwise} \end{cases} \tag{3.149}$$

Having noted that ω is a small quantity (an order of $1/E$), it can be proved easily that the series defined in equations (3.145) and (3.149) converge very rapidly for all realistic values of α and β. In general, only the first few terms in the series are sufficient for giving accurate results.

(e) Solution procedure

All the necessary equations for determining the complete pressure-expansion curve and the stress distributions have been derived. As the pressure-expansion relationship is not expressed in terms of a single equation, it is necessary to summarize the procedure which can be used to construct a complete pressure-expansion curve. Presented below is such a procedure:

(1) Choose input soil parameters $E, \nu, C, \phi, \psi, p_0$ and initial sphere geometry b_0/a_0.

(2) Calculate the derived parameters $G, Y, \alpha, \beta, \gamma, \delta$.

(3) If cavity pressure is less than p_{1y}, calculate the cavity radius from the small strain elastic solution (3.118).

(4) If cavity pressure is greater than p_{1y}, (3.127), (3.130) and (3.144) are needed:

(a) for a chosen value of c/b (less than 1 and greater than a_0/b_0), calculate c/a_0 and c/b_0 from equation (3.130), and c/a and a/a_0 from (3.144);

(b) calculate the pressure required to expand the sphere with a cavity expansion ratio of a/a_0 from equation (3.127).

(5) When $c/b = 1$, the entire hollow sphere becomes plastic and the pressure-expansion curve needs to be constructed from equations (3.128) and (3.148):

(a) choose a value of cavity pressure p which must be less than the value required to cause the whole sphere become plastic, and calculate b/a from equation (3.128);

(b) calculate $b/a_0, b/b_0$ and a/a_0 from equation (3.148).

Repetition of steps (1)–(4a,b) by varying c/b, and (5a,b) for varying cavity pressure p provides the data for the complete pressure-expansion curve. The distribution of stresses at any stage of expansion can be obtained from equations (3.121) to (3.124).

A simple computer code written in FORTRAN for performing the above calculation is available and readers of this book may request a copy of the code from the author.

3.3.2 Expansion of a cylindrical cavity in a finite medium

Initially the radii of inner and outer boundaries of a thick-walled cylinder are a_0, b_0 respectively and a hydrostatic pressure p_0 acts throughout the soil which is assumed to be homogeneous. This subsection is concerned with stress and strain fields around the cavity when the internal pressure p increases slowly from its initial value p_0. The solution presented here was developed by Yu (1992).

At any time in any position in the soil around the cavity with inner and outer radii of a and b, the stresses must satisfy the following equation of equilibrium:

$$(\sigma_\theta - \sigma_r) = r\frac{\partial \sigma_r}{\partial r} \tag{3.150}$$

which is subject to two boundary conditions (3.112) and (3.113).

(a) Elastic solution

For cylinder expansion, the elastic stress-strain relationship can be expressed as:

$$\dot{\varepsilon}_r = \frac{\partial \dot{u}}{\partial r} = \frac{1-\nu^2}{E}[\dot{\sigma}_r - \frac{\nu}{1-\nu}\dot{\sigma}_\theta] \tag{3.151}$$

$$\dot{\varepsilon}_\theta = \frac{\dot{u}}{r} = \frac{1-\nu^2}{E}[-\frac{\nu}{1-\nu}\dot{\sigma}_r + \dot{\sigma}_\theta] \tag{3.152}$$

The solution of equations (3.150)-(3.152), subject to the boundary conditions (3.112) and (3.113), is shown to be:

$$\sigma_r = -p_0 + (p-p_0)[\frac{1}{(\frac{b}{a})^2-1} - \frac{1}{(\frac{r}{a})^2-(\frac{r}{b})^2}] \tag{3.153}$$

$$\sigma_\theta = -p_0 + (p-p_0)[\frac{1}{(\frac{b}{a})^2-1} + \frac{1}{(\frac{r}{a})^2-(\frac{r}{b})^2}] \tag{3.154}$$

$$u = \frac{p-p_0}{2G(\frac{1}{a^2}-\frac{1}{b^2})}[\frac{1-2\nu}{b^2}r + \frac{1}{r}] \tag{3.155}$$

Like spherical cavity expansion, initial yielding occurs first at the inner wall of the cylinder when the stresses satisfy the yield condition (3.119) and that happens when the cavity pressure reaches the following value:

$$p = p_{1y} = p_0 + \frac{(b^2-a^2)[Y + (\alpha-1)p_0]}{(1+\alpha)b^2 + (\alpha-1)a^2} \tag{3.156}$$

(b) Elastic-plastic stress analysis

After initial yielding takes place at the cavity wall, a plastic zone will form around the inner cavity wall with an increase in the applied pressure p. The outer radius of the plastic zone is denoted by c.

Stresses in the plastic region

The stress components, which satisfy equilibrium (3.150) and the yield condition (3.119), are found to be:

$$\sigma_r = \frac{Y}{\alpha-1} + Ar^{-\frac{(\alpha-1)}{\alpha}} \tag{3.157}$$

$$\sigma_\theta = \frac{Y}{\alpha-1} + \frac{A}{\alpha} r^{-\frac{(\alpha-1)}{\alpha}} \tag{3.158}$$

where A is a constant of integration.

Stresses in the elastic region

The stress components in the elastic region can be obtained from the equilibrium equation and elastic stress-strain equations as follows:

$$\sigma_r = -p_0 + B\left(\frac{1}{b^2} - \frac{1}{r^2}\right) \tag{3.159}$$

$$\sigma_\theta = -p_0 + B\left(\frac{1}{b^2} + \frac{1}{r^2}\right) \tag{3.160}$$

where B is the second constant of integration.

The continuity of stress components at the elastic-plastic interface can be used to determine the constants A and B:

$$A = -[Y + (\alpha-1)p_0]$$

$$\times \left[\frac{1}{\alpha-1} - \frac{(\frac{c}{b})^2 - 1}{(\alpha-1)(\frac{c}{b})^2 + 1 + \alpha}\right] c^{\frac{(\alpha-1)}{\alpha}} \tag{3.161}$$

$$B = \frac{Y + (\alpha-1)p_0}{\frac{\alpha-1}{b^2} + \frac{1+\alpha}{c^2}} \tag{3.162}$$

Combining equations (3.112) and (3.157) gives the radius of elastic-plastic boundary:

$$\frac{c}{a} = \left\{ \frac{[(\frac{c}{b})^2 + 1 + \frac{2}{\alpha-1}](Y + (\alpha-1)p)}{(2 + \frac{2}{\alpha-1})(Y + (\alpha-1)p_0)} \right\}^{\frac{\alpha}{\alpha-1}} \tag{3.163}$$

The entire cylinder becomes plastic when the elastic-plastic interface reaches the outer boundary $c = b$. When this occurs, the internal applied cavity pressure can be obtained from equation (3.163):

$$p = \frac{Y + (\alpha-1)p_0}{\alpha-1}[(\frac{b}{a})^{\frac{\alpha-1}{\alpha}}-1] + p_0 \tag{3.164}$$

(c) Elastic-plastic displacement analysis

By substituting equations (3.159) and (3.160) into (3.152), the displacement in the elastic zone can be shown to be:

$$u = r-r_0 = \frac{1+\nu}{\delta}[\frac{1-2\nu}{\alpha-1+(1+\alpha)(\frac{b}{c})^2}r$$

$$+ \frac{1}{(\alpha-1)(\frac{r}{b})^2 + (1+\alpha)(\frac{r}{c})^2}r] \tag{3.165}$$

where $\delta = E/[Y + (\alpha - 1)p_0]$. In particular, for the outer boundary of the cylinder:

$$u(b) = b-b_0 = \frac{2(1-\nu^2)b}{\delta[\alpha-1+(1+\alpha)(\frac{b}{c})^2]} \tag{3.166}$$

The determination of the displacement field in the plastic zone requires the use of plastic flow rule. It is assumed that while yielding occurs, the total strain is decomposed additively into elastic and plastic components.

For the expansion of a cylinder, the non-associated Mohr-Coulomb flow rule can be expressed as:

$$\frac{\dot{\varepsilon}_r^p}{\dot{\varepsilon}_\theta^p} = \frac{\dot{\varepsilon}_r-\dot{\varepsilon}_r^e}{\dot{\varepsilon}_\theta-\dot{\varepsilon}_\theta^e} = -\frac{1}{\beta} \tag{3.167}$$

where $\beta = (1 + \sin\psi)/(1 - \sin\psi)$ and ψ is soil dilation angle.

Substituting the elastic strain solutions (3.151) and (3.152) into the plastic flow rule (3.167) results in:

$$\beta\dot{\varepsilon}_r + \dot{\varepsilon}_\theta = \frac{1-\nu^2}{E}[(\beta-\frac{\nu}{1-\nu})\dot{\sigma}_r + (1-\frac{\beta\nu}{1-\nu})\dot{\sigma}_\theta] \tag{3.168}$$

The distribution of stress and strain in the soil at the initiation of plastic yield is obtained from equations (3.151) to (3.155) by putting $p = p_{1y}$ defined in equation

(3.156). Subject to this initial condition, equation (3.168) can be integrated to give the following total stress-strain relationship:

$$\beta\varepsilon_r + \varepsilon_\theta = \frac{1-\nu^2}{E}[(\beta - \frac{\nu}{1-\nu})\sigma_r + (1 - \frac{\beta\nu}{1-\nu})\sigma_\theta$$

$$+ (\beta + 1 - \frac{\nu(1+\beta)}{1-\nu})p_0] \tag{3.169}$$

In order to account for the effect of large strain in the plastic zone, the logarithmic strains are adopted, namely:

$$\varepsilon_r = \ln(\frac{dr}{dr_0})$$

$$\varepsilon_\theta = \ln\frac{r}{r_0}$$

Substituting the above strain definitions and the stress equations in the plastic zone into (3.169) leads to:

$$\ln[(\frac{r}{r_0})^{\frac{1}{\beta}}\frac{dr}{dr_0}] = \ln\eta - \omega(\frac{c}{r})^{\frac{(\alpha-1)}{\alpha}} \tag{3.170}$$

where

$$\eta = \exp\left\{\frac{(\beta+1)(1-2\nu)(1+\nu)}{\delta(\alpha-1)\beta}\right\} \tag{3.171}$$

$$\omega = \frac{1+\nu}{\delta\beta}(\beta - \nu - \beta\nu + \frac{1-\nu-\beta\nu}{\alpha})$$

$$\times [\frac{1}{\alpha-1} - \frac{(\frac{c}{b})^2 - 1}{(\alpha-1)(\frac{c}{b})^2 + 1 + \alpha}] \tag{3.172}$$

By means of the following transformation:

$$\theta = \omega(\frac{c}{r})^{\frac{(\alpha-1)}{\alpha}} \tag{3.173}$$

$$\xi = (\frac{r_0}{c})^{\frac{\beta+1}{\beta}} \tag{3.174}$$

and use of equation (3.165) for $r = c$, equation (3.170) can be integrated over the interval [r, c], leading to:

$$\frac{\eta}{\gamma}\left\{[1-\frac{(1+\nu)g}{\delta}]^{\frac{\beta+1}{\beta}}-(\frac{r_0}{c})^{\frac{\beta+1}{\beta}}\right\} = -\omega^{\gamma}\int_{\theta}^{\omega} e^{\theta}\theta^{-\gamma-1}d\theta \tag{3.175}$$

where $\gamma = \dfrac{\alpha(1+\beta)}{(\alpha-1)\beta}$ and g is a function of b/c defined by:

$$g = \frac{1-2\nu}{\alpha-1+(1+\alpha)(\frac{b}{c})^2} + \frac{1}{(\alpha-1)(\frac{c}{b})^2+1+\alpha} \tag{3.176}$$

Introducing the infinite series

$$e^{\theta} = \sum_{n=0}^{\infty}\frac{\theta^n}{n!}$$

and putting $r = a$ and $r_0 = a_0$ in equation (3.175) results in:

$$\frac{\eta}{\gamma}\left\{[1-\frac{(1+\nu)g}{\delta}]^{\frac{\beta+1}{\beta}}-(\frac{a_0}{c})^{\frac{\beta+1}{\beta}}\right\} = \sum_{n=0}^{\infty}A_n^1 \tag{3.177}$$

in which A_n^1 is defined by:

$$A_n^1 = \begin{cases} \dfrac{(\alpha-1)\omega^n}{\alpha n!}\ln\dfrac{c}{a} & \text{if } n = \gamma \\ \dfrac{\omega^n}{n!(n-\gamma)}[(\dfrac{c}{a})^{\frac{(\alpha-1)}{\alpha}(n-\gamma)}-1] & \textit{otherwise} \end{cases} \tag{3.178}$$

(d) Fully plastic cylinders

The displacement equations obtained above for a partially plastic cylinder are no longer valid after the entire cylinder becomes plastic when $c = b$. In this case, equation (3.169) derived from the plastic flow rule still holds. However, it must be integrated over the interval [r,b], leading to:

$$\frac{\eta}{\gamma}\left\{(\frac{b_0}{b})^{\frac{\beta+1}{\beta}}-(\frac{r_0}{b})^{\frac{\beta+1}{\beta}}\right\} = -\omega^{\gamma}\int_{\theta}^{\omega} e^{\theta}\theta^{-\gamma-1}d\theta \tag{3.179}$$

where ω can be obtained from equation (3.172) by setting $c = b$ which gives:

$$\omega = \frac{1+\nu}{\delta\beta(\alpha-1)}(\beta-\nu - \beta\nu + \frac{1-\nu-\beta\nu}{\alpha}) \tag{3.180}$$

By putting $r = a$ and $r_0 = a_0$, we obtain the following displacement equation in the fully plastic sphere:

$$\frac{\eta}{\gamma}\left\{(\frac{b_0}{b})^{\frac{\beta+1}{\beta}}-(\frac{a_0}{b})^{\frac{\beta+1}{\beta}}\right\} = \sum_{n=0}^{\infty} A_n^2 \tag{3.181}$$

in which

$$A_n^2 = \begin{cases} \dfrac{(\alpha-1)\omega^n}{\alpha n!}\ln\dfrac{b}{a} & \text{if } n = \gamma \\ \dfrac{\omega^n}{n!(n-\gamma)}[(\dfrac{b}{a})^{\frac{(\alpha-1)}{\alpha}(n-\gamma)}-1] & \textit{otherwise} \end{cases} \tag{3.182}$$

As ω is a small quantity (an order of $1/E$), it can be easily proved that the series used in the solutions converge very rapidly for all realistic values of α and β.

(e) Solution procedure

All the necessary equations for determining the complete pressure-expansion curve and the stress distributions have been derived. The following procedure can be used to construct a complete pressure-expansion curve:

(1) Choose input soil parameters $E, \nu, C, \phi, \psi, p_0$ and initial cylinder geometry b_0/a_0.

(2) Calculate the derived parameters $G, Y, \alpha, \beta, \gamma, \delta$.

(3) If cavity pressure is less than p_{1y}, calculate the cavity radius from the small strain elastic solution (3.155).

(4) If cavity pressure is greater than p_{1y}, equations (3.163), (3.166) and (3.177) are needed:

(a) for a chosen value of c/b (less than 1 and greater than a_0/b_0), calculate c/a_0 and c/b_0 from equation (3.166), and c/a and a/a_0 from (3.177);

(b) calculate the pressure required to expand the sphere with a cavity expansion ratio of a/a_0 from equation (3.163).

(5) When $c/b = 1$, the entire cylinder becomes plastic and the pressure-expansion curve needs to be constructed from equations (3.164) and (3.181):

(a) choose a value of cavity pressure p which must be less than the value required to cause the whole cylinder become plastic, and calculate b/a from equation (3.164);

(b) Calculate $b/a_0, b/b_0$ and a/a_0 from equation (3.181).

Repetition of steps (1)–(4a,b) by varying c/b, and (5a,b) for varying cavity pressure p provides the data for the complete pressure-expansion curve. The distribution of stresses at any stage of expansion can be obtained from equations (3.157) to (3.160).

Once again, a simple computer code written in FORTRAN for performing the above calculation is available and readers of this book may request a copy of the code from the author.

(f) A special case: small strain solution

The solutions presented in the previous subsections are rigorous large strain solutions. If small strain assumptions can be made, the results can be simplified considerably.

The stress solutions in a small strain analysis are same as those for a large strain analysis. The main difference lies in the displacement equation in the plastic zone. For a small strain analysis, we can use the following definition for strains:

$$\varepsilon_r = \frac{du}{dr}$$

$$\varepsilon_\theta = \frac{u}{r}$$

Substituting the above equations for strains in the displacement equation (3.169) gives the following small strain solution for displacement fields:

$$u = Kr^{-\frac{1}{\beta}} + \frac{\beta \ln \eta}{1+\beta} r - \frac{\alpha\beta\omega}{\alpha+\beta} c^{\frac{\alpha-1}{\alpha}} r^{\frac{1}{\alpha}} \tag{3.183}$$

where

$$K = \frac{(1+\nu)[1+(1-2\nu)(\frac{c}{b})^2]c^{1+\frac{1}{\beta}}}{\delta[(\alpha-1)(\frac{c}{b})^2+1+\alpha]} + \frac{\alpha\beta\omega}{\alpha+\beta} c^{1+\frac{1}{\beta}} - \frac{\beta \ln \eta}{1+\beta} c^{1+\frac{1}{\beta}} \tag{3.184}$$

In particular, for a cavity wall when $r = a$, the displacement can be expressed in closed form:

$$\frac{u_a}{a} = \frac{\beta \ln \eta}{1+\beta}[1-(\frac{c}{a})^{1+\frac{1}{\beta}}] + \frac{(1+\nu)[1+(1-2\nu)(\frac{c}{b})^2](\frac{c}{a})^{1+\frac{1}{\beta}}}{\delta[(\alpha-1)(\frac{c}{b})^2+1+\alpha]}$$

$$+ \frac{\alpha\beta\omega}{\alpha+\beta}[(\frac{c}{a})^{1+\frac{1}{\beta}}-(\frac{c}{a})^{1-\frac{1}{\alpha}}] \tag{3.185}$$

in which the radius of the plastic zone c can be obtained from equation (3.163) for a given value of cavity pressure.

3.3.3 Expansion of cavities in an infinite medium

This section is concerned with the special case when outer radius of the cylinder or sphere is infinite, that is, the expansion of cavities in an infinite soil mass. For this case, the mathematics of the solution can be simplified. The solution presented here was derived by Yu (1990) and also presented by Yu and Houlsby (1991).

The properties of soil are defined by Young's modulus E and Poisson's ratio ν, cohesion C, angles of friction and dilation ϕ and ψ. The initial stress (assumed to be isotropic) is p_0. To simplify the presentation, it is possible to combine both the cylindrical and spherical analyses. The parameter k can be used to indicate cylindrical analysis ($k=1$) or spherical analysis ($k=2$).

Several functions of these variables recur throughout the analysis and to abbreviate the mathematics it is convenient to define the following quantities, all of which are constants in any given analysis:

$$G = \frac{E}{2(1+\nu)}$$

$$M = \frac{E}{1-\nu^2(2-k)}$$

$$Y = \frac{2C\cos\phi}{1-\sin\phi}$$

$$\alpha = \frac{1+\sin\phi}{1-\sin\phi}$$

$$\beta = \frac{1 + \sin\psi}{1 - \sin\psi}$$

$$\gamma = \frac{\alpha(\beta + k)}{k(\alpha - 1)\beta}$$

$$\delta = \frac{Y + (\alpha - 1)p_0}{2(k + \alpha)G}$$

$$\mu = \frac{(1 + k)\delta[1 - \nu^2(2 - k)]}{(1 + \nu)(\alpha - 1)\beta}$$

$$\times\ [\alpha\beta + k(1-2\nu) + 2\nu - \frac{k\nu(\alpha + \beta)}{1 - \nu(2-k)}]$$

$$\chi = \exp\left\{\frac{(\beta + k)(1-2\nu)(1 + (2-k)\nu)[Y + (\alpha-1)p_0]}{E(\alpha-1)\beta}\right\}$$

At any time in any position in the soil around the cavity with cavity radius of a, the stresses must satisfy the following equation of equilibrium:

$$(\sigma_\theta - \sigma_r) = \frac{r}{k}\frac{\partial \sigma_r}{\partial r} \tag{3.186}$$

which is subject to two boundary conditions:

$$\sigma_r|_{r=a} = -p$$

$$\sigma_r|_{r=\infty} = -p_0$$

(a) Elastic solution

As the cavity pressure increases from its initial value, the deformation of the soil is at first purely elastic. Under conditions of radial symmetry, the elastic stress-strain relationship can be expressed as:

$$\dot{\varepsilon}_r = \frac{\partial \dot{u}}{\partial r} = \frac{1}{M}[\dot{\sigma}_r - \frac{k\nu}{1 - \nu(2 - k)}\dot{\sigma}_\theta] \tag{3.187}$$

$$\dot{\varepsilon}_\theta = \frac{\dot{u}}{r} = \frac{1}{M}[-\frac{\nu}{1 - \nu(2 - k)}\dot{\sigma}_r + [1-\nu(k - 1)]\dot{\sigma}_\theta] \tag{3.188}$$

The solution of equations (3.186)–(3.188), subject to the stress boundary conditions, is shown to be:

$$\sigma_r = -p_0 - (p-p_0)(\frac{a}{r})^{1+k} \tag{3.189}$$

$$\sigma_\theta = -p_0 + \frac{p-p_0}{k}(\frac{a}{r})^{1+k} \tag{3.190}$$

$$u = \frac{p-p_0}{2kG}(\frac{a}{r})^{1+k} r \tag{3.191}$$

The initial yielding occurs at the inner cavity wall when the stresses satisfy the yield condition (3.119) and that happens when the cavity pressure reaches the following value:

$$p = p_{1y} = \frac{k[Y + (\alpha-1)p_0]}{k + \alpha} + p_0 = 2kG\delta + p_0 \tag{3.192}$$

(b) Elastic-plastic stress analysis

After initial yielding takes place at the cavity wall, a plastic zone will form around the inner cavity wall with an increase in the applied cavity pressure p. The outer radius of the plastic zone is denoted by c.

Stresses in the plastic region

The stress components in the plastic zone must satisfy the equilibrium equation (3.186) and the yield condition (3.119), and are shown to be in the form:

$$\sigma_r = \frac{Y}{\alpha-1} + Ar^{-\frac{k(\alpha-1)}{\alpha}} \tag{3.193}$$

$$\sigma_\theta = \frac{Y}{\alpha-1} + \frac{A}{\alpha} r^{-\frac{k(\alpha-1)}{\alpha}} \tag{3.194}$$

where A is a constant of integration.

Stresses in the elastic region

The stress components in the elastic region can be obtained from the equilibrium equation and elastic stress-strain equations as follows:

$$\sigma_r = -p_0 - Br^{-(1+k)} \tag{3.195}$$

$$\sigma_\theta = -p_0 + \frac{B}{k} r^{-(1+k)} \tag{3.196}$$

where B is the second integration constant.

The continuity of stress components at the elastic-plastic interface can be used to determine the constants A and B:

$$A = -\frac{(1+k)\alpha[Y+(\alpha-1)p_0]}{(\alpha-1)(k+\alpha)}c^{\frac{k(\alpha-1)}{\alpha}} \tag{3.197}$$

$$B = \frac{k[Y+(\alpha-1)p_0]}{k+\alpha}c^{1+k} \tag{3.198}$$

At the cavity wall $\sigma_r|_{r=a} = -p$ and this condition can be used, in conjunction with (3.193) and (3.197), to express the plastic radius c in terms of the current cavity radius and applied pressure

$$\frac{c}{a} = \left\{\frac{(k+\alpha)[Y+(\alpha-1)p]}{\alpha(1+k)[Y+(\alpha-1)p_0]}\right\}^{\frac{\alpha}{k(\alpha-1)}} \tag{3.199}$$

The stresses are now established in terms of a single unknown c. In the next subsection the displacements are examined, allowing for the determination of c and therefore the complete pressure-expansion relationship.

(c) Elastic-plastic displacement analysis

Substituting equations (3.195) and (3.196) into (3.188) gives the displacement in the elastic zone:

$$u = \delta\left(\frac{c}{r}\right)^{1+k} r \tag{3.200}$$

where δ is defined at the beginning of this section. The determination of the displacement field in the plastic zone requires the use of a plastic flow rule which indicates the relative magnitude of plastic strains in different directions.

For the expansion of cylindrical and spherical cavities, the non-associated Mohr-Coulomb flow rule can be expressed as:

$$\frac{\dot{\varepsilon}_r^p}{\dot{\varepsilon}_\theta^p} = \frac{\dot{\varepsilon}_r - \dot{\varepsilon}_r^e}{\dot{\varepsilon}_\theta - \dot{\varepsilon}_\theta^e} = -\frac{k}{\beta} \tag{3.201}$$

where β is a simple function of dilation angle given before. If $\beta = \alpha$ then the flow rule for the soil is said to be fully associated.

Substituting elastic stress equations (3.187) and (3.188) into the plastic flow rule (3.201) results in:

$$\beta \dot{\varepsilon}_r + k \dot{\varepsilon}_\theta = \frac{1}{M}[\beta - \frac{k\nu}{1-\nu(2-k)}]\dot{\sigma}_r$$

$$+ \frac{1}{M}[k(1-2\nu) + 2\nu - \frac{k\beta\nu}{1-\nu(2-k)}]\dot{\sigma}_\theta \qquad (3.202)$$

where M is defined at the beginning of this subsection. The distributions of stress and strain in the soil at the initiation of plastic yield are obtained from equations (3.187)-(3.191) by putting $p = p_{1y}$ defined in equation (3.192). The integral of equation (3.202) subject to this initial condition is found to be:

$$\beta \varepsilon_r + k \varepsilon_\theta = \frac{1}{M}[\beta - \frac{k\nu}{1-\nu(2-k)}]\sigma_r$$

$$+ \frac{1}{M}[k(1-2\nu) + 2\nu - \frac{k\beta\nu}{1-\nu(2-k)}]\sigma_\theta$$

$$+ \frac{1}{M}[\beta + k(1-2\nu) + 2\nu - \frac{k\nu(1+\beta)}{1-\nu(2-k)}]p_0 \qquad (3.203)$$

To account for effects of large strain in the plastic zone the logarithmic strains are adopted, namely

$$\varepsilon_r = \ln(\frac{dr}{dr_0})$$

$$\varepsilon_\theta = \ln \frac{r}{r_0}$$

Substituting the above large strain definitions and plastic stresses (3.193) and (3.194) into equation (3.203) leads to:

$$\ln[(\frac{r}{r_0})^{\frac{k}{\beta}} \frac{dr}{dr_0}] = \ln \chi - \mu (\frac{c}{r})^{\frac{k(\alpha-1)}{\alpha}} \qquad (3.204)$$

where χ and μ are defined at the start of this section.

By means of transformation:

$$\varrho = (\frac{c}{r})^{\frac{k(\alpha-1)}{\alpha}} \qquad (3.205)$$

$$\xi = (\frac{r_0}{c})^{\frac{\beta+k}{\beta}} \qquad (3.206)$$

and use of equation (3.200), equation (3.204) can be integrated over the interval [c,r], leading to:

$$\frac{\chi}{\gamma}\left\{(1-\delta)^{\frac{\beta+k}{\beta}}-\left(\frac{r_0}{c}\right)^{\frac{\beta+k}{\beta}}\right\} = \int_1^{\varrho} \exp(\mu\varrho)\varrho^{-\gamma-1}d\varrho \qquad (3.207)$$

By putting $r = r$ and $r_0 = a_0$ and making use of equation (3.199), we find:

$$\frac{\chi}{\gamma}\left\{(1-\delta)^{\frac{\beta+k}{\beta}}-R^{-\gamma}\left(\frac{a_0}{a}\right)^{\frac{\beta+k}{\beta}}\right\} = \int_1^{R} \exp(\mu\varrho)\varrho^{-\gamma-1}d\varrho \qquad (3.208)$$

in which R is a function of the current cavity pressure given by:

$$R = \frac{(k+\alpha)[Y+(\alpha-1)p]}{\alpha(1+k)[Y+(\alpha-1)p_0]} \qquad (3.209)$$

For the special case of a spherical cavity in Mohr-Coulomb materials with an associated flow rule (i.e. $k = 2$ and $\beta = \alpha$), equation (3.208) reduces to the solution derived by Chadwick (1959).

With the aid of the series expansion:

$$\exp(\mu\varrho) = \sum_{n=0}^{\infty}\frac{(\mu\varrho)^n}{n!} \qquad (3.210)$$

the following explicit expression for the pressure-expansion relationship is obtained:

$$\left(\frac{a}{a_0}\right)^{\frac{\beta+k}{\beta}} = \frac{R^{-\gamma}}{(1-\delta)^{\frac{\beta+k}{\beta}}-\frac{\gamma}{\chi}\sum_{n=0}^{\infty}A_n(R,\mu)} \qquad (3.211)$$

in which A_n is defined by:

$$A_n(R,\mu) = \begin{cases} \frac{\mu^n}{n!}\ln R & \text{if } n = \gamma \\ \frac{\mu^n}{n!(n-\gamma)}(R^{n-\gamma}-1) & \textit{otherwise} \end{cases} \qquad (3.212)$$

(d) Special cases

Limit pressure

When a cavity is expanded in a plastically deforming material the cavity pressure does not increase indefinitely, but a limit pressure is approached. By putting $a/a_0 \rightarrow \infty$ in equation (3.211), the limit cavity pressure $p_{\lim}$ can be obtained by finding $R_{\lim}$ from the following equation:

$$\sum_{n=0}^{\infty} A_n(R_{\lim}, \mu) = \frac{\chi}{\gamma}(1-\delta)^{\frac{\beta+k}{\beta}} \tag{3.213}$$

where A_n is related to $R_{\lim}$ by equation (3.212). Once $R_{\lim}$ is obtained, the limit cavity pressure $p_{\lim}$ can be readily derived from the following equation:

$$R_{\lim} = \frac{(k + \alpha)[Y + (\alpha-1)p_{\lim}]}{\alpha(1 + k)[Y + (\alpha-1)p_0]} \tag{3.214}$$

It is found that the cavity limit pressure depends strongly on the angles of friction and dilation, as well as the stiffness properties of the soil.

Frictionless case

The solution presented here can be applied to soils with friction and dilation. It does not reduce to the solution for a frictionless soil (i.e. Tresca soil) when friction angle $\phi = 0$. This is because in this case $\alpha = 1$ and the terms in $\alpha - 1$ which frequently appear in the denominator make the expressions indeterminate. However, it can be confirmed that at very small ϕ values the solution presented in this section approaches the solution (3.67) obtained in the previous section for Tresca materials.

Small strain case

Large strain theory is complex and wherever possible, small strain theory is used to model cavity expansion. In small strain theory the fact that displacements modify the position of material points is ignored, so the theory is only valid for small expansions. In particular, prediction of limit pressure is not possible with small strain assumptions. It can be shown that if the small strain assumption is made it is possible to obtain the following closed form expression for displacement in the plastic zone:

$$u = [\delta + \frac{\alpha\beta\mu}{k(\alpha + \beta) + \alpha\beta(1-k)} - \frac{\beta \ln\chi}{\beta + k}](\frac{c}{r})^{\frac{k}{\beta}}\, c$$

$$+ \frac{\beta \ln\chi}{\beta + k} r - \frac{\alpha\beta\mu}{k(\alpha + \beta) + \alpha\beta(1-k)}(\frac{c}{r})^{\frac{k(\alpha-1)}{\alpha}} r \tag{3.215}$$

Equation (3.215) is only applicable to the situation in which the maximum value of the cavity pressure is sufficiently small for the squares and higher powers of strains included in the large strain definitions to be negligible. This small strain solution is also given by Carter *et al.* (1986). For the special case of spherical and cylindrical cavities with an associated flow rule, the solution (3.215) reduces to the small strain equations obtained by Chadwick (1959) and Florence and Schwer (1978) respectively.

Neglecting elastic strain in plastic zone

A common assumption which considerably simplifies the analysis of plastic cavity expansion is to ignore the contribution of elastic strain within the plastic zone (e.g. Hughes *et al.*, 1977). While this may at first seem to be a reasonable assumption, as the elastic strains are much smaller than the plastic strains, it does have a significant effect on the predicted results.

Neglecting the elastic deformation in the plastically deforming region results in a relatively simple expression for the relationship between the cavity pressure and cavity displacement:

$$\frac{(k+\alpha)[Y+(\alpha-1)p]}{\alpha(1+k)[Y+(\alpha-1)p_0]} = \left\{ \frac{1-(\frac{a_0}{a})^{1+\frac{k}{\beta}}}{1-(1-\delta)^{1+\frac{k}{\beta}}} \right\}^{\frac{1}{\gamma}} \tag{3.216}$$

A comparison between the approximate large strain solution (3.216) and the exact large strain solution (3.211) indicates that the effects of elastic strain in the plastic zone are more important for soils with high friction and dilation angles and low elastic stiffness.

By putting $a/a_0 \to \infty$ and $\psi = 0$ and assuming $\delta = 0$, equation (3.216) reduces to the limit solution derived by Vesic (1972).

3.3.4 Contraction of cavities in an infinite medium

In this section, analytical solutions are developed for large strain contraction of both cylindrical and spherical cavities in Mohr-Coulomb soils with a non-associated flow rule.

(a) Contraction from an elastic-plastic stress state

It is assumed that in an unbound Mohr-Coulomb soil mass a single cylindrical or spherical cavity has been expanded from an initial radius a_0 to current radius a. It is assumed that during the loading stage a plastic zone is formed around the cavity wall, with c denoting the radius of the elastic-plastic interface. The stress and displacement fields in the soil at end of loading have been established in the previous section. The purpose of this subsection is to determine the stress and strain fields in the soil as the cavity pressure p decreases. The rigorous solution presented here was derived by Yu (1990) and also presented by Yu and Houlsby (1995). It is noted that a less rigorous and approximate solution for cavity unloading in Mohr-Coulomb materials has also been presented by Salencon (1969). The unloading solutions presented in this section have applications to the analysis of pressuremeter unloading tests in sand.

The analysis of the unloading cavities is rather complex as reference must be made to the configuration before loading, at the end of loading, and after unloading. As a result, the analysis involves a rather large number of variables and is difficult to present simply.

As in the loading analysis, the properties of soil are defined by Young's modulus E and Poisson's ratio ν, cohesion C, angles of friction and dilation ϕ and ψ. The initial stress (assumed to be isotropic) is p_0. To simplify the presentation, it is possible to combine both the cylindrical and spherical analyses. The parameter k will be used to indicate cylindrical analysis ($k=1$) or spherical analysis ($k=2$).

To abbreviate the mathematics in the development, it is convenient to define the following functions of soil properties. Note that some of these functions have also been used in the loading analysis presented in the previous section.

$$G = \frac{E}{2(1 + \nu)}$$

$$M = \frac{E}{1-\nu^2(2-k)}$$

$$Y = \frac{2C\cos\phi}{1-\sin\phi}$$

$$\alpha = \frac{1 + \sin\phi}{1-\sin\phi}$$

$$\beta = \frac{1 + \sin\psi}{1-\sin\psi}$$

$$\delta = \frac{Y + (\alpha-1)p_0}{2(k + \alpha)G}$$

$$m = \frac{2\delta G(1 + k)}{\alpha-1}$$

$$N = \frac{1 + k\beta}{k(\alpha - 1)}$$

$$l = \frac{(\alpha - 1)E}{[1 - \nu^2(2 - k)](1 + k)(Y + (\alpha - 1)p_0)}$$

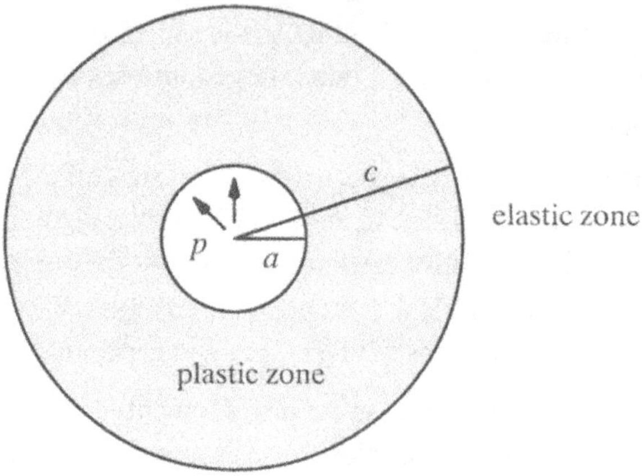

At end of cavity expansion

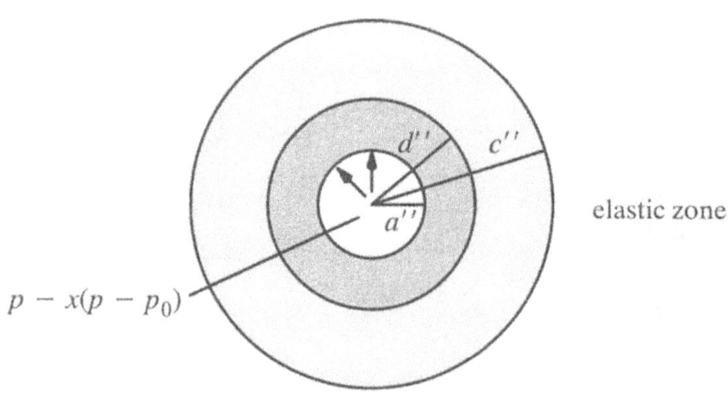

An instant during cavity contraction

Figure 3.3: Contraction of cavities from a plastic state

Elastic unloading and reverse yielding

Assuming that an elastic-plastic state of the soil around the cavity has been reached by monotonically increasing the cavity pressure from the initial stress p_0 to the current value p, the effect of monotonically reducing the cavity pressure from p to the value $p - x(p - p_0)$, where the unloading factor $0 \leq x \leq \frac{p}{p - p_0}$, is considered in detail. The restriction imposed on the value of x allows us only consider the case when all stresses in the soil remain compressive.

The configuration of the system on the completion of loading (i.e. the end of loading stage) is used as a reference state from which the stress and displacement of the unloading process are measured, Figure 3.3. Deviations from this state are denoted by a single prime and current values of field quantities by a double prime:

$$\sigma_r'' = \sigma_r + \sigma_r' \tag{3.217}$$

$$\sigma_\theta'' = \sigma_\theta + \sigma_\theta' \tag{3.218}$$

$$u'' = u + u' \tag{3.219}$$

where σ_r is the radial stress, σ_θ is the circumferential stress and u is the radial displacement.

As the unloading factor x increases from zero the soil at first unloads elastically. Superimposing upon the stress distribution at the completion of loading, equations (3.193)-(3.198), an elastic stress field produced by the application of an additional pressure $- x(p - p_0)$ at the cavity wall, it is found that:

$$\sigma_r'' = \frac{Y}{\alpha-1} - m\alpha(\frac{c}{r})^{\frac{k(\alpha-1)}{\alpha}} + x(p-p_0)(\frac{a''}{r''})^{1+k} \tag{3.220}$$

$$\sigma_\theta'' = \frac{Y}{\alpha-1} - m(\frac{c}{r})^{\frac{k(\alpha-1)}{\alpha}} - \frac{x}{k}(p-p_0)(\frac{a''}{r''})^{1+k} \tag{3.221}$$

for the plastic zone $a'' \leq r'' \leq c''$, and

$$\sigma_r'' = -p_0 - 2\delta Gk(\frac{c}{r})^{1+k} + x(p-p_0)(\frac{a''}{r''})^{1+k} \tag{3.222}$$

$$\sigma_\theta'' = -p_0 + 2\delta G(\frac{c}{r})^{1+k} - \frac{x}{k}(p-p_0)(\frac{a''}{r''})^{1+k} \tag{3.223}$$

for the outer elastic zone $r'' \geq c''$, where a and c denote the radius of the cavity and the radius of the elastic-plastic boundary at the end of loading phase respectively.

As x increases further, the reverse yielding will commence when the yield condition:

$$\alpha\sigma_r''-\sigma_\theta'' = Y \tag{3.224}$$

is satisfied. Using equations (3.220) and (3.221) it can be shown that the above yield condition is first satisfied at the cavity wall when the unloading factor reaches the following value:

$$x = x_1 = \frac{k(1+\alpha)[Y+(\alpha-1)p]}{\alpha(1+k\alpha)(p-p_0)} \tag{3.225}$$

Plastic unloading: stress analysis

Consider now the situation when a reverse plastic zone has formed. We denote the outer radius of the reverse plastic zone by d'' and assume, for the present, that the reverse plastic is always smaller than the loading plastic zone and that is $d'' \leq c''$. The distribution of stress in the soil is now obtained by considering the elastic and plastic regions separately and matching the solutions at the elastic-plastic interface $r'' = d''$.

The general results of this stress analysis, which is detailed in Yu (1990) and Yu and Houlsby (1995), are as follows:

$$\sigma_r'' = \frac{Y}{\alpha-1} - m(\frac{c}{d})^{\frac{k(\alpha-1)}{\alpha}}(\frac{r''}{d''})^{k(\alpha-1)} \tag{3.226}$$

$$\sigma_\theta'' = \frac{Y}{\alpha-1} - m\alpha(\frac{c}{d})^{\frac{k(\alpha-1)}{\alpha}}(\frac{r''}{d''})^{k(\alpha-1)} \tag{3.227}$$

for the reverse plastic zone $a'' \leq r'' \leq d''$, and

$$\sigma_r' = mk(1+\alpha)(\frac{c}{d})^{\frac{k(\alpha-1)}{\alpha}}(\frac{d''}{r''})^{1+k} \tag{3.228}$$

$$\sigma_\theta' = -m(1+\alpha)(\frac{c}{d})^{\frac{k(\alpha-1)}{\alpha}}(\frac{d''}{r''})^{1+k} \tag{3.229}$$

for the reverse elastic but loading plastic zone $r'' \geq d''$.

The stress condition at the cavity wall

$$\sigma_r''|_{r''=a''} = -[p-x(p-p_0)] \tag{3.230}$$

can be used to provide an additional relation:

$$(\frac{a''}{d''})^{k(\alpha-1)}(\frac{a}{d})^{\frac{k(\alpha-1)}{\alpha}} =$$

$$\frac{a(1+ka)[Y+(a-1)(p_0+(1-x)(p-p_0))]}{(k+a)(Y+(a-1)p)} \tag{3.231}$$

Plastic unloading: displacement analysis

The unloading displacement in the region $r'' \geq d''$ is shown to be:

$$u' = -\frac{(1+k)(1+a)\delta}{1+ka}(\frac{c}{d})^{\frac{k(a-1)}{a}}(\frac{d''}{r''})^{1+k}r'' \tag{3.232}$$

and in particular for the elastic-plastic interface:

$$d = d''-u'|_{r''=d''}$$

$$= d''[1+\frac{(1+k)(1+a)\delta}{1+ka}(\frac{c}{d})^{\frac{k(a-1)}{a}}] \tag{3.233}$$

As shown by Yu (1990) and Yu and Houlsby (1995), an earlier assumption $d'' \leq c''$ is always true if a simple condition is imposed on the unloading factor:

$$x \leq 1 + \frac{k[Y+(a-1)p_0]}{(1+ka)(p-p_0)} \tag{3.234}$$

The non-associated Mohr-Coulomb flow rule for reverse plasticity is defined by:

$$\frac{\dot{\varepsilon}_r^{p''}}{\dot{\varepsilon}_\theta^{p''}} = -k\beta \tag{3.235}$$

and can be used for the evaluation of the displacement in the reverse plastic zone $a'' \leq r'' \leq d''$.

Substituting the elastic solution for strains (3.187) and (3.188) into the reverse plastic flow rule (3.235) leads to:

$$\dot{\varepsilon}_r'' + k\beta\dot{\varepsilon}_\theta'' = \frac{1}{M}[1-\frac{k\beta\nu}{1-\nu(2-k)}]\dot{\sigma}_r''$$

$$+\frac{1}{M}[k\beta(1+\nu-k\nu)-\frac{k\nu}{1-\nu(2-k)}]\dot{\sigma}_\theta'' \tag{3.236}$$

The differential relation (3.236) can be integrated subject to the condition for the initial yielding to occur. The total stress-strain relation is found to be:

$$\varepsilon_r' + k\beta\varepsilon_\theta' = \frac{1}{M}[1-\frac{k\beta\nu}{1-\nu(2-k)}]\dot{\sigma}_r''$$

$$+ \frac{1}{M}[k\beta(1 + \nu - k\nu) - \frac{k\nu}{1-\nu(2-k)}]\dot{\sigma}_\theta''$$

$$- \frac{Y}{M(\alpha - 1)}[1 + k\beta(1 + \nu - k\nu) - \frac{k\nu(1 + \beta)}{1-\nu(2-k)}]$$

$$+ \frac{m}{M}[\alpha(1 - \frac{k\nu\beta}{1 - \nu(2 - k)}) + k\beta(1 + \nu - k\nu)$$

$$- \frac{k\nu}{1-\nu(2-k)}](\frac{c}{r})^{\frac{k(\alpha-1)}{\alpha}} \tag{3.237}$$

The logarithmic strain components measured from the fully loaded state are:

$$\varepsilon_r' = \ln(\frac{dr''}{dr}) \tag{3.238}$$

$$\varepsilon_\theta' = \ln(\frac{r''}{r}) \tag{3.239}$$

and it should be noted that the reference state for strains used in (3.238) and (3.239) is the end of loading phase.

Substituting the stresses in the reverse plastic region (3.226) and (3.227) into (3.237) results in the differential equation:

$$\ln\left\{(\frac{r''}{r})^{k\beta}\frac{dr''}{dr}\right\} = -\lambda(\frac{r''}{d''})^{k(\alpha-1)} + \varkappa(\frac{d}{r})^{\frac{k(\alpha-1)}{\alpha}} \tag{3.240}$$

where

$$\lambda = \frac{1}{(1 + k\alpha)l}\{1 - \frac{k\beta\nu}{1-\nu(2-k)}$$

$$+ \alpha[k\beta\ (1 + \nu - k\nu) - \frac{k\nu}{1-\nu(2-k)}]\}(\frac{c}{d})^{\frac{k(\alpha-1)}{\alpha}} \tag{3.241}$$

$$\varkappa = \frac{1}{(k + \alpha)l}\{\alpha[1 - \frac{k\beta\nu}{1-\nu(2-k)}]$$

$$+ k\beta\ (1 + \nu - k\nu) - \frac{k\nu}{1-\nu(2-k)}\}(\frac{c}{d})^{\frac{k(\alpha-1)}{\alpha}} \tag{3.242}$$

By means of the transformation

$$\varrho = (\frac{d}{r})^{\frac{k(a-1)}{a}} \tag{3.243}$$

$$\xi = (\frac{r''}{d''})^{k(a-1)} \tag{3.244}$$

The displacement equation (3.240) can be integrated over the interval $[d'', r'']$ to give the solution:

$$a \int_1^{\varrho} \varrho^{-aN-1} \exp(\varkappa \varrho) d\varrho + (\frac{d''}{d})^{1+k\beta} \int_1^{\xi} \xi^{N-1} \exp(\lambda \xi) d\xi = 0 \tag{3.245}$$

By putting $r = a$ and $r'' = a''$ and with the aid of a series expansion, equation (3.245) can be integrated to give a closed form solution for displacement:

$$a \sum_{n=0}^{\infty} A_n^1 + (\frac{d''}{d})^{1+k\beta} \sum_{n=0}^{\infty} A_n^2 = 0 \tag{3.246}$$

in which the series functions are defined by:

$$A_n^1 = \begin{cases} \dfrac{k(a-1)\varkappa^n}{an!} \ln \dfrac{d}{a} & \text{if } n = aN \\ \dfrac{\varkappa^n}{n!(n-aN)} [(\dfrac{d}{a})^{\frac{k(a-1)}{a}(n-aN)} - 1] & \textit{otherwise} \end{cases} \tag{3.247}$$

$$A_n^2 = \frac{\lambda^n}{n!(n+N)} [(\frac{a''}{d''})^{k(a-1)(n+N)} - 1] \tag{3.248}$$

This analysis is an exact large strain solution because no approximation is used in the development. It is worth noting that the exact displacement equation in the plastic region (3.246) can be simplified considerably by ignoring the contribution of elastic strain within the plastically deforming zone as the right hand side of equation (3.236) will become zero under this assumption. Using this assumption, an approximate small strain unloading solution has been presented by Houlsby *et al.* (1986) and Withers *et al.* (1989) for both cylindrical and spherical cavities.

Small strain solutions for the special case of loading and unloading of a cylindrical cavity in the Mohr-Coulomb medium with an associated flow rule were also obtained by Senseny *et al.* (1989).

Solution procedure

All the equations necessary for determining a cavity pressure-contraction curve and the stress distributions have been derived. The following procedure may be used to construct a complete pressure-contraction curve:

(1) Choose input soil parameters $E, \nu, C, \phi, \psi, p_0$.

(2) Calculate the derived parameters $G, M, Y, \alpha, \beta, \delta, \gamma, m, N, l$.

(3) For a given pressure p at the end of loading phase, the size of the elastic-plastic boundary *c/a* before unloading starts can be calculated from the loading solution (3.199):

$$\frac{c}{a} = \left\{ \frac{(k+\alpha)[Y+(\alpha-1)p]}{\alpha(1+k)[Y+(\alpha-1)p_0]} \right\}^{\frac{\alpha}{k(\alpha-1)}}$$

(4) For a given value of unloading factor x less than x_1 given by equation (3.225), the unloading is purely elastic and the cavity pressure and the displacement at the cavity wall are related by the elastic solution.

$$\frac{a''-a}{a} = \frac{-x(p-p_0)}{2kG}$$

(5) For x greater than x_1, the unloading is elastic-plastic; choose a value of *d/a* (greater than 1), calculate *c/d* from *c/a* and *d/a*.

(6) Evaluate d/d'' from equation (3.233) using the known value of *c/d*.

(7) Calculate a''/d'' from equation (3.246) using the known values of *d/a* and d/d''; as equation (3.246) is a non-linear equation in terms of a single unknown a''/d'', an iterative procedure such as the Newton-Raphson scheme must be used to solve for the unknown.

(8) Calculate the current cavity pressure from equation (3.231) using the known values of a''/d'' and d/a.

Repetition of steps (5)-(8) by varying d/a provides the data for the complete pressure-contraction curve. A simple computer code written in FORTRAN for performing the above calculation is available and readers of this book may request a copy of the code from the author.

(b) Contraction from an *in-situ* stress state

It is assumed that an unbound Mohr-Coulomb medium contains a single cylindrical or spherical cavity. Initially the radius of the cavity is a_0 and a hydrostatic pressure p_0 acts throughout the soil which is assumed to be homogeneous. This subsection is concerned with the distribution of stress and displacement in the soil as the cavity pressure p reduces from its initial or *in-situ* value. The symbol k is used to denote cylindrical ($k=1$) and spherical cavity ($k=2$). The solution presented here was derived by Yu and Rowe (1999) for predicting soil behaviour around tunnels. Note that the definitions used in the previous section for Y, α and β are retained in this section.

Elastic response and initial yielding

As the pressure p decreases from p_0, the deformation of the soil is at first purely elastic. Under conditions of radial symmetry the elastic stress-strain relationship may be expressed as:

$$\dot{\varepsilon}_r = \frac{\partial \dot{u}}{\partial r} = \frac{1}{M}[\dot{\sigma}_r - \frac{k\nu}{1-\nu(2-k)}\dot{\sigma}_\theta] \tag{3.249}$$

$$\dot{\varepsilon}_\theta = \frac{\dot{u}}{r} = \frac{1}{M}[-\frac{\nu}{1-\nu(2-k)}\dot{\sigma}_r + [1-\nu(k-1)]\dot{\sigma}_\theta] \tag{3.250}$$

where $M = E/(1 - \nu^2(2 - k))$.

The elastic solutions for stresses and displacement are as follows:

$$\sigma_r = -p_0 - (p - p_0)(\frac{a}{r})^{1+k} \tag{3.251}$$

$$\sigma_\theta = -p_0 + \frac{p - p_0}{k}(\frac{a}{r})^{1+k} \tag{3.252}$$

$$u = \frac{p - p_0}{2kG}(\frac{a}{r})^{1+k} r \tag{3.253}$$

For unloading of cavities, the Mohr-Coulomb yield equation takes the form:

$$\alpha\sigma_r - \sigma_\theta = Y \tag{3.254}$$

As the internal pressure decreases further, yielding starts to occur at the cavity wall when the condition

$$p = p_{1y} = \frac{1 + k}{1 + \alpha k}p_0 - \frac{kY}{1 + \alpha k} \tag{3.255}$$

is satisfied.

Elastic-plastic stress analysis

After initial yielding at the cavity wall a plastic zone within the region $a \leq r \leq c$ forms around the inner wall of the cavity with a decrease in the cavity pressure p.

The stresses in the elastic zone are shown to be as follows:

$$\sigma_r = -p_0 - Br^{-(1+k)} \tag{3.256}$$

$$\sigma_\theta = -p_0 + \frac{B}{k} r^{-(1+k)} \tag{3.257}$$

On the other hand, the stresses in the plastic zone must satisfy equilibrium and yield condition and are shown to be of the form:

$$\sigma_r = \frac{Y}{\alpha - 1} + Ar^{k(\alpha - 1)} \tag{3.258}$$

$$\sigma_\theta = \frac{Y}{\alpha - 1} + A\alpha r^{k(\alpha - 1)} \tag{3.259}$$

The continuity of stress components at the elastic-plastic interface can be used to determine the constants A and B in terms of the plastic radius c:

$$A = -\frac{(1+k)[Y+(\alpha-1)p_0]}{(\alpha-1)(1+k\alpha)} c^{(1-\alpha)k} \tag{3.260}$$

$$B = \frac{k[(1-\alpha)p_0 - Y]}{1+k\alpha} c^{1+k} \tag{3.261}$$

Applying $\sigma_r = -p$ at the cavity wall leads to a relationship between the cavity pressure p and the plastic radius c:

$$\frac{c}{a} = \left\{ \frac{(1+k\alpha)[Y+(\alpha-1)p]}{(1+k)[Y+(\alpha-1)p_0]} \right\}^{\frac{1}{k(1-\alpha)}} \tag{3.262}$$

Elastic-plastic displacement analysis

The results obtained above can not be used to calculate the distribution of stresses until the displacement field is known. The displacement in the elastic zone can be shown to be:

$$u = \frac{(1-\alpha)p_0 - Y}{2G(1+k\alpha)} \left(\frac{c}{r}\right)^{1+k} r \tag{3.263}$$

Therefore, at the elastic-plastic interface the displacement should be:

$$u|_{r=c} = c - c_0 = -\frac{[(1-\alpha)p_0 + Y]c}{2(1+k\alpha)G} \tag{3.264}$$

For unloading cavities, the non-associated Mohr-Coulomb flow rule can be expressed as:

$$\frac{\dot{\varepsilon}_r^p}{\dot{\varepsilon}_\theta^p} = \frac{\dot{\varepsilon}_r - \dot{\varepsilon}_r^e}{\dot{\varepsilon}_\theta - \dot{\varepsilon}_\theta^e} = -k\beta \tag{3.265}$$

where β is a simple function of dilation angle. If $\beta = \alpha$ then the flow rule for the soil is associated.

If the elastic strains are included it is difficult to solve equation (3.265) analytically. However, if we ignore the elastic contribution in the plastically deforming region the solution is considerably simplified. By adopting the logarithmic strains, the flow rule (3.265) can be integrated to give the following relation:

$$r^{k\beta} dr = r_0^{k\beta} dr_0 \tag{3.266}$$

which can be further integrated over the interval $[c, r]$ to give:

$$r^{1+k\beta} - c^{1+k\beta} = r_0^{1+k\beta} - c_0^{1+k\beta} \tag{3.267}$$

As c can be linked to c_0 by the elastic solution (3.264), equation (3.267) defines the displacement field in the plastic region. At the cavity wall, equation (3.267) reduces to:

$$a^{1+k\beta} - a_0^{1+k\beta} = c^{1+k\beta} - c_0^{1+k\beta} \tag{3.268}$$

With the aid of equation (3.262), the displacement at the cavity wall can be linked to the cavity pressure as follows:

$$\frac{1-(\frac{a_0}{a})^{1+k\beta}}{1-(\frac{c_0}{c})^{1+k\beta}} = \left\{\frac{(1+k\alpha)[Y+(\alpha-1)p]}{(1+k)[Y+(\alpha-1)p_0]}\right\}^{\frac{1+k\beta}{k(1-\alpha)}} \tag{3.269}$$

Note that c_0/c can be determined from equation (3.264).

Small strain displacement analysis

With the small strain assumption, the displacement field in the plastic region is easily shown to be:

$$u = r - r_0 = \frac{(1-\alpha)p_0 - Y}{2G(1+\alpha k)} (\frac{c}{r})^{1+k\beta} r \tag{3.270}$$

In particular, the displacement at the cavity wall can be obtained by putting $r = a$ in equation (3.270):

$$\frac{u_a}{a} = \frac{[(1-\alpha)p_0-Y]}{2G(1+\alpha k)} \left\{ \frac{(1+k\alpha)[Y+(\alpha-1)p]}{(1+k)[Y+(\alpha-1)p_0]} \right\}^{\frac{1+k\beta}{k(1-\alpha)}} \tag{3.271}$$

3.3.5 Expansion of cavities from zero initial radius

This section deals with the special case of cavity expansion from zero initial radius in an infinite soil mass. As noted by Hill (1950), this problem has no characteristic length and hence will possess a similarity solution, in which the cavity pressure is constant and the continuing deformation is geometrically self-similar. As a result, the velocity approach used by Hill (1950) to analyse cavity expansion in Tresca materials may be conveniently used to obtain a solution for limiting pressure of a cavity expansion in Mohr-Coulomb materials.

One of the first attempts to obtain an analytical solution for cavity limit pressure in Mohr-Coulomb soils was made by Carter *et al.* (1986). However, the solution of Carter *et al.* (1986) may only be regarded as approximate because the convected part of the stress rate was ignored in their derivation. Later Collins and Wang (1990) derived a rigorous solution for purely frictional soils by including the convected part of the stress rate. The solution of Collins and Wang (1990) was however obtained using numerical integration as they could not express the solution in explicit form. Simple comparisons reported by Collins and Wang (1990) and Wang (1992) indicated that the difference introduced by ignoring the convected part of the stress rate was generally small for low values of dilation angle, but may increase to as much as 20% when the dilation angle becomes very large.

By following the solution procedures used in Section 3.2.1, this section derives a rigorous analytical solution for the expansion of cavities from zero initial radius in an infinite cohesive-frictional soil mass. The solution procedure adopted here is slightly different from that used by Collins and Wang (1990) as the plastic radius c is used as the time scale. Unlike Collins and Wang's solution, it is shown that by using a series expansion the solution can be expressed in closed form.

The properties of soil are defined by Young's modulus E and Poisson's ratio ν, and the cohesion C, angles of friction and dilation ϕ and ψ. The initial stress (assumed to be isotropic) is p_0. To simplify the presentation, the parameter k is used to indicate cylindrical analysis (k=*1*) or spherical analysis (k=2).

To abbreviate the mathematics, it is convenient to define the following quantities, all of which are constants in any given analysis:

$$G = \frac{E}{2(1+\nu)}$$

$$M = \frac{E}{1-\nu^2(2-k)}$$

$$Y = \frac{2C\cos\phi}{1-\sin\phi}$$

$$\alpha = \frac{1+\sin\phi}{1-\sin\phi}$$

$$\beta = \frac{1+\sin\psi}{1-\sin\psi}$$

$$\delta = \frac{Y+(\alpha-1)p_0}{2(k+\alpha)G}$$

(a) Elastic solution in the outer elastic zone

The stress-strain relations for soils in the outer elastic zone can be expressed as:

$$d\varepsilon_r^e = \frac{\partial \dot{u}}{\partial r} = \frac{1}{M}[d\sigma_r - \frac{k\nu}{1-\nu(2-k)}d\sigma_\theta] \qquad (3.272)$$

$$d\varepsilon_\theta^e = \frac{\dot{u}}{r} = \frac{1}{M}[-\frac{\nu}{1-\nu(2-k)}d\sigma_r + [1-\nu(k-1)]d\sigma_\theta] \qquad (3.273)$$

The solution of stresses is shown to be:

$$\sigma_r = -p_0 - (p_{1y}-p_0)(\frac{c}{r})^{1+k} \qquad (3.274)$$

$$\sigma_\theta = -p_0 + \frac{p_{1y}-p_0}{k}(\frac{c}{r})^{1+k} \qquad (3.275)$$

$$u = \frac{p_{1y}-p_0}{2kG}(\frac{c}{r})^{1+k}r \qquad (3.276)$$

where

$$p_{1y} = \frac{k[Y+(\alpha-1)p_0]}{k+\alpha} + p_0 = 2kG\delta + p_0 \qquad (3.277)$$

(b) Stress solution in the plastic zone

Stresses in the plastic region

The stress components in the plastic zone must satisfy the equilibrium equation (3.186) and the yield condition (3.119), and can be shown to be:

$$\sigma_r = \frac{Y}{\alpha-1} + Ar^{-\frac{k(\alpha-1)}{\alpha}} \tag{3.278}$$

$$\sigma_\theta = \frac{Y}{\alpha-1} + \frac{A}{\alpha}r^{-\frac{k(\alpha-1)}{\alpha}} \tag{3.279}$$

where A is a constant of integration.

Stresses in the elastic region

The stress components in the elastic region can be solved using the equilibrium equation and elastic stress-strain equations as follows:

$$\sigma_r = -p_0 - Br^{-(1+k)} \tag{3.280}$$

$$\sigma_\theta = -p_0 + \frac{B}{k}r^{-(1+k)} \tag{3.281}$$

where B is the second constant of integration.

The continuity of stress components at the elastic-plastic interface can be used to determine the constants A and B:

$$A = -\frac{(1+k)\alpha[Y+(\alpha-1)p_0]}{(\alpha-1)(k+\alpha)}c^{\frac{k(\alpha-1)}{\alpha}} \tag{3.282}$$

$$B = \frac{k[Y+(\alpha-1)p_0]}{k+\alpha}c^{1+k} \tag{3.283}$$

At the cavity wall we have $\sigma_r|_{r=a} = -p$ and this condition can be used to express the plastic radius c in terms of the current cavity pressure and radius:

$$\frac{c}{a} = \left\{\frac{(k+\alpha)[Y+(\alpha-1)p]}{\alpha(1+k)[Y+(\alpha-1)p_0]}\right\}^{\frac{\alpha}{k(\alpha-1)}} \tag{3.284}$$

The stresses are now established in terms of a single unknown c. In the next subsection the displacements are examined which allows for the determination of c and therefore the cavity pressure.

(c) Elastic-plastic displacement analysis

Substituting equations (3.277) into (3.276) gives the displacement in the elastic zone:

$$u = \delta\left(\frac{c}{r}\right)^{1+k} r \tag{3.285}$$

where δ is defined previously in this section. The determination of the displacement field in the plastic zone requires the use of a plastic flow rule which indicates the relative magnitude of plastic strains in different directions.

For the expansion of cavities, the non-associated Mohr-Coulomb flow rule can be expressed as:

$$\frac{d\varepsilon_r^p}{d\varepsilon_\theta^p} = \frac{d\varepsilon_r - d\varepsilon_r^e}{d\varepsilon_\theta - d\varepsilon_\theta^e} = -\frac{k}{\beta} \tag{3.286}$$

where β is a simple function of dilation angle as given before. If $\beta = \alpha$ then the flow rule for the soil is said to be associated.

Substituting equations (3.272) and (3.273) into the plastic flow rule (3.286) results in:

$$\beta d\varepsilon_r + k d\varepsilon_\theta = \frac{1}{M}[\beta - \frac{k\nu}{1 - \nu(2 - k)}]d\sigma_r$$

$$+ \frac{1}{M}[k(1-2\nu) + 2\nu - \frac{k\beta\nu}{1-\nu(2-k)}]d\sigma_\theta \tag{3.287}$$

where M is defined at the beginning of this section. In the plastic zone, from the yield equation $d\sigma_\theta = \frac{1}{\alpha}d\sigma_r$ and as a result equation (3.287) reduces to:

$$d\varepsilon_r + \frac{k}{\beta}d\varepsilon_\theta = \frac{\chi}{\beta}d\sigma_r \tag{3.288}$$

where

$$\chi = \frac{1}{M}[\beta - \frac{k\nu}{1-\nu(2-k)}]$$

$$+ \frac{1}{M\alpha}[k(1-2\nu) + 2\nu - \frac{k\beta\nu}{1-\nu(2-k)}] \tag{3.289}$$

In calculating the displacement of any individual particle it is convenient to take the movement of the plastic boundary as the scale of 'time' or progress of the expansion, since the parameter c appears in the formulae for the stresses. We can speak of the velocity V of a particle, meaning that the particle is displaced by an amount Vdc when the plastic boundary moves outwards a further distance dc. V can be expressed directly in terms of the total displacement u, which is a function of both the current radius r and plastic radius c so that:

$$du = \frac{\partial u}{\partial c}dc + \frac{\partial u}{\partial r}dr = \left(\frac{\partial u}{\partial c} + V\frac{\partial u}{\partial r}\right)dc$$

where r and c are taken as the independent variables. Equating the above expression to Vdc we obtain the expression for the particle velocity:

$$V = \frac{\frac{\partial u}{\partial c}}{1-\frac{\partial u}{\partial r}}$$

To evaluate the increments of stress and strain, we must follow a given element and therefore:

$$d\varepsilon_r = \frac{\partial(du)}{\partial r} = \frac{\partial V}{\partial r}dc$$

$$d\varepsilon_\theta = \frac{du}{r} = \frac{Vdc}{r}$$

$$d\sigma_r = \left(\frac{\partial \sigma_r}{\partial c} + V\frac{\partial \sigma_r}{\partial r}\right)dc$$

Equation (3.288) can be written in terms of velocity:

$$\frac{\partial V}{\partial r} + \frac{k}{\beta}\frac{V}{r} = \frac{\chi}{\beta}\left(\frac{\partial \sigma_r}{\partial c} + V\frac{\partial \sigma_r}{\partial r}\right) \tag{3.290}$$

Substituting the expressions for radial stress in the plastic zone (3.278) and (3.282) into equation (3.290) provides the following differential equation for velocity:

$$\frac{\partial V}{\partial r} + P(r)V = Q(r) \tag{3.291}$$

in which

$$P(r) = \frac{k}{\beta r} - \frac{\chi q k(\alpha-1)}{\alpha\beta}\left(\frac{c}{r}\right)^{\frac{k(\alpha-1)}{\alpha}}\frac{1}{r} \tag{3.292}$$

$$Q(r) = \frac{s}{c}\left(\frac{c}{r}\right)^{\frac{k(a-1)}{a}} \tag{3.293}$$

and q and s are defined by:

$$q = \frac{(1+k)a[Y+(a-1)p_0]}{(a-1)(k+a)}$$

$$s = -\frac{\chi q k(a-1)}{a\beta}$$

It is noted that the velocity is known on the plastic boundary from the solution for the displacement in the elastic region. Thus from (3.285):

$$V_{r=c} = \delta(1+k) \tag{3.294}$$

With the above boundary condition, equation (3.291) can be solved to give the following solution for velocity V:

$$V = \exp\left[-\frac{\chi q}{\beta}\left(\frac{c}{r}\right)^{\frac{k(a-1)}{a}}\right]\left\{\sum_{n=0}^{\infty} A_n\left(\frac{c}{r}\right)^{\frac{k(a-1)(1+n)}{a}-1} + \left[\delta(1+k)\exp\left(\frac{\chi q}{\beta}\right) - \sum_{n=0}^{\infty} A_n\right]\left(\frac{c}{r}\right)^{\frac{k}{\beta}}\right\} \tag{3.295}$$

in which A_n is defined by:

$$A_n = \frac{1}{n!}\left(\frac{\chi q}{\beta}\right)^n \frac{a\beta s}{ka - k\beta(a-1)(1+n) + a\beta} \tag{3.296}$$

and n is an integer ranging from zero to infinity.

At the cavity wall $r=a$, $V=da/dc$, so that:

$$\frac{da}{dc} = \exp\left[-\frac{\chi q}{\beta}\left(\frac{c}{a}\right)^{\frac{k(a-1)}{a}}\right]\left\{\sum_{n=0}^{\infty} A_n\left(\frac{c}{a}\right)^{\frac{k(a-1)(1+n)}{a}-1} + \left[\delta(1+k)\exp\left(\frac{\chi q}{\beta}\right) - \sum_{n=0}^{\infty} A_n\right]\left(\frac{c}{a}\right)^{\frac{k}{\beta}}\right\} \tag{3.297}$$

For cavity expansion from zero initial radius, the deformation is assumed to be geometrically similar in the plastic zone such that the ratio of the radius of the elastic-plastic boundary to that of cavity wall is constant, namely:

$$\frac{da}{dc} = \frac{a}{c} \tag{3.298}$$

With the above relation, equation (3.297) reduces to a nonlinear equation of c/a:

$$\frac{a}{c} = \exp[-\frac{\chi q}{\beta}(\frac{c}{a})^{\frac{k(\alpha-1)}{\alpha}}]\left\{\sum_{n=0}^{\infty} A_n(\frac{c}{a})^{\frac{k(\alpha-1)(1+n)}{\alpha}-1}\right.$$

$$\left. + [\delta(1+k)\exp(\frac{\chi q}{\beta}) - \sum_{n=0}^{\infty} A_n](\frac{c}{a})^{\frac{k}{\beta}}\right\} \tag{3.299}$$

which can be solved for the value of c/a. Once c/a is determined, equation (3.284) can be used to determine the limit pressure p.

Neglecting the convected part of the stress rate

If the convected part of the stress rate is ignored, equation (3.290) reduces to:

$$\frac{\partial V}{\partial r} + \frac{k}{\beta}\frac{V}{r} = \frac{\chi}{\beta}\frac{\partial \sigma_r}{\partial c} \tag{3.300}$$

which can be simplified to:

$$\frac{\partial V}{\partial r} + \frac{k}{\beta}\frac{V}{r} = Q(r) \tag{3.301}$$

in which $Q(r)$ is given by equation (3.293). With the boundary condition (3.294), it is possible to solve this equation and find the following solution for velocity V:

$$V = \gamma(\frac{c}{r})^{\frac{k(\alpha-1)}{\alpha}-1} + [\delta(1+k)-\gamma](\frac{c}{r})^{\frac{k}{\beta}} \tag{3.302}$$

where

$$\gamma = \frac{\alpha\beta s}{k\alpha - k\beta(\alpha-1) + \alpha\beta} \tag{3.303}$$

Following the same argument as used before, equation (3.302) can be used to obtain a nonlinear equation on c/a:

$$1 = \gamma(\frac{c}{a})^{\frac{k(\alpha-1)}{\alpha}} + [\delta(1+k)-\gamma](\frac{c}{a})^{1+\frac{k}{\beta}} \tag{3.304}$$

The above equation is the same as that derived by Carter *et al.* (1986).

3.4 SUMMARY

1. For spherical cavities, the von Mises yield criterion is identical to the Tresca criterion. For the expansion of an infinitely long, plane strain cylindrical cavity, the von Mises criterion can be approximated by the Tresca criterion if the shear strength is multiplied by a factor of 1.15.

2. The large strain cavity pressure-displacement relationship for the expansion of a spherical cavity from a non-zero initial radius in a Tresca material with a finite boundary cannot be expressed in a single equation. However the complete cavity expansion curves can be derived by combining equations (3.19) and (3.32). For the special case of cavity expansion from zero radius in an infinite incompressible medium, the cavity pressure is a constant that is defined by equation (3.34).

3. The large strain cavity pressure-displacement relationship for the expansion of a cylindrical cavity from a non-zero initial radius in a Tresca material with a finite boundary can be derived by combining equations (3.54) and (3.60). Two special cases are considered. First, for cavity expansion from zero radius in an infinite incompressible medium, the cavity pressure remains a constant and is defined by equation (3.34). Second, for finite cavity expansion in an infinite incompressible material, the large strain cavity pressure-displacement relationship can be expressed by the single equation (3.67). As will be discussed in Chapter 8, this solution serves as the theoretical basis for many interpretation methods for pressuremeter tests in clay.

4. The large strain relationship between cavity pressure and displacement for the unloading of cylindrical and spherical cavities in Tresca materials from an *in-situ* stress state is defined by equation (3.86). With the assumption of small strains, this solution reduces to equation (3.88). These solutions will be used in Chapter 10 as a theoretical tool for predicting ground settlements caused by underground tunnelling in clay.

5. The large strain relationship between cavity pressure and displacement for the unloading of cylindrical and spherical cavities in Tresca materials from a limiting plastic state is defined by equation (3.109). This solution will be used in Chapter 8 as a theoretical basis for interpreting cone pressuremeter tests in clay.

6. The large strain cavity pressure-displacement relationship for the expansion of spherical and cylindrical cavities from a non-zero initial radius in a Mohr-Coulomb material with a finite boundary cannot be expressed in a single equation. However the complete cavity expansion curves can be derived by combining a

number of equations. Simple procedures are given that can be used to compute such a complete pressure-expansion curve. With the assumption of small strains, it is possible to derive a closed form solution. For example, with a cylindrical cavity, the cavity pressure expansion curve is defined by equations (3.185) and (3.163). The cylinder expansion solutions presented in this chapter can be used to assess the effect of chamber size on the measured cone tip resistance.

7. As in the case of finite boundary, the large strain cavity pressure-displacement relationship for the expansion of spherical and cylindrical cavities from a non-zero initial radius in an infinite Mohr-Coulomb material cannot be expressed in a single equation. However the complete cavity expansion curves can be derived by combining a number of equations. Simple procedures are presented for computing such a complete pressure-expansion curve. With the assumption of small strains, the solution for displacement in the plastic zone can be expressed in closed form (equation (3.215)). If we make a further assumption that the elastic deformation in the plastically deforming zone can be ignored, the relationship between cavity pressure and displacement is defined by equation (3.216). This assumption has a considerable effect on the solution for soils with high friction and dilation angles.

8. The large strain cavity pressure-displacement relationship for the unloading of spherical and cylindrical cavities from an elastic-plastic stress state in an infinite Mohr-Coulomb material cannot be expressed in a single equation. However we presented all the equations needed to construct a complete pressure-expansion curve. A simple procedure is given that can be used for conducting such a computation. This unloading solution will be used in Chapter 8 in relation to the development of an interpretation method for pressuremeter unloading tests in sand.

9. By ignoring the elastic deformation in the plastically deforming zone, a closed form, large strain solution is derived for the unloading of cavities from an *in-situ* stress state in an infinite Mohr-Coulomb material. This simplified large strain solution is defined by equation (3.269). This solution will be used in Chapter 10 in relation to the development of an predictive method for soil displacement around a tunnel in cohesive-frictional soils.

10. As a special case, cavity expansion from zero initial radius in an infinite Mohr-Coulomb material possess a similarity solution, in which the cavity pressure is a constant and the continuing deformation is geometrically self-similar. This constant cavity pressure is equal to the limiting pressure achieved at very large strains for finite cavity expansion. The rigorous similarity solution for the cavity pressure can be determined by combining equations (3.284) and (3.299). By neglecting the convected part of the stress rate, the approximate limit solution of

Carter *et al.* (1986) is recovered, that is defined by equations (3.284) and (3.299). The errors introduced by this assumption are generally small for low values of dilation angle but may increase to as much as 20% when soil dilation angle becomes large. The solutions for cavity expansion limit pressures will be used in Chapters 8 and 9 for estimating cone tip resistance and end bearing capacity of driven piles in sand.

REFERENCES

Bigoni, D. and Laudiero, F. (1989). The quasi-static finite cavity expansion in a non-standard elasto-plastic medium. *International Journal of Mechanical Sciences*, 31, 825–837.

Carter, J.P., Booker, J.R. and Yeung, S.K. (1986). Cavity expansion in cohesive frictional soils. *Geotechnique*, 36(3), 349–353.

Chadwick, P. (1959). The quasi-static expansion of a spherical cavity in metals and ideal soils. *Quarterly Journal of Mechanics and Applied Mathematics*, 12, 52–71.

Collins, I.F. and Wang, Y. (1990). Similarity solutions for the quasi-static expansion of cavities in frictional materials. *Research report No.* 489, Department of Engineering Science, University of Auckland, New Zealand.

Davis, E.H. (1968). Theories of plasticity and the failure of soil masses. In: *Soil Mechanics* (Editor: I.K. Lee). London, Butterworths.

Florence, A.L. and Schwer, L.E. (1978). Axisymmetric compression of a Mohr-Coulomb medium around a circular hole. *International Journal for Numerical and Analytical Methods in Geomechanics*, 2, 367–379.

Gibson, R.E. and Anderson, W.F. (1961). In situ measurement of soil properties with the pressuremeter. *Civil Engineering and Public Works Review*, 56, 615–618.

Hill, R. (1950). *The Mathematical Theory of Plasticity.* Oxford University Press.

Houlsby, G.T. and Withers, N.J. (1988). Analysis of the cone pressuremeter test in clay. *Geotechnique*, 38(4), 575–587.

Houlsby, G.T., Clarke, B.G. and Wroth, C.P. (1986). Analysis of unloading of a pressuremeter in sand. *Proceedings of the 2nd International Symposium on Pressuremeters,* ASTM STP 950, 245–264.

Hughes, J.M.O., Wroth, C.P. and Windle, D. (1977). Pressuremeter tests in sands. *Geotechnique*, 27(4), 455–477.

Jefferies, M.G. (1988). Determination of horizontal geostatic stress in clay with self-bored pressuremeter. *Canadian Geotechnical Journal*, 25, 559–573.

Ladanyi, B. (1963). Evaluation of pressuremeter tests in granular soils. *Proc. 2nd Pan American Conference on Soil Mechanics*, San Paulo, 1, 3–20.

Salencon, J. (1969). Contraction quasi-statique d'une cavite a symetrie spherique ou cylindrique dans un milieu elastoplastique. *Annales des Ponts et Chaussees*, 4, 231–236.

Senseny, P.E., Lindberg, H.E. and Schwer, L.E. (1989). Elastic-plastic response of a circular hole to repeated loading. *International Journal for Numerical and Analytical Methods in Geomechanics*, 13, 459–476.

Vesic, A.S. (1972). Expansion of cavities in infinite soil mass. *Journal of the Soil Mechanics and Foundations Division,* ASCE, 98, 265–290.

Wang, Y. (1992). *Cavity Expansion in Sands with Applications to Cone Penetrometer Tests*. PhD Thesis, Auckland University .

Withers, N.J., Howie, J., Hughes, J.M.O. and Robertson, P.K. (1989). Performance and analysis of cone pressuremeter tests in sands. *Geotechnique*, 39, 433–454.

Yu, H.S. (1990). *Cavity Expansion Theory and Its Application to the Analysis of Pressuremeters*. DPhil Thesis, Oxford University .

Yu, H.S. (1992). Expansion of a thick cylinder of soils. *Computers and Geotechnics*, 14, 21–41.

Yu, H.S. (1993). Finite elastoplastic deformation of an internally pressurized hollow sphere. *Acta Mechanica Solida Sinica*, 6(1), 81–97.

Yu, H.S. and Houlsby, G.T. (1991). Finite cavity expansion in dilatant soils: loading analysis. *Geotechnique*, 41(2), 173–183.

Yu, H.S. and Houlsby, G.T. (1995). A large strain analytical solution for cavity contraction in dilatant soils. *International Journal for Numerical and Analytical Methods in Geomechanics*, 19(11), 793–811.

Yu, H.S. and Rowe, R.K. (1999). Plasticity solutions for soil behaviour around contracting cavities and tunnels. *International Journal for Numerical and Analytical Methods in Geomechanics*, 23, 1245–1279.

CHAPTER 4

CRITICAL STATE SOLUTIONS

4.1 INTRODUCTION

In Chapter 3, cavity expansion solutions were developed by assuming that soil can be modelled as an elastic-perfectly plastic material. The key feature of perfect plasticity is that the strength of material remains constant during the loading and unloading process. In reality, however, the strength of soil may vary considerably with deformation history. To account for this dependence in cavity expansion analysis, a strain-hardening/softening plasticity model must be used.

Perhaps the most widely used strain-hardening/softening models in soil mechanics are those based on the critical state concept developed at the University of Cambridge (Schofield and Wroth, 1968; Roscoe and Burland, 1968). Due to the complex nature of the cavity expansion problems in critical state soils, most existing critical state solutions have been developed only in recent years. This chapter will present some of the currently available analytical solutions for cavity expansion in soils modelled by critical state based plasticity models. Both clay and sand are considered.

4.2 CAVITY EXPANSION FROM A FINITE INITIAL RADIUS

4.2.1 Undrained expansion of cavities in clays

This section is concerned with the undrained cavity expansion in both normally and overconsolidated clays. The soil is modelled by a variety of Cam clay critical state models (Schofield and Wroth, 1968; Muir Wood, 1990). The analytical solutions presented in this section were developed by Collins and Yu (1996).

In undrained deformations it is common to work with total stresses. However, this is no longer appropriate in models where the strength of the soil is a variable since the strength is a function of the effective stresses rather than the total stresses. In particular, unlike the effective stress approach, the total stress analysis can not take account of the influence of stress history on soil behaviour.

Four fundamental equations that need to be considered in cavity expansion analysis are: *(a)* conservation of mass or "continuity", *(b)* quasi-static equilibrium, *(c)* the yield condition and *(d)* the elastic/plastic flow rule. In an undrained expansion *(a)* is automatically satisfied as the total volume of each soil element remains constant. This results in a simple relation between the finite shear strain and the position coordinates of the element which is valid in both the elastic and plastic regions. The equilibrium equation *(b)* serves only to determine the excess pore pressure distribution at the end of the calculation after the effective stress distributions have been found by integrating the constitutive equations *(c)* and *(d)*. Since for isotropic materials these two equations are most naturally expressed in terms of the effective mean and deviatoric stresses, the distributions of the effective stress components are most conveniently found by first solving for these two stress invariants.

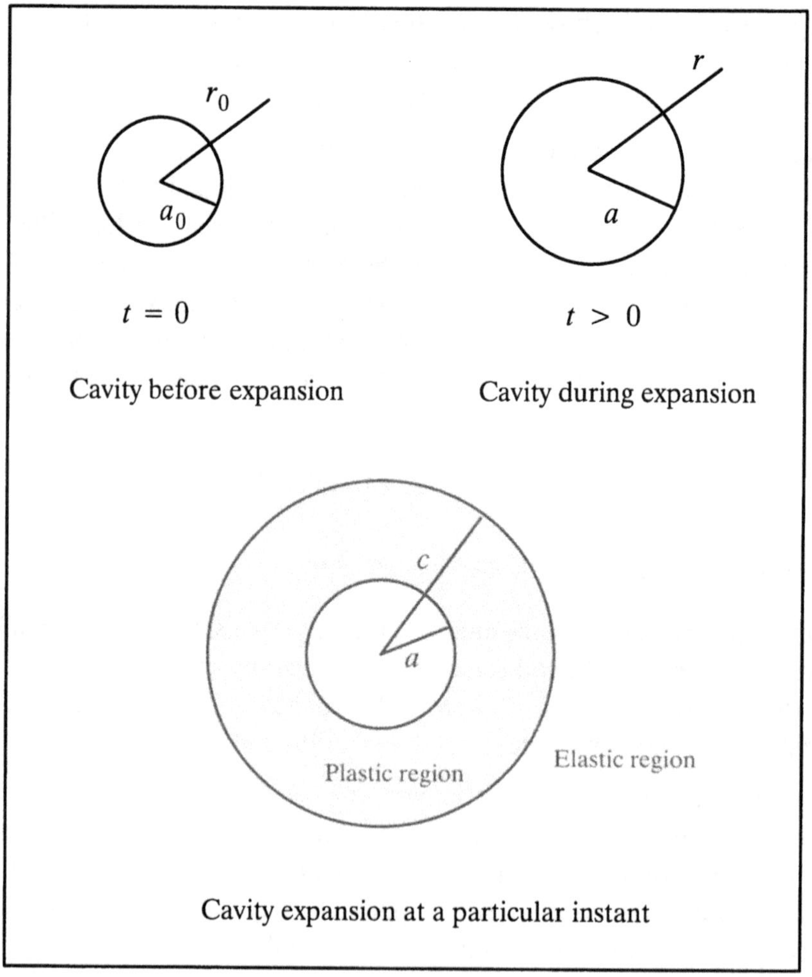

Figure 4.1: Kinematics of cavity expansion

(a) Kinematics of cavity expansion

Since an undrained deformation is necessarily isochoric, the conservation of volume condition gives the following relation between r, the current radius of a material element which was initially at r_0 and the current and initial radii of the cavity – a and a_0 respectively (Figure 4.1):

$$r^{k+1} - r_0^{k+1} = a^{k+1} - a_0^{k+1} \tag{4.1}$$

where k is an index parameter used to denote cylindrical (k=*1*) and spherical cavities (k=2). The radial speed of the element is hence related to the speed of expansion of the cavity by:

$$w = \dot{r} = (\tfrac{a}{r})^k \, \dot{a} \tag{4.2}$$

Hence the radial, circumferential, shear and volumetric strain rates can be expressed as follows:

$$e_r = -\frac{\partial w}{\partial r} = [\frac{ka^k}{r^{k+1}}] \, \dot{a} \tag{4.3}$$

$$e_\theta = -\frac{w}{r} = -[\frac{a^k}{r^{k+1}}] \, \dot{a} \tag{4.4}$$

$$\dot{\gamma} = e_r - e_\theta = [(k+1)\frac{a^k}{r^{k+1}}] \, \dot{a} \tag{4.5}$$

$$\dot{\delta} = e_r + ke_\theta = 0 \tag{4.6}$$

Using equation (4.1) the shear strain rate can also be written in terms of the initial position of the particle r_0:

$$\dot{\gamma} = [\frac{(k+1)\, a^k}{(a^{k+1} + r_0^{k+1} - a_0^{k+1})}] \, \dot{a} \tag{4.7}$$

Since r_0 is fixed for a given particle, equation (4.7) can be integrated to give the finite Lagrangean shear strain:

$$\gamma = \ln(\frac{a^{k+1} + r_0^{k+1} - a_0^{k+1}}{r_0^{k+1}}) = (k+1)\ln\frac{r}{r_0} \tag{4.8}$$

associated with the particle originally located at r_0. This relation can now be rewritten in terms of r, the current coordinate of the particle:

$$\gamma = -\ln[1 - \frac{(a^{k+1} - a_0^{k+1})}{r^{k+1}}] \tag{4.9}$$

or inversely :

$$r^{k+1} = \frac{a^{k+1} - a_0^{k+1}}{1 - \exp(-\gamma)} \tag{4.10}$$

by using equation (4.1) to eliminate r_0. Equation (4.9) gives the distribution of shear strain with radius r at the instant when the current cavity radius is a. From equations (4.8)-(4.10) it follows that the relations between radial and shear strain increments *(a)* for a given particle and *(b)* at a fixed instant of time are respectively:

$$(k + 1)\frac{dr}{r} = d\gamma \quad \text{and} \quad (k + 1)\frac{dr}{r} = -\frac{d\gamma}{\exp(\gamma) - 1} \tag{4.11}$$

Note that at the cavity wall the shear strain is:

$$\gamma_c = (k + 1)\ \ln\frac{a}{a_0} \tag{4.12}$$

which is infinite when the initial cavity radius is zero. It is emphasised that these results are kinematic and apply in both the elastic and elastic/plastic phases of the expansion.

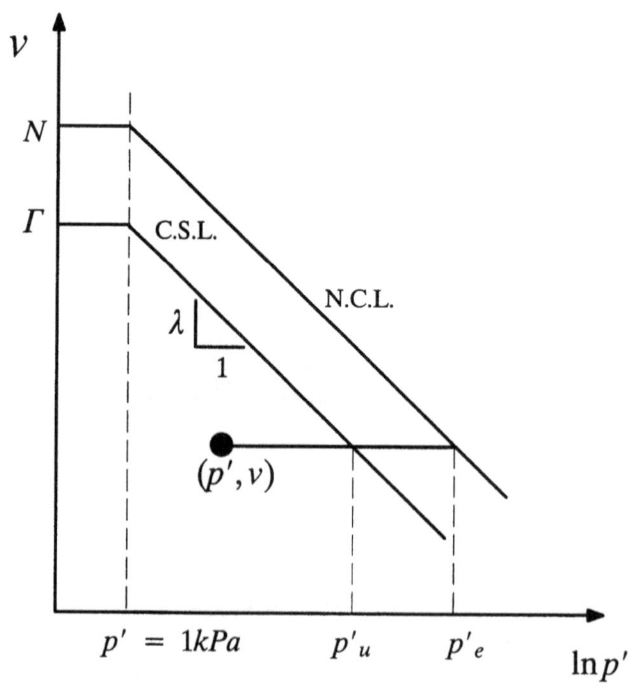

Figure 4.2: Definition of critical state constants

(b) Elastic phase of expansion

Following Collins and Stimpson (1994), the two effective stress invariants defined below are used in the analysis of cavity expansion problems:

$$q = \sigma'_r - \sigma'_\theta \quad \text{and} \quad p' = \frac{\sigma'_r + k\sigma'_\theta}{1+k} \tag{4.13}$$

where σ'_r and σ'_θ are the effective radial and hoop stresses respectively. In order to fully take account of the scaling laws it is convenient to non-dimensionalise all stresses and moduli by some representative stress. Such nondimensional variables will be denoted by superposed bars. The usual convention in critical state soil mechanics is adopted and the equivalent consolidation pressure p'_e as this representative pressure is used (see Figure 4.2). The elastic constitutive law is most conveniently expressed in rate form:

$$\dot{\delta}^e = \frac{\overset{\circ}{\bar{p}'}}{\bar{K}(\bar{p}', v)} \quad \text{and} \quad \dot{\gamma}^e = \frac{\overset{\circ}{\bar{q}}}{2\bar{G}(\bar{p}', v)} \tag{4.14}$$

in which $\dot{\delta}^e$ and $\dot{\gamma}^e$ represent the elastic volumetric and shear strain rates respectively; $\overset{\circ}{\bar{p}'}$ and $\overset{\circ}{\bar{q}}$ are the material rates of change of the non-dimensional, effective mean and shear stress invariants. The instantaneous bulk and shear moduli are both functions of the specific volume v and mean effective pressure p' in general, so that the elastic stress-strain relation obtained by integration will be nonlinear. The symbol $\overset{\circ}{()}$ denotes the material time derivative associated with a given material particle and it is related to the local time derivative $\dot{()}$, evaluated at fixed position r, by

$$\overset{\circ}{()} = \dot{()} + w\frac{\partial()}{\partial r} \tag{4.15}$$

where w is the radial speed of a material element.

In the initial purely elastic phase of an undrained expansion the elastic volumetric strain rate $\dot{\delta}^e = 0$, so that from equation (4.14) the effective mean pressure remains constant and is equal to its initial value p'_0. The instantaneous elastic bulk and shear moduli hence also remain constant and equal to their initial values K_0 and G_0 respectively. The second part of equation (4.14) for the elastic shear strain rate can be integrated along a particle path, so that the shear stress invariant q is just twice the initial elastic shear modulus times the finite shear strain γ, namely:

$$\gamma = \frac{\bar{q}}{2\bar{G}_0} \tag{4.16}$$

The radial and circumferential components of effective stress are therefore given by:

$$\overline{\sigma}_r' = \overline{p}'_0 + \frac{2\overline{G}_0\gamma k}{k+1} \quad \text{and} \quad \overline{\sigma}'_\theta = \overline{p}'_0 - \frac{2\overline{G}_0\gamma}{k+1} \tag{4.17}$$

These stresses can now be expressed in terms of the radial position coordinates by eliminating γ using either equation (4.8) or equation (4.9). It is noted that these effective stress distributions have been found without reference to the equilibrium equations and without the need to make any small strain assumptions. Since the effective mean pressure is constant the stress path of a material element in the elastic phase of the expansion is a vertical line in the $\overline{q} - \overline{p}'$ diagram (Figure 4.3).

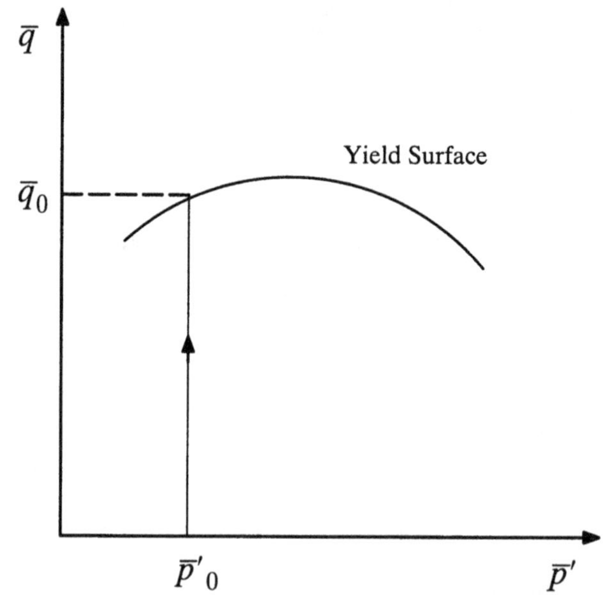

Figure 4.3: Stress path for elastic phase of cavity expansion

The soil first goes plastic at the cavity wall when the shear stress invariant reaches the yield value q_0, depending upon the particular yield criterion. The corresponding shear strain is:

$$\gamma_0 = (k+1)\ln\frac{a_1}{a_0} = \frac{\overline{q}_0}{2\overline{G}_0} = \frac{q_0}{2G_0} \tag{4.18}$$

where a_1 is the cavity radius at the onset of yielding, γ_0 is the shear strain to yield and is a measure of the compliance of the material. In a perfectly plastic model q_0 is twice the undrained shear strength so that γ_0 is the reciprocal of the rigidity index. This is not true in a more general model.

These results are also valid in the outer elastic region during the elastic-plastic phase of the expansion. The radius of the elastic/plastic boundary c at the instant when the cavity has radius a is given by:

$$c^{k+1} = \frac{a^{k+1} - a_0^{k+1}}{1 - \exp(-\gamma_0)} \tag{4.19}$$

using equation (4.9) or (4.10). This relation between the shear strain and radial coordinate can alternatively be written as:

$$\left(\frac{r}{c}\right)^{k+1} = \frac{1 - \exp(-\gamma_0)}{1 - \exp(-\gamma)} \tag{4.20}$$

(c) Elastic-plastic phase of expansion

Effective stress distributions

Here the basic solution is developed in a general form appropriate to a wide class of materials, where the yield condition and plastic flow rule can be written in the form:

$$\bar{q} = f(\bar{p}') \quad \text{and} \quad \frac{\dot{\delta}^p}{\dot{\gamma}^p} = g(\bar{p}') \tag{4.21}$$

In an undrained deformation the total volumetric strain-rate is zero, so that $\dot{\delta}^e = -\dot{\delta}^p$. It follows from equations (4.14) and (4.21) that the total strain-rate is:

$$\dot{\gamma} = \dot{\gamma}^e + \dot{\gamma}^p = L(\bar{p}')\overset{\circ}{\bar{p}} \tag{4.22}$$

where

$$L(\bar{p}') = \frac{f'(\bar{p}')}{2\bar{G}(\bar{p}')} - \frac{1}{\bar{K}(\bar{p}')\, g(\bar{p}')} \tag{4.23}$$

Integrating equation (4.22) along a particle path starting at the elastic-plastic boundary gives a relation between the finite shear strain and the effective mean pressure:

$$\gamma = \gamma_0 + I(\bar{p}') - I(\bar{p}'_0) \tag{4.24}$$

where

$$I(\bar{p}') = \int^{\bar{p}'} L(\bar{p}')\, d\bar{p}' \tag{4.25}$$

As a special case, equation (4.24) describes the relationship between the cavity pressure and the cavity shear strain after the cavity wall becomes plastic. As will be shown in a later section the integral (4.25) can be evaluated analytically for the original Cam clay model and is readily evaluated numerically in other cases. The variation of $\overline{p}'$ with radius r can be obtained implicitly by eliminating γ between equations (4.24) and (4.8), (4.9) or (4.20).

Calculation of excess pore pressures

The distribution of pore pressure $U(r)$ can be calculated from the quasi-static radial equilibrium equation:

$$\frac{d\,\overline{\sigma}_r}{d\,r} + k\,\frac{\overline{\sigma}_r - \overline{\sigma}_\theta}{r} = 0 \tag{4.26}$$

Since $\overline{\sigma}_r = \overline{p} + (k/(k+1))\,\overline{q}$ and $\overline{p} = \overline{U} + \overline{p}'$, the non-dimensional pore pressure gradient is given by:

$$\frac{d\,\overline{U}}{d\,r} = -\frac{d\,\overline{p}'}{d\,r} - \frac{k}{k+1}\frac{d\,\overline{q}}{d\,r} - \frac{k\overline{q}}{r} \tag{4.27}$$

As the effective mean stress distribution is constant in the elastic region, the change in the pore pressure (i.e. excess pore pressure) in the elastic zone is provided by:

$$\Delta\,\overline{U} = -\frac{k}{k+1}\,\overline{q} - k\int \overline{q}\,\frac{d\,r}{r} \tag{4.28}$$

But $\overline{q}$ and (dr/r) can both be expressed in terms of γ from equation (4.16) and the second equation of (4.11), so that equation (4.28) becomes:

$$\Delta\,\overline{U} = -\frac{2k\overline{G}_0}{k+1}\left[\,\gamma - \int_0^{\gamma} \frac{\gamma}{\exp(\gamma) - 1}\,d\gamma\,\right] \doteq -\frac{k\overline{G}_0\gamma^2}{2(k+1)} \tag{4.29}$$

to the second order in γ.

The excess pore pressure is hence constant in the elastic region to first order in the shear strain as already noted by Collins and Stimpson (1994). To second order in strain, the excess pore pressure at the elastic-plastic boundary is:

$$\Delta\,\overline{U}_0 = -\frac{k\overline{G}_0\gamma_0^2}{2(k+1)} \tag{4.30}$$

Excess pore pressures in plastic regions

Integrating equation (4.27) through the plastic region from the elastic/plastic boundary yields a relationship between the excess pore pressure and the finite shear strain:

$$\Delta\,\bar{U} = \Delta\bar{U}_0 - (\bar{p}' - \bar{p}'_0) - \frac{k}{k+1}\,(\bar{q} - \bar{q}_0 - (J(\gamma) - J(\gamma_0))) \qquad (4.31)$$

where the integral J is most conveniently evaluated numerically by expressing both $\bar{q}$ and r in terms of $\bar{p}$ using equations (4.11), (4.21) and (4.22):

$$J(\gamma) = \int^{\gamma} \frac{\bar{q}}{(\exp(\gamma) - 1)}\, d\gamma = \int^{\gamma} \frac{f(\bar{p}')\,L(\bar{p}')}{\exp(\gamma) - 1}\, d\bar{p}' \qquad (4.32)$$

Again, as a special case, equation (4.31) represents the plastic relationship between the excess pore pressure and the finite shear strain for the cavity wall. Once the effective stress state has essentially reached the critical state, the value of $\bar{q}$ is effectively constant and equation (4.32) can then be integrated analytically to give:

$$J(\gamma) \doteq \bar{q}_{cs}\,\ln(1 - \exp(-\gamma)) \qquad (4.33)$$

(d) A special case: perfectly plastic model

Before discussing the solutions for various critical state models it is instructive to firstly review the solution for a perfectly plastic model when yielding is governed by the Tresca criterion. This corresponds to the situation where under undrained loading the in situ soil behave purely elastically before reaching the critical state. Since the shear stress and effective mean pressure are now constant through the plastic annulus the J integral in (4.32) can be evaluated analytically, giving:

$$\Delta\,\bar{U} = \Delta\bar{U}_0 + \frac{k}{k+1}\,\bar{q}_0\,\ln\frac{1 - \exp(-\gamma)}{1 - \exp(-\gamma_0)} \qquad (4.34)$$

or, in terms of the radial coordinate:

$$\Delta\,\bar{U} = \Delta\bar{U}_0 + k\bar{q}_0\,\ln\tfrac{c}{r} = \Delta\bar{U}_0 + \frac{k}{k+1}\,\bar{q}_0\,\ln\frac{(\frac{a}{r})^{k+1} - (\frac{a_0}{r})^{k+1}}{1 - \exp(-\gamma_0)} \qquad (4.35)$$

using equations (4.19) and (4.20). To first order in the elastic limit strain γ_0 the excess pore pressure at the cavity wall is:

$$\Delta\,\bar{U}_c \doteq \frac{k}{k+1}\,\bar{q}_0\,[\,\ln(1 - (\tfrac{a_0}{a})^{k+1}) + \ln I_r\,] \qquad (4.36)$$

where $I_r = 1/\gamma_0$ is the well known rigidity index. Using the above equation, the well known total radial stress solution at the cavity wall derived by Gibson and Anderson (1961) is recovered:

$$\bar{\sigma}_r|_c = \bar{p}_0 + \frac{k}{k+1}\,\bar{q}_0\,[\,1 + \ln(1 - (\tfrac{a_0}{a})^{k+1}) + \ln I_r\,] \qquad (4.37)$$

In the initial stages of the deformation where γ is small, equation (4.36) reduces to:

$$\Delta\,\overline{U}_c \doteq \frac{k}{k+1}\,\overline{q}_0\,[\,\ln\gamma_c + \ln I_r\,] \tag{4.38}$$

showing that in the early phases of the expansion the excess pore pressure is proportional to the logarithm of the cavity shear strain. On the other hand for large expansions $a_0/a \to 0$:

$$\Delta\,\overline{U}_c \doteq \frac{k}{k+1}\,\overline{q}_0\,\ln I_r \tag{4.39}$$

$$\overline{\sigma}_r|_c = \overline{p}_0 + \frac{k}{k+1}\,\overline{q}_0\,(\,1 + \ln I_r\,) \tag{4.40}$$

which are the well known limiting solutions for the excess pore pressure and the total cavity pressure in a cavity expanding from zero initial radius in the perfectly plastic material.

(e) Critical state plasticity models

The original Cam clay model for both normally and overconsolidated clays

The yield function in the original Cam clay model proposed by Schofield and Wroth (1968) is:

$$\overline{q} = f(\overline{p}') = -\frac{M}{\Lambda}\,\overline{p}'\ln\overline{p}' \tag{4.41}$$

where the stresses have been non-dimensionalised by the equivalent consolidation pressure at the same specific volume v:

$$p'_e = \exp\left(\frac{N-v}{\lambda}\right) \tag{4.42}$$

The constant $\Lambda = 1 - \varkappa/\lambda$, where $\varkappa$ and λ are the slopes of the elastic swelling line and normal consolidation line respectively, in $\ln p' - v$ space and N is the value of v on the normal consolidation line when $p' = 1$kPa. The final critical state constant M is the slope of the critical state line in $\overline{p}' - \overline{q}$ space.

In this model the elastic moduli are given by:

$$\overline{K} = \frac{v\overline{p}'}{\varkappa} \quad \text{and} \quad \overline{G} = a\overline{K} \tag{4.43}$$

where $a = (1 + k)(1 - 2\mu)/(2(1 + (k - 1)\mu))$ and μ denotes Poisson's ratio. Some researchers assume μ to be constant whilst others keep G fixed and use (4.43) to calculate Poisson's ratio (Atkinson and Bransby, 1978; Muir Wood, 1990). In the present context it proves advantageous to assume a constant value of Poisson's ratio, since then all the material parameters defining the model are dimensionless, which results in a number of simplifying scaling laws. Nevertheless, it will be demonstrated in the next section that a constant shear modulus can also be included in the present cavity expansion solution procedure.

The overconsolidation ratio (OCR) in terms of the mean effective stress is:

$$n_p = (\bar{p}'_0)^{-\frac{1}{\Lambda}} \tag{4.44}$$

At the critical state $\bar{q}/\bar{p}' = M$, $n_p = e$ and $\bar{p}' = e^{-\Lambda}$, whilst the undrained stress path in $\bar{p}' - \bar{q}$ space has a maximum value when $\bar{p}' = 1/e$ and $\bar{q}/\bar{p}' = M/\Lambda$ as shown in Figure 4.4.

The ratio of the plastic volumetric and shear strain rates calculated from the normal flow rule is:

$$\frac{\dot{\delta}^p}{\dot{\gamma}^p} = g(\bar{p}') = \frac{kM}{(k+1)\Lambda}(\Lambda + \ln\bar{p}') \tag{4.45}$$

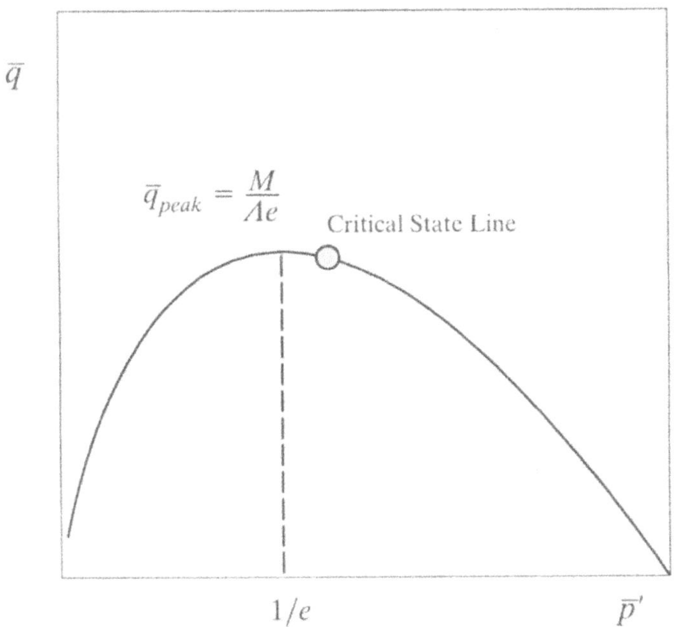

Figure 4.4: The original Cam clay yield surface

Hence, the function $L(\bar{p}')$ needed to calculate the effective pressure distributions in equation (4.23) is:

$$L(\bar{p}') = -\frac{A(1 + \ln\bar{p}')}{\bar{p}'} - \frac{B}{\bar{p}'(\Lambda + \ln\bar{p}')} \tag{4.46}$$

and upon integration the function needed in equation (4.24) to calculate the shear strain is:

$$I(\bar{p}') = -A(\ln\bar{p}' + \frac{1}{2}(\ln\bar{p}')^2) - B\ln|(\Lambda + \ln\bar{p}')| \tag{4.47}$$

where $A = M\kappa/(2\Lambda\alpha v)$ and $B = (k+1)\Lambda\kappa/(kMv)$ are constants. Note that the value of the integral I, and hence the shear strain, is inversely proportional to the specific volume v.

Original Cam clay for normally consolidated and lightly overconsolidated clays – The Hvorslev yield surface for heavily overconsolidated clays

It is well known that for overconsolidated clays, the original Cam clay yield criterion tends to significantly overpredict the soil strength. In this region, the Hvorslev surface has often been used as the yield function. The Hvorslev yield surface is a straight line in $\bar{p}' - \bar{q}$ space (Atkinson and Bransby, 1978):

$$\bar{q} = h\bar{p}' + (M - h)\exp(-\Lambda) \tag{4.48}$$

where h is the slope of the Hvorslev yield surface, Figure 4.5.

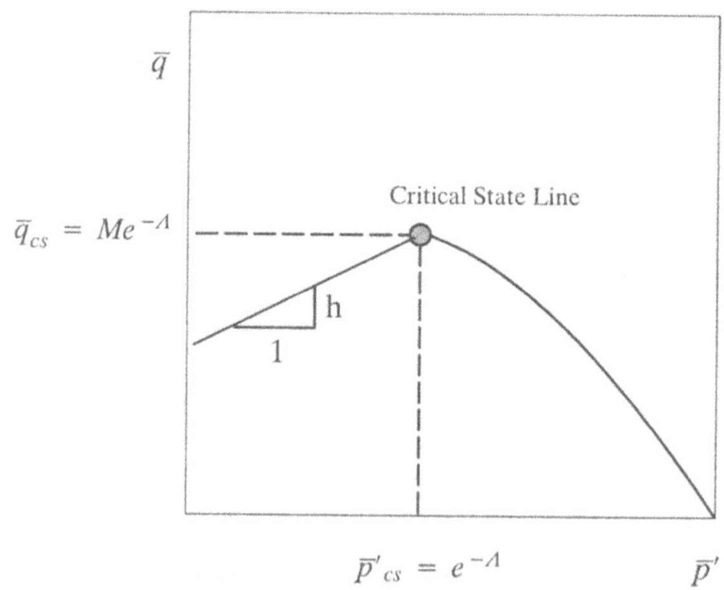

Figure 4.5: The original Cam clay–Hvorslev yield surface

It has been found that the use of equation (4.43) for heavily overconsolidated clays results in unrealistically low values of the elastic moduli. To overcome this shortcoming, Randolph *et al.* (1979) proposed a more realistic hypothesis by selecting G as half of the maximum value of the elastic bulk modulus, K_{max} that was ever reached during the loading history of the soil. In their proposal, the bulk modulus is still assumed to be pressure dependent, and so the resulting model is conservative for elastic behaviour (Zytynski *et al.*, 1978). Equation (4.43) can be replaced by the following expression for heavily overconsolidated clays:

$$\overline{K} = \frac{v\overline{p}'}{\varkappa} \quad \text{and} \quad \overline{G} = \frac{v + \lambda(\Lambda - 1)\ \ln n_p}{2\varkappa}\ (n_p)^{1-\Lambda} \tag{4.49}$$

Using equation (4.48) as the yield function, equation (4.41) as the plastic potential and equation (4.49) for the elastic moduli, we obtain:

$$L(\overline{p}') = \frac{h}{2\overline{G}} - \frac{(1+k)\varkappa}{kv(M-h)}\ \frac{1}{(\overline{p}' - \exp(-\Lambda))} \tag{4.50}$$

and

$$I(\overline{p}') = \frac{h\overline{p}'}{2\overline{G}} - \frac{(1+k)\varkappa}{kv(M-h)}\ \ln(\overline{p}' - \exp(-\Lambda)) \tag{4.51}$$

where the constant $\overline{G}$ is given by equation (4.49).

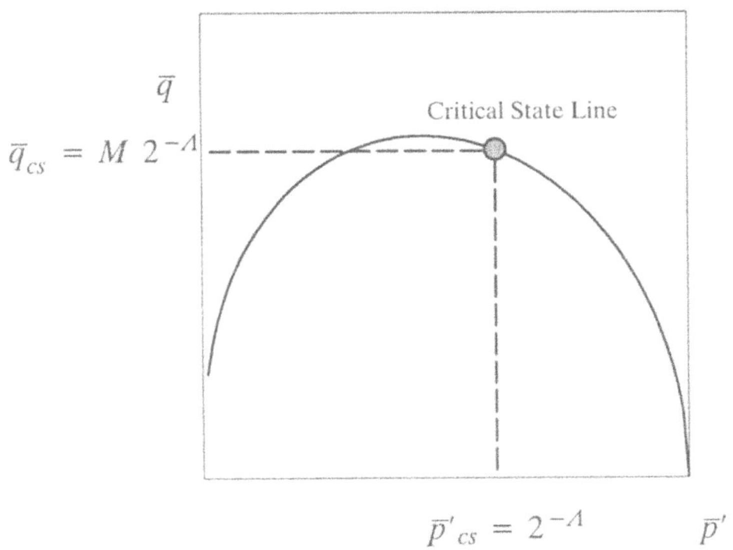

Figure 4.6: The modified Cam clay yield surface

Modified Cam clay for both normally and overconsolidated clays

The yield function in the modified Cam clay model (Muir Wood, 1990) is:

$$\bar{q} = f(\bar{p}') = M\bar{p}' \sqrt{(\bar{p}'^{-\frac{1}{\Lambda}} - 1)} \tag{4.52}$$

At the critical state $\bar{q}/\bar{p}' = M$, $n_p = 2$ and $\bar{p}' = 2^{-\Lambda}$ as shown in Figure 4.6.

The ratio of the plastic volumetric and shear strain rates calculated from the normal flow rule is:

$$\frac{\dot{\delta}^p}{\dot{\gamma}^p} = g(\bar{p}') = -\frac{kM}{2(k+1)} \frac{\bar{p}'^{-1/\Lambda}}{(\bar{p}'^{-1/\Lambda} - 1)^{1/2}} \tag{4.53}$$

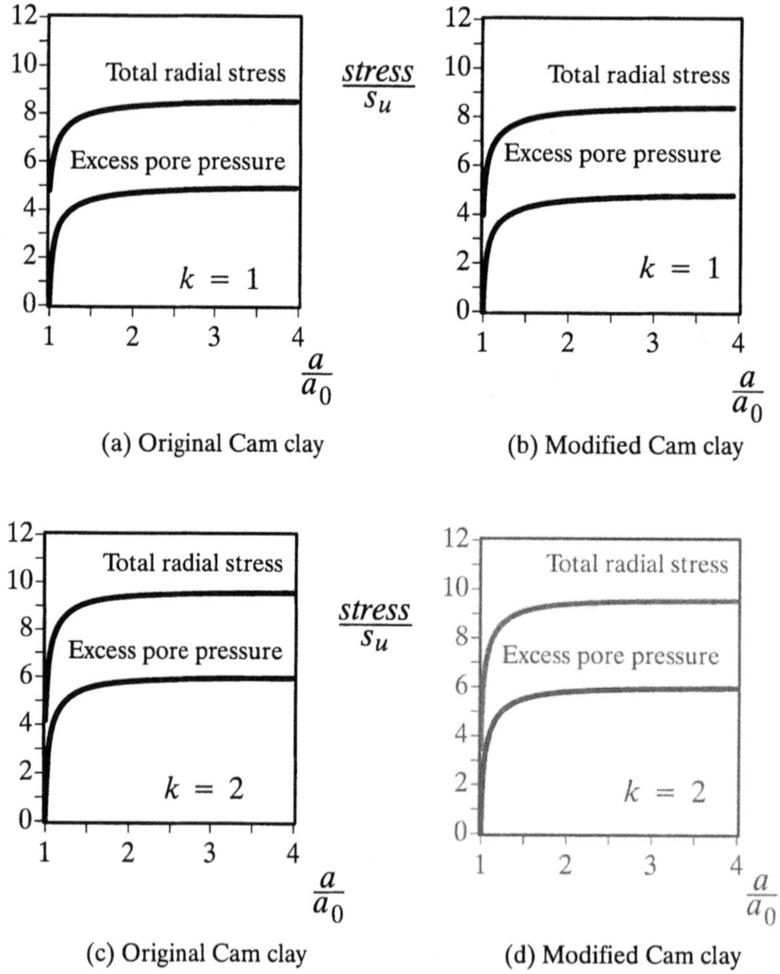

Figure 4.7: Cavity expansion curves for OCR of $n_p = 1.001$

The function $L(\bar{p}')$ needed to calculate the effective pressure distributions in equation (4.23) can be obtained by using (4.52) and (4.53). Unlike the original Cam clay model, the function needed in equation (4.24) to calculate the shear strain for the modified Cam clay model can not be obtained in a closed form and instead a simple numerical integration must be used.

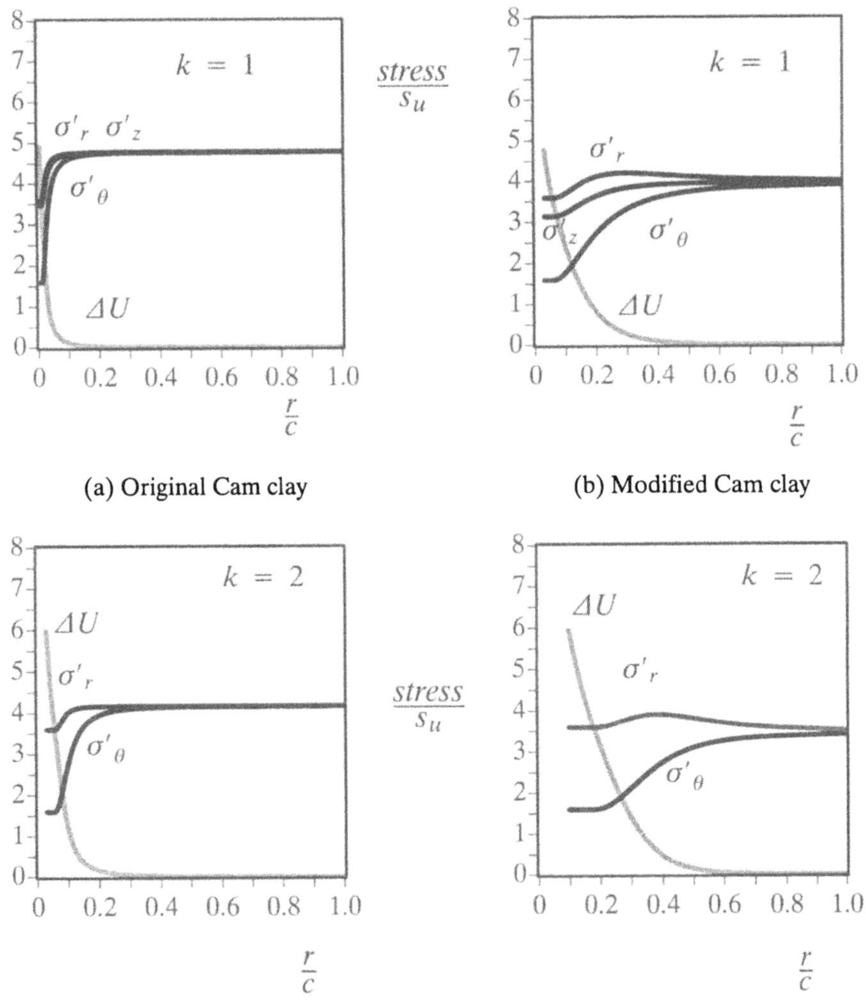

(a) Original Cam clay

(b) Modified Cam clay

(c) Original Cam clay

(d) Modified Cam clay

Figure 4.8: Stress distributions in plastic zone for OCR of $n_p = 1.001$

(f) Sample results

In this section sample results are presented for the cavity expansion curves and stress distributions for the undrained cavity expansion in critical state soils. The values of the critical state parameters chosen for the examples presented here those relevant to London clay: $\Gamma = 2.759$, $\lambda = 0.161$, $\varkappa = 0.062$, the critical state friction angle $\phi'_{cs} = 22.75^{\circ}$ and the Hvorslev friction angle $\phi'_{hc} = 19.7^{\circ}$ (Muir Wood, 1990). All results presented relate to the case when the specific volume of the soil v is equal to 2.0. The Poisson ratio μ is assumed to be 0.3 for all the calculations when the original and modified Cam clay models are used. If the critical state friction angle of the soil is assumed to be the same for both the triaxial and plane strain loading conditions (see Muir Wood, 1990 on page 178 for experimental evidence and further discussion), the values of M for both spherical and cylindrical cavities can be determined using $M = 6\sin\phi'_{cs}/(3 - \sin\phi'_{cs})$ and $M = 2\sin\phi'_{cs}$ respectively. Similar expressions can be used to determine h from ϕ'_{hc}. Following Yu and Houlsby (1991), the vertical effective stress σ'_z for the cylindrical case can be obtained from the other two stress components by using the plane strain condition in the vertical direction. This condition therefore requires the elastic vertical strain rate to be zero since equations (4.13) and (4.21) together with the plastic flow rule imply that the plastic vertical strain rates are zero.

The numerical results are obtained with the original Cam clay and the modified Cam clay models. Cavity expansion curve and stress distribution in the plastically deformed region at the instant of $a/a_0 = 4$ for the overconsolidation ratios of $n_p = 1.001$ and 8 are presented in Figure 4.7 to Figure 4.10. Both spherical and cylindrical cases have been included. The reason for using $n_p = 1.001$ to represent a normally consolidated clay is that when $n_p = 1$ the shear strain required to reach the yield surface is zero and the radius of the elastic-plastic boundary becomes indeterminate. All the stresses and pressures have been normalised by the undrained shear strength of the soil which is defined by $s_u = 0.5M\exp((\Gamma - v)/\lambda)$. It should also be stressed that the total radial stress values presented in this section do not include ambient pore pressure. As far as the cavity expansion curves are concerned, it is interesting to note that the responses of the modified Cam clay and the original Cam clay are similar for normally consolidated clays. For overconsolidated soils, the cavity expansion curves predicted by the original Cam clay model are slightly stiffer than those from the modified Cam clay model. In particular, the limit solutions of the total cavity pressure and the excess pore pressure predicted by the modified Cam clay model are typically 10–20% smaller than those obtained with the original Cam clay model. With reference to the stress distributions, it is found that the radius of the elastic-plastic boundary calculated with the original Cam clay model is much greater than that from the modified Cam clay model. This is because the shear strain required to reach the modified Cam clay yield surface γ_0 is greater than that required to reach the original Cam clay yield surface. It is also found that when the overconsolidation ratio increases, the excess pore pressure could become negative and this is mainly due to the fact that very low elastic moduli are obtained from equation (4.43) for heavily overconsolidated clays.

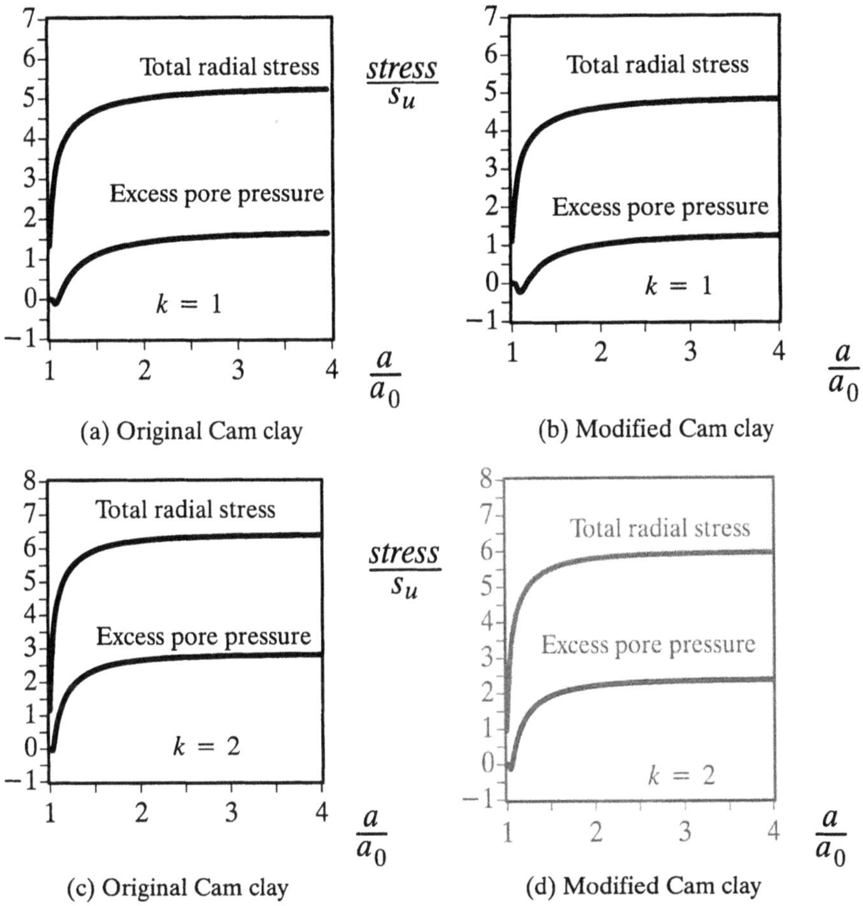

Figure 4.9: Cavity expansion curves for OCR of $n_p = 8$

A comparison between the results from the Hvorslev yield criterion and those of the original and modified Cam clay models detailed in Collins and Yu (1996) indicates that the limit solutions of the total cavity pressure and the excess pore pressure derived from the Hvorslev yield criterion are slightly higher than those from the original and modified Cam clay models. However, this difference tends to decrease with the increase of the overconsolidation ratio. It is also noted that the radius of the elastic-plastic boundary calculated with the Hvorslev model is much greater than those from both the original and modified Cam clay models. Unlike the original and modified Cam clay models, no negative excess pore pressures have been observed in the results of the Hvorslev yield criterion when the overconsolidation ratio is less than 16. This is because a constant shear modulus has been used in the Hvorslev model.

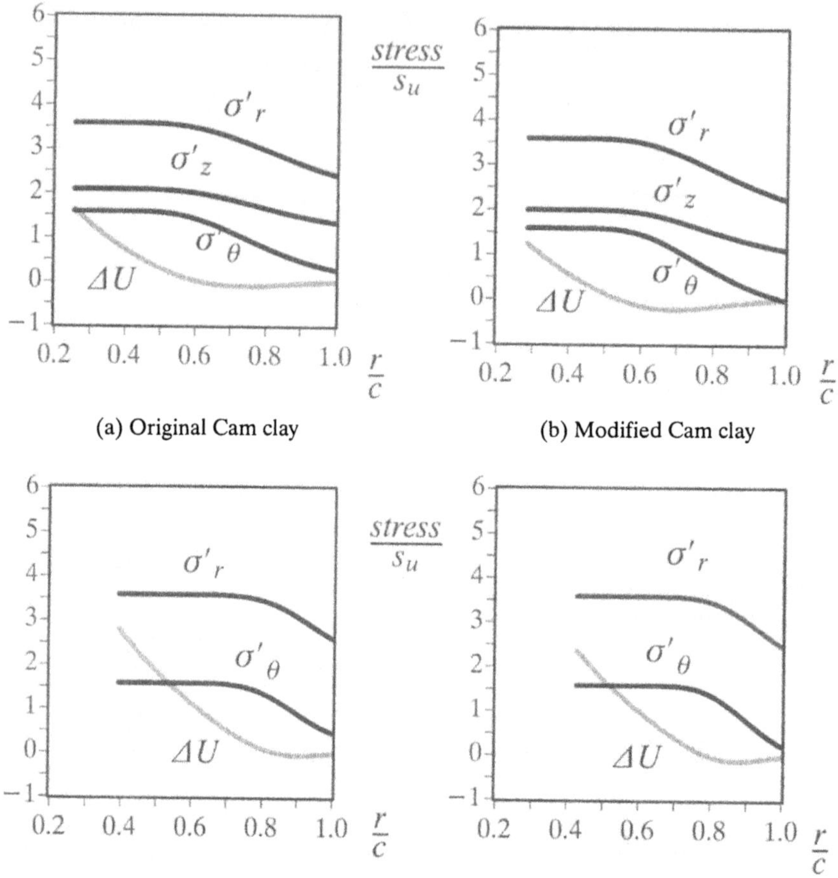

(a) Original Cam clay

(b) Modified Cam clay

(c) Original Cam clay

(d) Modified Cam clay

Figure 4.10: Stress distributions in plastic zone for OCR of $n_p = 8$

In summary, it may be concluded that the critical state model has a greater effect on the results for excess pore pressures, effective stress distributions and the size of the plastically deformed region as compared to the total cavity pressures.

4.2.2 Undrained contraction of cavities in clays

Yu and Rowe (1999) extended the loading solution presented in the previous section to the case when a cavity is unloaded from its initial stress. While some key solutions are presented here, a complete description of these solutions will be given in Chapter 10 when they are applied to predict soil deformation around a tunnel.

Equations (4.1)-(4.33) established for cavity expansion are still valid for the unloading of cavities in critical state soils. The difference lies in the fact that the yield functions for unloading cavities are different from those for loading cavities. Shown in Figure 4.11 is the stress path for the unloading of cavities in critical state soils.

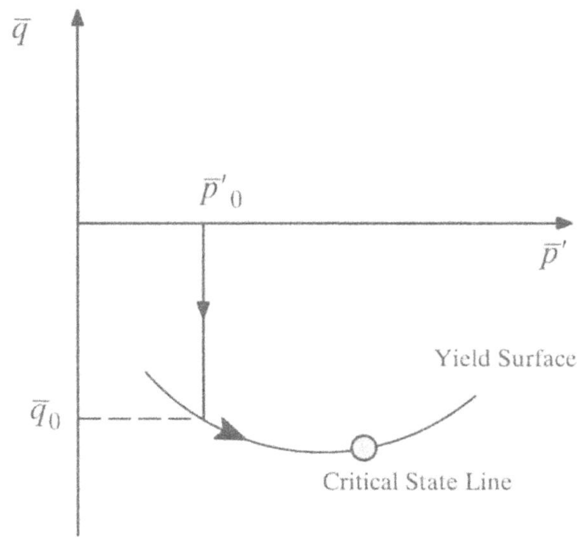

Figure 4.11: Stress path for elastic-plastic cavity unloading

(a) Linear elastic-perfectly plastic Tresca model

As in the loading cavities, before discussing the unloading solutions for various critical state models, it is useful to present the unloading solution for a perfectly plastic Tresca model. This corresponds to the situation where the in situ soil under undrained loading conditions behaves purely elastically before reaching the critical state. Since the shear stress and effective mean pressure are now constant through the plastic annulus the J integral in equation (4.32) can be evaluated analytically, giving:

$$\Delta \bar{U} = \Delta \bar{U}_0 + \frac{k}{k+1} \bar{q}_{cs} \ln \frac{\exp(-\gamma) - 1}{\exp(-\gamma_0) - 1} \tag{4.54}$$

or, in terms of the radial coordinate:

$$\Delta\ \overline{U} = \Delta\overline{U}_0 + k\overline{q}_{cs}\ \ln\frac{c}{r} = \Delta\overline{U}_0 + \frac{k}{k+1}\ \overline{q}_{cs}\ \ln\frac{(\frac{a_0}{r})^{k+1} - (\frac{a}{r})^{k+1}}{\exp(-\gamma_0) - 1} \tag{4.55}$$

using equations (4.19) and (4.20). To first order in the elastic limit strain γ_0 the excess pore pressure at the cavity wall is:

$$\Delta\ \overline{U}_c \doteq \frac{k}{k+1}\ \overline{q}_{cs}\ [\ \ln((\frac{a_0}{a})^{k+1} - 1) + \ln I_r\ ] \tag{4.56}$$

where $I_r = G/s_u$ is the well known rigidity index and s_u is undrained shear strength. Using the above equation, the total radial stress solution at the cavity wall can be derived:

$$\overline{\sigma}_r|_c = \overline{p}_0 + \frac{k}{k+1}\ \overline{q}_{cs}\ [\ 1 + \ \ln((\frac{a_0}{a})^{k+1} - 1) + \ln\frac{G}{s_u}\ ] \tag{4.57}$$

With $q_{cs} = -2s_u$ the above equation reduces to the following cavity pressure-contraction relationship:

$$p = p_0 - \frac{2ks_u}{1+k}[1 + \ln\frac{G}{s_u}] - \frac{2ks_u}{1+k}\ln[(\frac{a_0}{a})^{k+1} - 1] \tag{4.58}$$

where p is the total cavity pressure and p_0 is the initial total stress in the soil. It is noted that the above solution is the same as that obtained in Chapter 3 for unloading cavity in Tresca soils.

For a complete unloading cavity (relevant to unlined tunnels), the maximum displacement at the cavity wall is obtained by setting $p = 0$ in equation (4.58):

$$\frac{a_0}{a} = [1 + \exp(\frac{1+k}{2k} \times \frac{p_0}{s_u} - 1 - \ln\frac{G}{s_u})]^{\frac{1}{k+1}} \tag{4.59}$$

(b) Critical state soil plasticity models

The original Cam clay model for both normally and overconsolidated clays

The yield function of the original Cam clay model for cavity unloading problems can be shown to be:

$$\overline{q} = f(\overline{p}') = \frac{M}{\Lambda}\ \overline{p}'\ln\overline{p}' \tag{4.60}$$

where the stresses have been non-dimensionalised by the equivalent consolidation pressure at the same specific volume v.

The ratio of the plastic volumetric and shear strain rates calculated from the normal flow rule is:

$$\frac{\dot{\delta}^p}{\dot{\gamma}^p} = g(\overline{p}') = -\frac{kM}{(k+1)\Lambda}\ (\Lambda + \ln\overline{p}') \tag{4.61}$$

The function $L(\bar{p}')$ needed to calculate the effective pressure distributions in equation (4.23) is hence:

$$L(\bar{p}') = \frac{A(1 + \ln\bar{p}')}{\bar{p}'} + \frac{B}{\bar{p}'(\Lambda + \ln\bar{p}')} \tag{4.62}$$

and upon integration the function needed in equation (4.24) to calculate the shear strain is:

$$I(\bar{p}') = A(\ln\bar{p}' + \frac{1}{2}(\ln\bar{p}')^2) + B\ln|(\Lambda + \ln\bar{p}')| \tag{4.63}$$

where $A = M\varkappa/(2\Lambda\alpha v)$ and $B = (k+1)\Lambda\varkappa/(kMv)$ are constants. The value of the integral I, and hence the shear strain, is inversely proportional to the specific volume v.

The original Cam clay model for normally consolidated and lightly overconsolidated clays – The Hvorslev yield surface for heavily overconsolidated clays

The Hvorslev yield surface is a straight line in $\bar{p}' - \bar{q}$ space:

$$\bar{q} = -h\bar{p}' - (M - h)\exp(-\Lambda) \tag{4.64}$$

where h is the slope of the Hvorslev yield surface.

Using equation (4.64) as the yield function, equation (4.60) as the plastic potential and equation (4.49) for the elastic moduli, we can obtain:

$$L(\bar{p}') = -\frac{h}{2\overline{G}} + \frac{(1+k)\varkappa}{kv(M-h)}\frac{1}{(\bar{p}' - \exp(-\Lambda))} \tag{4.65}$$

and

$$I(\bar{p}') = -\frac{h\bar{p}'}{2\overline{G}} + \frac{(1+k)\varkappa}{kv(M-h)}\ln(\bar{p}' - \exp(-\Lambda)) \tag{4.66}$$

where the constant $\overline{G}$ is given by equation (4.49).

The modified Cam clay model for both normally and overconsolidated clays

To improve the performance of the original Cam clay model for normally consolidated clays, the modified Cam clay model can be used. The modified Cam clay yield surface for cavity unloading problems is:

$$\bar{q} = f(\bar{p}') = -M\bar{p}'\sqrt{(\bar{p}'^{-\frac{1}{\Lambda}} - 1)} \tag{4.67}$$

At the critical state $\bar{q}/\bar{p}' = M$, $n_p = 2$ and $\bar{p}' = 2^{-\Lambda}$.

The ratio of the plastic volumetric and shear strain rates calculated from the normal flow rule is:

$$\frac{\dot{\delta}^p}{\dot{\gamma}^p} = g(\bar{p}') = \frac{k}{(k+1)}(\frac{M^2 - \eta^2}{2\eta}) \qquad (4.68)$$

where η can be expressed as a function of mean effective stress using equation (4.67). The function $L(\bar{p}')$ needed to calculate the effective pressure distributions in equation (4.23) can be obtained by using equations (4.67) and (4.68). Unlike the original Cam clay model, the function needed in equation (4.24) to calculate the shear strain for the modified Cam clay model cannot be obtained in closed form and instead a simple numerical integration must be used.

4.2.3 Drained expansion of a cylindrical cavity in NC clays

Using a very simple critical state model, Palmer and Mitchell (1970) were among the first to derive an approximate small strain analytical solution for fully drained expansion of a cylindrical cavity in normally consolidated clay. In the solution of Palmer and Mitchell elastic deformation was ignored for simplicity.

(a) The simplified critical state model

Figure 4.12 shows a constant specific volume cross-section of the yield surface for normally consolidated clays. Unlike the yield surfaces of Cam clay and Modified Cam clay where yield surfaces are curved, the model used here is a straight line, an idealisation suggested by Palmer (1967). Note that since elastic deformation is neglected swelling lines are parallel to the mean stress axis.

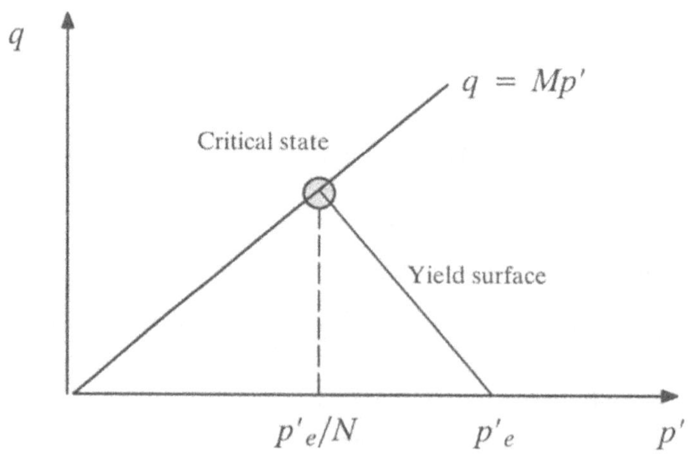

Figure 4.12: Yield surface adopted by Palmer and Mitchell (1970)

The yield surface for a given specific volume can be written as:

$$\chi\sigma'_r - \sigma'_\theta = (\chi - 1)p'_e \qquad (4.69)$$

where the size of yield surface is controlled by the equivalent consolidation pressure p'_e, which is linked to the initial mean effective stress p'_0, the initial specific volume v_0, and the current specific volume v by the following equation:

$$v_0 - v = \lambda \ln \frac{p'_e}{p'_0} \tag{4.70}$$

Substituting the expression for p'_e from equation (4.70) into equation (4.69) leads to the equation for the state boundary surface:

$$\chi \sigma'_r - \sigma'_\theta = (\chi - 1) p'_0 \exp\left(\frac{v_0 - v}{\lambda}\right) \tag{4.71}$$

The value of χ depends on the assumption for the vertical stress. Two possibilities are presented below:

$$\chi = \begin{cases} \dfrac{M + 3N - 3}{3N - 2M - 3} & \text{if } \sigma'_z = \sigma'_\theta \\ \dfrac{M + 2N - 2}{2N - M - 2} & \text{if } \sigma'_z = \frac{1}{2}(\sigma'_r + \sigma'_\theta) \end{cases} \tag{4.72}$$

The study of Palmer and Mitchell (1970) only considered the first possibility (i.e., $\sigma'_z = \sigma'_\theta$).

(b) Solution procedure

As noted by Palmer and Mitchell (1970), it is convenient in cavity expansion analysis to non-dimensionalise all lengths and displacements with respect to the initial radius of the cavity, so at the cavity wall r=1.

When the cavity is expanded by internal pressure, the radial strain is compressive, the axial strain zero, and the circumferential strain tensile. The order of the stresses is: $\sigma'_r > \sigma'_z > \sigma'_\theta$. For a normally consolidated clay, the soil becomes plastic as soon as the cavity pressure ψ is increased from its initial value p'_0. In other words, there is no elastic zone in the soil surrounding the cavity for this particular case.

Displacement analysis

Considering a small radial displacement of u, we can obtain the strains:

$$\varepsilon_r = -\frac{du}{dr} \quad \text{and} \quad \varepsilon_\theta = -\frac{u}{r} \tag{4.73}$$

The use of an associated flow rule and the yield function (4.71) gives the following relationship between strains:

$$\frac{\varepsilon_r}{\varepsilon_\theta} = -\chi \tag{4.74}$$

Substituting equation (4.73) into the above plastic flow rule leads to:

$$\frac{du}{dr} + \chi \frac{u}{r} = 0 \tag{4.75}$$

which can be integrated to give the following relationship between the displacement at cavity wall u_1 and displacement u at any radius r:

$$u = u_1 r^{-\chi} \tag{4.76}$$

As a result, the change in specific volume can be determined as follows:

$$v_0 - v = -v_0\left(\frac{u}{r} + \frac{du}{dr}\right) = (\chi-1)v_0 u_1 r^{-(\chi+1)} \tag{4.77}$$

Stress analysis

The stresses in the soil need to satisfy the equation of equilibrium:

$$r\frac{d\sigma'_r}{dr} + \sigma'_r - \sigma'_\theta = 0 \tag{4.78}$$

and the yield function (4.71).

Combining the yield function (4.71), the specific volume change (4.77), and the equilibrium equation (4.78) gives the following equation on the radial stress:

$$r\frac{d\sigma'_r}{dr} + (1-\chi)\sigma'_r = (1-\chi)p'_0 \exp(ar^{-(1+\chi)}) \tag{4.79}$$

where a is a function of the cavity displacement defined by:

$$a = \frac{(\chi-1)v_0}{\lambda} u_1 \tag{4.80}$$

With the following boundary condition:

$$\sigma'_r|_{r\to\infty} = p'_0 \tag{4.81}$$

The governing equation (4.79) can be integrated to give the following solution for the radial stress:

$$\frac{\sigma'_r}{p'_0} = 1 + (\chi-1)\sum_{n=1}^{\infty} \frac{(ar^{-(1+\chi)})^n}{n![(n+1)\chi + n-1]} \tag{4.82}$$

From the yield equation, the hoop stress can be expressed in terms of the radial stress as follows:

$$\frac{\sigma'_\theta}{p'_0} = \chi\frac{\sigma'_r}{p'_0} - (\chi-1)\exp(ar^{-(1+\chi)}) \tag{4.83}$$

It is important to note that the series in equation (4.82) converges very rapidly for all realistic values of α.

In particular, at the cavity wall r=1 equation (4.82) reduces to:

$$\frac{\psi}{p'_0} = 1 + (\chi-1)\sum_{n=1}^{\infty}\frac{\alpha^n}{n![(n+1)\chi + n-1]} \tag{4.84}$$

which defines the relationship between cavity pressure and cavity displacement.

Analysis with the critical state being approached

The solution presented so far is valid provided that the soil has nowhere reached the critical state. Initially the stress state is assumed to be isotropic and the stress ratio $\sigma'_r/\sigma'_\theta = 1.0$. As loading continues, the stress ratio increases and it is always the greatest at the inner boundary. At some point in time, the critical state is reached, first at the cavity wall and then in an enlarging zone around the cavity wall.

The critical state is reached when the following equation is satisfied by the stress components:

$$\frac{\sigma'_r}{\sigma'_\theta} = \beta \tag{4.85}$$

The value of β depends on the assumption for the vertical stress, namely:

$$\beta = \begin{cases} \dfrac{3+2M}{3-M} & \text{if } \sigma'_z = \sigma'_\theta \\ \dfrac{2+M}{2-M} & \text{if } \sigma'_z = \dfrac{1}{2}(\sigma'_r + \sigma'_\theta) \end{cases} \tag{4.86}$$

The value of the cavity displacement $u_1{}^*$ at the moment the critical state is reached at the cavity wall can be found by substituting equations (4.82) and (4.83) with $r=1$ into equation (4.85). This gives the following nonlinear equation for $\alpha^* = (\chi - 1)v_0 u_1{}^*/\lambda$:

$$\frac{1 + (\chi-1)\sum_{n=1}^{\infty} A_n(\alpha^*)^n}{\chi + (\chi^2-\chi)\sum_{n=1}^{\infty} A_n(\alpha^*)^n - (\chi-1)\exp(\alpha^*)} = \beta \tag{4.87}$$

in which

$$A_n = \frac{1}{n![(n+1)\chi + n-1]} \tag{4.88}$$

The equation (4.87) can be used to determine the critical cavity displacement $u_1{}^*$ required for cavity wall to reach the critical state.

If loading continues, there will be two separate regions (see Figure 4.13), one within a radius r_1 is the critical state zone, but outside that radius the stress solution of the previous plastic analysis is still valid.

Let us assume that in the outer region which has not yet reached the critical state the displacement can be expressed in the form of (4.76):

$$u = \delta r^{-\chi} \tag{4.89}$$

It can be easily established that the stresses in the outer plastic region can still be expressed by equations (4.82) and (4.83) but α should be expressed in terms of δ by $\alpha = (\chi - 1)v_0\delta/\lambda$.

At the interface $r = r_1$, the stresses must satisfy the critical state condition (4.85). As a result, it can be shown that:

$$\frac{(\chi-1)v_0}{\lambda}\delta r_1{}^{-(1+\chi)} = \alpha^* \tag{4.90}$$

which can be used to express the quantity δ in terms of the radius of the critical state zone r_1 as follows:

$$\delta = \frac{\lambda\alpha^*}{(\chi-1)v_0} r_1{}^{1+\chi} \tag{4.91}$$

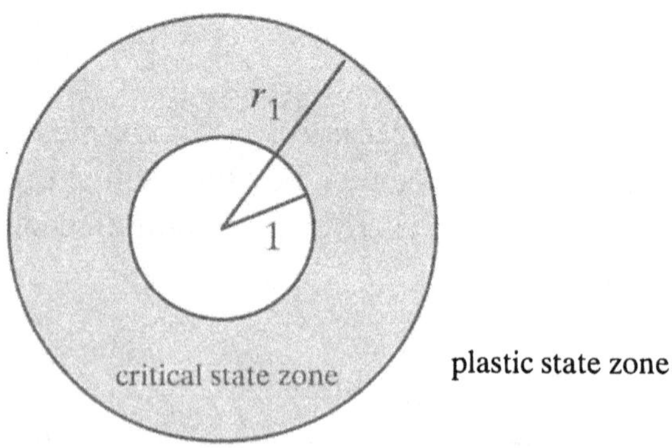

Figure 4.13: Drained cavity expansion in clays

Hence the displacement outside the critical state zone is:

$$u = \frac{\lambda\alpha^*}{(\chi-1)v_0}\left(\frac{r_1}{r}\right)^{1+\chi} r \tag{4.92}$$

Now let us turn our attention to stresses in the critical state zone. The stresses in the critical state zone must satisfy both equilibrium (4.78) and the yield condition (4.85). Combining these two equations gives:

$$r\frac{d\sigma'_r}{dr} + (1-\frac{1}{\beta})\sigma'_r = 0 \tag{4.93}$$

Integration of the above equation leads to:

$$\sigma'_r \, r^{(1-\frac{1}{\beta})} = constant \tag{4.94}$$

After using the continuity for the radial stress at $r = r_1$, we obtain the expression for radial stress in the critical state zone:

$$\frac{\sigma'_r}{p_0} = \left\{1 + (\chi-1)\sum_{n=1}^{\infty}\frac{(a^*)^n}{n![(n+1)\chi + n-1]}\right\}\left(\frac{r}{r_1}\right)^{\frac{1}{\beta}-1} \tag{4.95}$$

In particular, at the cavity wall $r = 1$, the cavity pressure ψ is linked to the radius of the critical state zone by:

$$\frac{\psi}{p_0} = \left\{1 + (\chi-1)\sum_{n=1}^{\infty}\frac{(a^*)^n}{n![(n+1)\chi + n-1]}\right\}(r_1)^{1-\frac{1}{\beta}} \tag{4.96}$$

In the critical state zone it is often assumed that the volume does not change so that material is incompressible with which we can link the displacement at the cavity wall to the radius of the critical zone:

$$u_1 = \delta r_1^{(1-\chi)} = \frac{\lambda a^*}{(\chi-1)v_0}(r_1)^{2\chi} \tag{4.97}$$

By substituting equation (4.97) into equation (4.96), the relationship between the cavity pressure ψ and the cavity displacement u_1 is determined.

4.2.4 Drained expansion of cavities in heavily OC clays

This section shows that it is also possible to derive a small strain analytical solution for cavity expansion in heavily overconsolidated (OC) clays. As discussed previously, the Hvorslev surface is often used as the yield function for heavily OC clays. The cavity expansion solution in OC clays presented here was derived by Yu (1993).

(a) Yield function, plastic potential and flow rule

As shown in Figure 4.5, the Hvorslev yield surface is a straight line in $\bar{p}' - \bar{q}$ space (Atkinson and Bransby, 1978):

$$q = hp' + (M - h)\ \exp(\frac{\Gamma - v}{\lambda}) \tag{4.98}$$

where h is the slope of the Hvorslev yield surface.

As in Collins and Stimpson (1994), the following two effective stress invariants are used for the analysis of cavity expansion problems:

$$q = \sigma'_r - \sigma'_\theta \quad \text{and} \quad p' = \frac{\sigma'_r + k\sigma'_\theta}{1 + k}$$

where σ'_r and σ'_θ are the effective radial and hoop stresses respectively. Note that the effective stresses are identical to the total stresses for fully drained loading conditions.

The Hvorslev yield function (4.98) can be rewritten in terms of radial and hoop stresses as follows:

$$\alpha\sigma'_r - \sigma'_\theta = Y \exp(\frac{\Gamma - v}{\lambda}) \tag{4.99}$$

in which

$$\alpha = \frac{1 + k - h}{1 + k + kh} \tag{4.100}$$

$$Y = \frac{(1 + k)(M - h)}{1 + k + kh} \tag{4.101}$$

For generality, a non-associated flow rule with a plastic potential function similar to the yield function (4.99) is used:

$$\beta\sigma'_r - \sigma'_\theta = constant \tag{4.102}$$

where β is a measure of soil dilatancy as defined by:

$$\beta = \frac{1 + k - h_1}{1 + k + kh_1} \tag{4.103}$$

and h_1 is the slope of the plastic potential function which is in the range of 0 and h. If $h_1 = h$, the flow rule is associated.

At any time in any position in the soil around the cavity with cavity radius of a, the stresses must satisfy the following equation of equilibrium:

$$(\sigma'_\theta - \sigma'_r) = \frac{r}{k} \frac{\partial \sigma'_r}{\partial r} \tag{4.104}$$

that is subject to two boundary conditions:

$$\sigma'_r|_{r=a} = p'$$

$$\sigma'_r|_{r=\infty} = p'_0$$

(b) Elastic solution

As the cavity pressure increases from its initial value, the deformation of the soil is at first purely elastic. The solutions for stresses and displacement can be shown to be as follows:

$$\sigma'_r = p'_0 + (p'-p'_0)(\frac{a}{r})^{1+k} \tag{4.105}$$

$$\sigma'_\theta = p'_0 - \frac{p'-p'_0}{k}(\frac{a}{r})^{1+k} \tag{4.106}$$

$$u = \frac{p'-p'_0}{2kG}(\frac{a}{r})^{1+k} r \tag{4.107}$$

The initial yielding occurs at the inner wall of the cavity when the stresses satisfy the yield condition (4.99) and that is when the cavity pressure reaches the following value:

$$p' = p'_{1y} = \frac{k[Y \exp(\frac{\Gamma - v}{\lambda}) + (1 - a)p'_0]}{1 + ka} + p'_0 \tag{4.108}$$

(c) Elastic-plastic analysis

After initial yielding takes place at the cavity wall, a plastic zone will form around the inner wall of the cavity with an increase in the applied pressure p'. The outer radius of the plastic zone is denoted by c.

Displacements in the plastic region

If the elastic deformation is neglected, the use of the plastic potential function (4.102) gives the following relationship between strains:

$$\frac{\varepsilon_r}{\varepsilon_\theta} = -k\beta \tag{4.109}$$

which can be integrated to give the following relationship between the cavity wall displacement u_1 and displacement u at any radius r:

$$u = u_1(\frac{a}{r})^{k\beta} \tag{4.110}$$

As a result, the change in specific volume can be determined from :

$$v_0 - v = -v_0(\frac{u}{r} + k\frac{du}{dr}) \tag{4.111}$$

which in turn provides the following relationship:

$$v = v_0[1 + \frac{(k-k\beta)u_1}{a}(\frac{a}{r})^{1+k\beta}] \tag{4.112}$$

Using the displacement in the outer elastic zone and the continuity of the displacement at the elastic-plastic interface $r=c$, it is possible to link the cavity displacement u_1 and the radius of the plastic zone c:

$$\frac{u_1}{a} = \frac{[Y\exp(\frac{\Gamma-v_0}{\lambda}) + (1-\alpha)p'_0]}{2G(1 + k\alpha)}(\frac{c}{a})^{1+k\beta} \tag{4.113}$$

so for a given cavity displacement we can use the above equation to determine the radius of plastic zone.

Stresses in the plastic region

The stress components in the plastic zone must satisfy equilibrium (4.104) and the yield condition (4.99) which can be combined to give:

$$\frac{d\sigma'_r}{dr} + \frac{k(1-\alpha)}{r}\sigma'_r = -\frac{kY\exp(\frac{\Gamma-v}{\lambda})}{r} \tag{4.114}$$

With the aid of equation (4.112), the above equation can be used to give the radial stress in the plastic zone:

$$\sigma'_r = Ar^{-k(1-\alpha)} - kY\exp(\frac{\Gamma-v_0}{\lambda})r^{-k(1-\alpha)}\sum_{n=0}^{\infty} A_n(r) \tag{4.115}$$

where A is a constant of integration and the function $A_n(r)$ is given by:

$$A_n(r) = \frac{m^n\, r^{k-k\alpha-nk\beta-n}}{n!(k-k\alpha-nk\beta-n)} \tag{4.116}$$

and

$$m = -\frac{v_0(k-k\beta)u_1}{\lambda}a^{k\beta} \tag{4.117}$$

Note that the hoop stress can be obtained from the radial stress by using the yield function (4.99).

Stresses in the elastic region

The stress components in the elastic region are obtained from the equilibrium equation and elastic stress-strain equations as follows:

$$\sigma'_r = p'_0 + Br^{-(1+k)} \tag{4.118}$$

$$\sigma'_\theta = p'_0 - \frac{B}{k}r^{-(1+k)} \tag{4.119}$$

where B is the second constant of integration.

The continuity of stress components at the elastic-plastic interface can be used to determine the constant A:

$$A = \frac{(1+k)p'_0}{1+ak} c^{k(1-a)} + kY \exp(\frac{\Gamma - v_0}{\lambda})[\sum_{n=0}^{\infty} A_n(c) + \frac{c^{k(1-a)}}{1+ak}] \tag{4.120}$$

At the cavity wall we have $\sigma_r|_{r=a} = p'$ and this condition can be used to link the plastic radius c to the current cavity radius and applied pressure:

$$p' = Aa^{-k(1-a)} - kY \exp(\frac{\Gamma - v_0}{\lambda}) a^{-k(1-a)} \sum_{n=0}^{\infty} A_n(a) \tag{4.121}$$

The above equation determines the cavity pressure from the radius of the plastic zone which is calculated for a given cavity displacement. In this way, a complete cavity expansion curve can be obtained.

4.3 CAVITY EXPANSION FROM ZERO INITIAL RADIUS

In this section, some analytical studies on the expansion of cavities from zero initial radius in soils modelled with critical state models are presented. As discussed in Chapter 3, the pressure in the cavity is constant when the cavity is expanded from zero radius. The ratio of the radii of the elastic-plastic interface and cavity wall will also remain constant. This is true because the problem of cavity expansion from zero radius has no characteristic length scale. The constant cavity pressure can be viewed as the limit pressure attained asymptotically as a finite cavity is expanded to infinity.

4.3.1 Drained expansion of cavities in sands

A semi-analytical solution was developed by Collins *et al.* (1992) for the expansion of cavities from zero initial radius in cohesionless sands using a simple state parameter-based critical state model. A summary of this solution procedure is briefly described.

(a) The state parameter-based critical state model for sands

The general formulations of the state parameter models are discussed by Collins (1990), Jefferies (1993) and Yu (1994, 1998). The basic assumption of the state parameter model is the existence of a critical state at which the sand deforms without any plastic volume change so that the dilation angle is zero. The material behaviour prior to the achievement of the critical state is controlled by the state parameter as defined by:

$$\xi = v + \lambda \ln(\frac{p'}{p'_1}) - \Gamma \tag{4.122}$$

where v is specific volume, p' denotes the mean effective stress and p'_1 is the reference mean pressure which is often taken as unity in soil mechanics. The state parameter ξ is zero at the critical state, positive at loose side and negative at dense side.

The state parameter model used by Collins *et al.* (1992) is an elastic–plastic strain hardening (or softening) model which postulates that a sand specimen can be considered as an isotropic continuum.

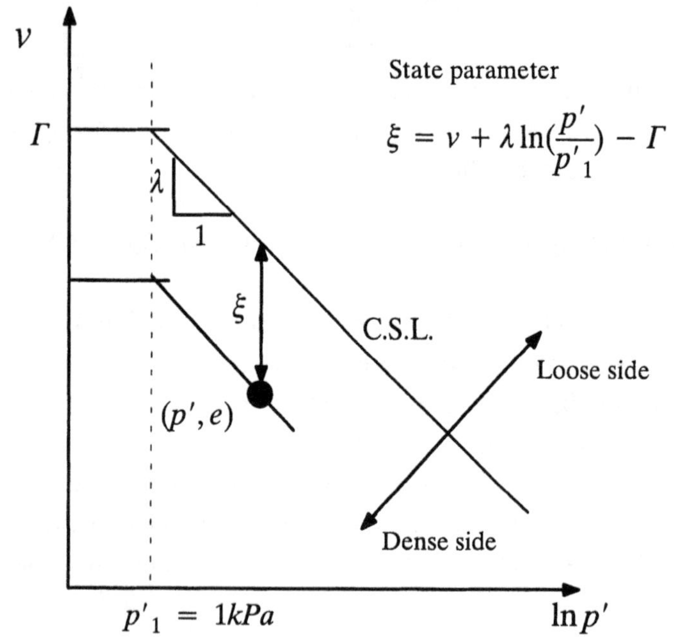

Figure 4.14: Definition of state parameter

The choice of plastic potential and yield function is an important step in constitutive modelling since strain rates essentially depend on their derivatives with respect to stresses.

In the solution of Collins *et al.* (1992), the Mohr-Coulomb yield criterion was adopted to describe sand behaviour, namely:

$$\sigma'_r - \alpha\sigma'_\theta = 0 \tag{4.123}$$

where α is a function of the mobilised friction angle ϕ_m:

$$\alpha = \frac{1 + \sin\phi_m}{1 - \sin\phi_m} \tag{4.124}$$

To derive the plastic strain rates Collins *et al.* (1992) proposed the use of the following plastic potential:

$$\sigma'_r - \beta\sigma'_\theta = 0 \tag{4.125}$$

where β is a function of the mobilised dilation angle ψ_m defined by:

$$\beta = \frac{1 + \sin\psi_m}{1 - \sin\psi_m} \tag{4.126}$$

As discussed by Collins *et al.* (1992), it is reasonable to use the following exponential type of empirical relationship to fit experimental data compiled by Been and Jefferies (1985):

$$\phi_m - \phi_{cv} = A[\exp(-\xi) - 1] \tag{4.127}$$

where ϕ_{cv} is the angle of internal friction at the critical state, and A is a curve fitting parameter ranging from 0.6–0.95 depending on the type of sands. The friction angle is measured in radians. According to the above relationship, the mechanical properties (i.e. yield function) of a sand prior to the achievement of the critical state are therefore dependent on both the void ratio and the mean effective stress through a single composite state parameter. The critical state friction angle ϕ_{cv} may be determined by conventional laboratory tests and normally ranges from 30–33 degrees.

To link the plastic potential and the state parameter, a stress–dilatancy equation is used which defines the relationship between the angles of friction and dilation. Perhaps the most successful stress–dilatancy model is that developed by Rowe (1962), which has been further simplified by Bolton (1986) as follows:

$$\psi_m = 1.25(\phi_m - \phi_{cv}) = 1.25A[\exp(-\xi) - 1] \tag{4.128}$$

The above assumption implies a non–associated flow rule as the angles of friction and dilation are not the same.

(b) The solution in the elastic region

In the outer elastic region, the elastic stress-strain relations and equilibrium equation can be used to derive the solution for both displacement and stresses:

$$u = \varepsilon_c\left(\frac{c}{r}\right)^k c \tag{4.129}$$

$$\sigma'_r = p'_0 + 2Gk\varepsilon_c\left(\frac{c}{r}\right)^{1+k} \tag{4.130}$$

$$\sigma'_\theta = p'_0 - 2G\varepsilon_c\left(\frac{c}{r}\right)^{1+k} \tag{4.131}$$

where ε_c is the hoop strain at the elastic-plastic interface r=c.

It is noted that the mean effective stress, and hence the state parameter ξ, are both unchanged by the elastic deformation. At the elastic-plastic boundary:

$$\frac{\sigma'_r}{\sigma'_\theta} = \alpha_0 = \alpha(\xi_0) = \frac{1+\sin\phi_0}{1-\sin\phi_0} \tag{4.132}$$

where ξ_0 and ϕ_0 denote the initial state parameter and friction angle. By combining equations (4.130)-(4.132), we obtain:

$$\varepsilon_c = \frac{(\alpha_0-1)p'_0}{2(k+\alpha_0)G} \tag{4.133}$$

and the radial stress at the elastic-plastic interface is defined by:

$$\sigma'_c = \sigma'_r|_{r=c} = \frac{(1+k)\alpha_0 p'_0}{\alpha_0+k} \tag{4.134}$$

These elastic solutions for the elastic-plastic interface provide sufficient boundary conditions for the solution in the inner plastic zone.

(c) The solution in the plastic region

In the plastic region, stresses must satisfy the equilibrium equation:

$$(\sigma'_\theta-\sigma'_r) = \frac{r}{k}\frac{\partial\sigma'_r}{\partial r} \tag{4.135}$$

The plastic flow rule can be used to obtain the following governing equation in terms of material velocity V:

$$\frac{\partial V}{\partial r} + \frac{k}{\beta}\frac{V}{r} = -\frac{1}{2G}[A(\beta)\overset{0}{\sigma}'_r + B(\beta)\overset{0}{\sigma}'_\theta] \tag{4.136}$$

As before, the symbol $\overset{0}{()}$ denotes the material time derivative associated with a given material particle and it is related to the local time derivative $\dot{()}$, evaluated at fixed position r, by:

$$\overset{0}{()} = \dot{()} + V\frac{\partial()}{\partial r} \tag{4.137}$$

where V is the radial speed or velocity of a solid material element. $A(\beta)$ and $B(\beta)$ are defined by:

$$A(\beta) = \frac{(1-2\mu) + k\mu(\beta-1)/\beta}{1+(k-1)\mu} \tag{4.138}$$

$$B(\beta) = \frac{k(1-\mu)/\beta-k\mu}{1+(k-1)\mu} \tag{4.139}$$

where μ is Poisson's ratio. The rate form of the yield function (4.123) is:

$$\overset{\circ}{\sigma}'_r = \alpha\overset{\circ}{\sigma}'_\theta + \frac{\alpha'}{\alpha}\overset{\circ}{\xi}\sigma'_r \tag{4.140}$$

where $\alpha' = d\alpha/d\xi$. Since the state parameter depends on the specific volume and mean effective pressure, its material derivative is given by:

$$\overset{\circ}{\xi} = \xi_{,v}\overset{\circ}{v} + \xi_{,p}\overset{\circ}{p} \tag{4.141}$$

where the comma denotes a partial derivative (e.g. $\xi_{,v} = \partial/\partial v$).

The material derivatives of specific volume and mean effective stress can be expressed as follows:

$$\overset{\circ}{v} = -v\left(\frac{\partial V}{\partial r} + k\frac{V}{r}\right) \tag{4.142}$$

$$\overset{\circ}{p}' = \frac{\overset{\circ}{\sigma}'_r + k\overset{\circ}{\sigma}'_\theta}{1 + k} \tag{4.143}$$

Equations (4.140)-(4.143) can now be used to express both $\overset{\circ}{\sigma}'_\theta$ and $\overset{\circ}{\xi}$ in terms of $\overset{\circ}{\sigma}'_r$ and V, v, ξ, so that the velocity equation (4.136) and state evolutionary equation (4.141) become:

$$m\frac{\partial V}{\partial r} + n\frac{V}{r} = -\frac{q}{2G}\overset{\circ}{\sigma}'_r \tag{4.144}$$

and

$$h\overset{\circ}{\xi} = v\xi_{,v}\left(\frac{\partial V}{\partial r} + k\frac{V}{r}\right) + l\overset{\circ}{\sigma}'_r \tag{4.145}$$

in which

$$m = 1 - \frac{\alpha'\sigma'_r}{\alpha^2}[\xi_{,v}vB(\beta)/(2G) - \xi_{,p}k/(1 + k)] \tag{4.146}$$

$$n = \frac{k}{\beta}[1 + \frac{k\alpha'}{\alpha^2}\sigma'_r\xi_{,p}/(1 + k)] - \frac{k\alpha'}{\alpha^2}[\sigma'_r\xi_{,v}vB(\beta)/(2G)] \tag{4.147}$$

$$q = A(\beta) + \frac{B(\beta)}{\alpha} + [\alpha'\xi_{,p}\sigma_r/(1 + k)\alpha^2][kA(\beta) - B(\beta)] \tag{4.148}$$

$$h = 1 + k\xi_{,p}\alpha'\sigma'_r/(1 + k)\alpha^2 \tag{4.149}$$

$$l = \xi_{,p}(\alpha + k)/(1 + k)\alpha \tag{4.150}$$

For the problem of cavity expansion from zero radius, the solution for velocity, stress and state parameter depends on radius r and time t through the dimensionless radial coordinate:

$$\eta = \frac{r}{c} = \frac{r}{V_c t} \tag{4.151}$$

where $V_c = \partial c/\partial t = \dot{c}$ is the speed or velocity at the elastic-plastic interface. Note that $c = V_c t$ is valid because *c=0* when *t=0.*

It is also convenient to non-dimensionlise velocity and stress components as follows:

$$\overline{V} = \frac{V}{V_c}, \quad \text{and} \quad \overline{\sigma}' = \frac{\sigma'}{p'_0} \tag{4.152}$$

Using equation (4.151), various derivatives which appear in the governing equations can now be expressed in terms of $\partial/\partial\eta$:

$$\frac{\partial}{\partial r} = \frac{1}{c}\frac{\partial}{\partial \eta} \tag{4.153}$$

$$\frac{\partial}{\partial t} = \dot{()} = -\frac{V_c\eta}{c}\frac{\partial}{\partial \eta} \tag{4.154}$$

$$\frac{d}{dt} = \overset{0}{()} = \frac{V_c(\overline{V}-\eta)}{c}\frac{\partial}{\partial \eta} \tag{4.155}$$

A system of ordinary differential equations can now be solved. In terms of these dimensionless variables the equation becomes:

$$\frac{d\overline{\sigma}'_r}{d\eta} + k\left(1-\frac{1}{\alpha}\right)\frac{\overline{\sigma}'_r}{\eta} = 0 \tag{4.156}$$

and the constitutive equation (4.144) and state parameter evolutionary equation (4.145) can be written:

$$\frac{d\overline{V}}{d\eta} = \frac{k(1-\frac{1}{\alpha})(\overline{V}-\eta)(E(\xi)A(\beta) + C(\xi)B(\beta))(\sigma'_c/2G)(\overline{\sigma}'_r/\eta)}{E(\xi)-D(\xi,\nu)B(\beta)\overline{\sigma}'_r(\sigma'_c/2G)}$$
$$-\frac{k(E(\xi)/\beta-D(\xi,\nu)B(\beta)(\sigma'_c/2G)\overline{\sigma}'_r](\frac{\overline{V}}{\eta})}{E(\xi)-D(\xi,\nu)B(\beta)\overline{\sigma}'_r(\sigma'_c/2G)} \tag{4.157}$$

and

$$\frac{d\xi}{d\eta} = \frac{\nu(d\overline{V}/d\eta + k\overline{V}/\eta)}{E(\overline{V}-\eta)} + \frac{\lambda}{E\overline{\sigma}'_r}\frac{d\overline{\sigma}'_r}{d\eta} \tag{4.158}$$

where C, D and E are defined by:

$$C(\xi) = \frac{1-\lambda\alpha'/(\alpha + k)}{\alpha} \tag{4.159}$$

$$D(\xi, v) = \frac{v a'}{a^2} \qquad (4.160)$$

$$E(\xi) = 1 + \frac{\lambda k a'}{a(a + k)} \qquad (4.161)$$

Equations (4.156)-(4.158) can be used to solve for stress, velocity and state parameter by a standard NAG library differential equation solver. At each stage of the solution the specific volume in (4.158) is calculated from:

$$v = \xi + \Gamma - \lambda \ln[\frac{\bar{\sigma}'_r(a + k)p'_0}{a(1 + k)}] \qquad (4.162)$$

Boundary conditions

At the elastic-plastic interface, equation (4.134) can be used to give:

$$\bar{\sigma}'_r(\eta)|_{\eta=1} = \frac{(1 + k)a_0}{a_0 + k} \qquad (4.163)$$

Differentiating equation (4.129) with respect to time t and for $r=c$, the normalised velocity at the elastic-plastic interface is:

$$\bar{V}(\eta)|_{\eta=1} = (1 + k)\varepsilon_c = \frac{(1 + k)(a_0 - 1)p'_0}{2(k + a_0)G} \qquad (4.164)$$

On the cavity wall the normalised material velocity is:

$$\bar{V}(\eta)|_{\eta=a/c} = \frac{\dot{a}}{\dot{c}} = \frac{a}{c} = \eta \qquad (4.165)$$

and the initial value of the state parameter is:

$$\xi_0 = v_0 - \Gamma + \lambda \ln(\frac{p'_0}{p'_1}) \qquad (4.166)$$

where the reference pressure p'_1 can be taken to be 1. The solution starts from the elastic-plastic boundary to the cavity wall. From equation (4.165), it is known that the cavity wall is reached when $\bar{V} = \eta$.

The semi-analytical solution procedure as outlined above for sand was also used by Collins and Stimpson (1994) to analyse drained and undrained expansion of cavities from zero initial radius in clays.

4.3.2 Undrained expansion of a cylindrical cavity in a rate-type clay

The use of a simple rate-type (or hypoelastic) constitutive model to represent soil stress-strain behaviour has been considered by Davis *et al.* (1984) in deriving an analytical solution for the undrained expansion of a cylindrical cavity from zero initial radius. Although the rate type model may be easier to deal with than the conventional elastic-plastic constitutive models, its main limitation is that it is not yet clear how loading and unloading processes should be differentiated in actual calculations.

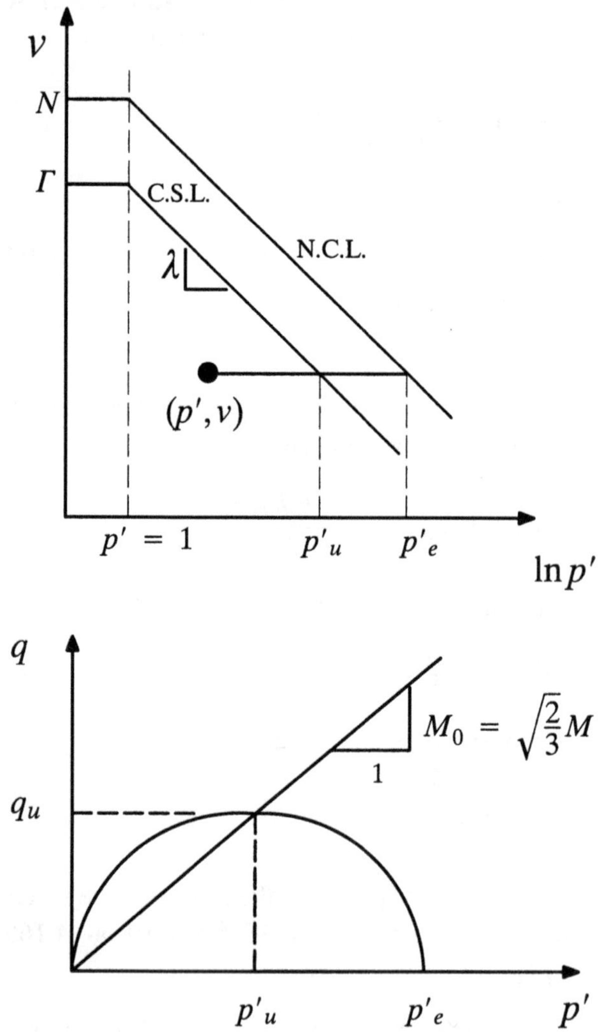

Figure 4.15: Failure surface implied by the rate-type model used by Davis *et al.* (1984)

The simple rate-type model used by Davis *et al.* (1984) includes strain hardening and softening effects. The failure surface implied in the model is similar to the yield surface of the modified Cam clay model proposed by Roscoe and Burland (1968).

(a) Kinematics of cavity expansion

Consider a cylindrical cavity that is expanded from zero initial radius under fully undrained loading conditions. The cavity radius becomes a at time t, and let r denote the position of the material particle which was initially located at r_0. Due to the zero soil volume change during undrained loading:

$$r^2 - a^2 = r_0^2 \tag{4.167}$$

in which a is a function of time t and $r = r(r_0, t)$. Differentiating the above equation gives:

$$\dot{r} = \frac{a}{r}\dot{a} \tag{4.168}$$

where $\dot{r} = \partial r/\partial t$ and $\dot{a} = \partial a/\partial t$ are local time derivatives.

The non-zero strain rates can then be expressed as follows:

$$\dot{\varepsilon}_r = -\frac{\partial \dot{r}}{\partial r} \tag{4.169}$$

$$\dot{\varepsilon}_\theta = -\frac{\dot{r}}{r} \tag{4.170}$$

Let us define a cavity strain rate $\dot{e}_c$ as negative of the hoop strain rate and using the incompressibility condition we have:

$$\dot{e}_c = -\dot{\varepsilon}_c = \frac{\dot{r}}{r} = -\frac{\partial \dot{r}}{\partial r} \tag{4.171}$$

Integrating the above equation leads to:

$$e_c = e_c(r,t) = \ln\left(\frac{r}{r_0}\right) = -\frac{1}{2}\ln\left(1-\frac{a^2}{r^2}\right) \tag{4.172}$$

Note that the cavity strain becomes infinity when r approaches a.

(b) Solution procedure

The simple rate-type stress-strain relations

The rate-type stress-strain relation used by Davis *et al.* (1984) adopts a failure surface similar (not same as) to the modified Cam clay yield surface (Figure 4.15). The quantities q and p' are defined as follows:

$$q = \left[\frac{(\sigma'_r-\sigma'_\theta)^2 + (\sigma'_\theta-\sigma'_z)^2 + (\sigma'_z-\sigma'_r)^2}{3}\right]^{\frac{1}{2}} \tag{4.173}$$

$$p' = \frac{\sigma'_r + \sigma'_\theta + \sigma'_z}{3} \tag{4.174}$$

Note that the definition q used by Davis *et al.* (1984) is $\sqrt{2/3}$ times the definition conventionally adopted for q in critical state soil mechanics.

In cylindrical cavity problems the stress-strain relation takes the following form:

$$\dot{\sigma}'_r = \frac{2G}{q_u^2}[(\sigma'_r - \sigma'_\theta)(p'_u - \sigma'_r) + q_u^2]\dot{e}_c \tag{4.175}$$

$$\dot{\sigma}'_\theta = \frac{2G}{q_u^2}[(\sigma'_r - \sigma'_\theta)(p'_u - \sigma'_\theta) - q_u^2]\dot{e}_c \tag{4.176}$$

$$\dot{\sigma}'_z = \frac{2G}{q_u^2}(\sigma'_r - \sigma'_\theta)(p'_u - \sigma'_\theta)\dot{e}_c \tag{4.177}$$

in which p'_u is the equivalent critical state pressure at a given specific volume and $q_u = M_0 p'_u = \sqrt{2/3}\, M p'_u$ where M is the usual critical state constant used in critical state soil mechanics.

Effective stresses

By introducing the following dimensionless stress variables:

$$\bar{\sigma}'_r = \frac{\sigma'_r - p'_u}{\sqrt{2}\, q_u} \tag{4.178}$$

$$\bar{\sigma}'_\theta = \frac{\sigma'_\theta - p'_u}{\sqrt{2}\, q_u} \tag{4.179}$$

$$\bar{\sigma}'_z = \frac{\sigma'_z - p'_u}{\sqrt{2}\, q_u} \tag{4.180}$$

$$\beta = \frac{\sqrt{2}\, G}{q_u} \tag{4.181}$$

The constitutive equations (4.175)-(4.177) become:

$$\dot{\bar{\sigma}}'_r = 2\beta[\tfrac{1}{2} - \bar{\sigma}'_r(\bar{\sigma}'_r - \bar{\sigma}'_\theta)]\dot{e}_c \tag{4.182}$$

$$\dot{\bar{\sigma}}'_\theta = 2\beta[-\tfrac{1}{2} - \bar{\sigma}'_\theta(\bar{\sigma}'_r - \bar{\sigma}'_\theta)]\dot{e}_c \tag{4.183}$$

$$\dot{\bar{\sigma}}'_z = -2\beta\bar{\sigma}'_z(\bar{\sigma}'_r - \bar{\sigma}'_\theta)\dot{e}_c \tag{4.184}$$

Integrating the above equations gives:

$$\bar{\sigma}'_r = \frac{1}{2}\left\{\frac{\exp(4\beta e_c) + 4\bar{\sigma}'_{r0}\exp(2\beta e_c) - 1}{\exp(4\beta e_c) + 1}\right\} \tag{4.185}$$

$$\bar{\sigma}'_\theta = \frac{1}{2}\left\{\frac{-\exp(4\beta e_c) + 4\bar{\sigma}'_{r0}\exp(2\beta e_c) + 1}{\exp(4\beta e_c) + 1}\right\} \tag{4.186}$$

$$\bar{\sigma}'_z = \frac{2\bar{\sigma}'_{z0}\exp(2\beta e_c)}{\exp(4\beta e_c) + 1} \tag{4.187}$$

in which $\bar{\sigma}'_{r0} = \bar{\sigma}'_{\theta 0}$ and $\bar{\sigma}'_{z0}$ are initial values of the dimensionless stresses.

By using the expression of the cavity strain (4.172), the above stress solutions can be further expressed in terms of radius r as follows:

$$\bar{\sigma}'_r = \frac{1}{2}\left\{\frac{-[1-(\frac{a}{r})^2]^{2\beta} + 4\bar{\sigma}'_{r0}[1-(\frac{a}{r})^2]^{\beta} + 1}{[1-(\frac{a}{r})^2]^{2\beta} + 1}\right\} \tag{4.188}$$

$$\bar{\sigma}'_\theta = \frac{1}{2}\left\{\frac{[1-(\frac{a}{r})^2]^{2\beta} + 4\bar{\sigma}'_{r0}[1-(\frac{a}{r})^2]^{\beta}-1}{[1-(\frac{a}{r})^2]^{2\beta} + 1}\right\} \tag{4.189}$$

$$\bar{\sigma}'_z = \frac{2\bar{\sigma}'_{z0}[1-(\frac{a}{r})^2]^{\beta}}{[1-(\frac{a}{r})^2]^{2\beta} + 1} \tag{4.190}$$

Total stresses and pore pressure

The determination of total stresses, and hence pore pressure, requires use of the equilibrium equation:

$$(\sigma_\theta - \sigma_r) = r\frac{\partial \sigma_r}{\partial r} \tag{4.191}$$

From equations (4.178)-(4.179), it can be shown that:

$$\sigma_r - \sigma_\theta = \sigma'_r - \sigma'_\theta = \sqrt{2}q_u(\bar{\sigma}'_r - \bar{\sigma}'_\theta) \tag{4.192}$$

With the above equation, the equilibrium equation (4.191) gives:

$$\frac{\partial \sigma_r}{\partial r} + \frac{\sqrt{2}q_u}{r}\left\{\frac{-(1-\frac{a^2}{r^2})^{2\beta} + 1}{(1-\frac{a^2}{r^2})^{2\beta} + 1}\right\} = 0 \tag{4.193}$$

which can be integrated to provide the following solution for total radial stress:

$$\sigma_r = \sigma_{r0} + \frac{\sqrt{2}q_u}{2}\int_1^{1-a^2/r^2} (\frac{x^{2\beta}-1}{x^{2\beta} + 1})\frac{dx}{1-x} \tag{4.194}$$

where σ_{r0} is the initial total horizontal stress. Although the right side of equation (4.194) may not be evaluated in closed form, the integration can be easily performed numerically. Once the total radial stress is determined, the pore pressure can be calculated by:

$$U = \sigma_r - \sigma'_r \tag{4.195}$$

which can then be used to determine the other two components of the total stress:

$$\sigma_\theta = U + \sigma'_\theta \tag{4.196}$$

$$\sigma_z = U + \sigma'_z \tag{4.197}$$

4.4 SUMMARY

1. It is common to formulate undrained cavity expansion problems in terms of total stresses. In Chapter 3, the expansion or contraction of cavities in undrained clay was studied using a total stress formulation with the Tresca yield criterion. However this is not appropriate if critical state soil models are used, primarily because soil strength is a function of effective stresses rather than total stresses. A major drawback of the total stress analysis is that it does not account for the effect of stress history.

2. Analytical solutions were derived by Collins and Yu (1996) for large strain cavity expansion in undrained clays modelled by a variety of critical state theories. The soil models used include Cam clay, modified Cam clay, and a combination of Cam clay and Hvorslev yield surface. The soil stress history (i.e. overconsolidation ratio OCR) has a significant effect on the cavity expansion solutions. The cavity expansion solutions presented in this chapter will be used in Chapter 9 for modelling pore pressure changes around driven piles in clay.

3. Analytical solutions were developed by Yu and Rowe (1999) for the contraction of cavities from an *in-situ* stress state in undrained clays modelled by critical state theories. As in the loading case, soil stress history (i.e. overconsolidation ratio OCR) has a significant effect on the cavity unloading solutions. These unloading solutions will be used in Chapter 10 in relation to the prediction of soil displacement around a tunnel.

4. Using a very simple critical state model, Palmer and Mitchell (1970) derived one of the first critical state solutions for cavity expansion problems. This is a small strain solution for drained expansion of a cylindrical cavity in normally consolidated (NC) soils.

5. Using the Hvorslev yield surface, Yu (1993) developed a small strain solution for drained expansion of a cylindrical cavity in overconsolidated (OC) soils. This solution is defined by equations (4.113) and (4.121). The drained solution has

potential applications in a drained analysis of self-boring pressuremeter tests in overconsolidated (OC) clays.

6. Semi-analytical solutions were obtained by Collins *et al.* (1992) for the expansion of cavities from zero initial radius in sand using a state parameter-based critical state model. The pressure in the cavity is constant when the cavity is expanded from zero radius, as this problem has no characteristic length scale. The constant cavity pressure can be viewed as the limit pressure attained asymptotically as a finite cavity is expanded to infinity. The effect of variable friction and dilation angles can be simply accounted for by using average values of strength parameters (friction and dilation angles). This finding will be discussed further in Chapter 9 when cavity limit pressure is used to estimate the end bearing capacity of driven piles in sand.
7. Using a simple rate-type soil model, Davis *et al.* (1984) derived an analytical solution for undrained cavity expansion from zero radius. Davis *et al.* (1984) also showed that the solution can be applied to predict the behaviour of driven piles in clay. The main limitation of using a rate-type model is that it is not yet clear how loading and unloading processes should be differentiated in actual calculations.

REFERENCES

Atkinson, J.H. and Bransby, P.L. (1978). *The Mechanics of Soils*. McGraw-Hill, England.

Been, K. and Jefferies, M.G. (1985). A state parameter for sands. *Geotechnique*, 35, 99–112.

Bolton, M.D. (1986). The strength and dilatancy of sands. *Geotechnique*, 36, 65–78.

Carter, J.P., Randolph, M.F. and Wroth, C.P. (1979). Stress and pore pressure changes in clay during and after the expansion of a cylindrical cavity. *International Journal for Numerical and Analytical Methods in Geomechanics*, 3, 305–323.

Collins, I.F. (1991). On the mechanics of state parameter models for sands. *Proc. of 7th International Conference on Computer Methods and Advances in Geomechanics,* Cairns, Australia, 1, 593–599.

Collins, I.F., Pender, M.J. and Wang, Y. (1992). Cavity expansion in sands under drained loading conditions. *International Journal for Numerical and Analytical Methods in Geomechanics*, 16(1), 3–23.

Collins, I.F. and Stimpson, J.R. (1994). Similarity solutions for drained and undrained cavity expansions in soils. *Geotechnique*, 44(1), 21–34.

Collins, I.F. and Yu, H.S. (1996). Undrained expansions of cavities in critical state soils. *International Journal for Numerical and Analytical Methods in Geomechanics*, 20(7), 489–516.

Davis, R.O., Scott, R.F. and Mullenger, G. (1984). Rapid expansion of a cylindrical cavity in a rate type soil. *International Journal for Numerical and Analytical Methods in Geomechanics*, 8, 3–23.

Jefferies, M.G. (1993). Nor-Sand: a simple critical state model for sand. *Geotechnique,* 43(1), 91–103.

Muir Wood, D. (1990). *Soil Behaviour and Critical State Soil Mechanics*. Cambridge University Press.

Palmer, A.C. and Mitchell, R.J. (1971). Plane strain expansion of a cylindrical cavity in clay. *Proceedings of the Roscoe Memorial Symposium,* Cambridge, 588–599.

Palmer, A.C. (1967). Stress strain relations for clays: an energy approach. *Geotechnique,* 17(4), 348–358.

Prevost, J.H. and Hoeg, K. (1975). Analysis of pressuremeter in strain softening soil. *Journal of the Geotechnical Engineering Division*, ASCE, 101(GT8), 717–732.

Randolph, M.F., Carter, J.P. and Wroth, C.P. (1979). Driven piles in clay – the effects of installation and subsequent consolidation. *Geotechnique,* 29, 361–393.

Roscoe, K.H. and Burland, J.B. (1968). On generalised stress strain behaviour of wet clay. In: *Engineering Plasticity* (edited by Heyman and Leckie), 535–609.

Rowe, P. W. (1962). The stress-dilatancy relation for static equilibrium of an assembly of particles in contact. *Proceedings of Royal Society,* 267, 500–527.

Schofield, A.N. and Wroth, C.P. (1968). *Critical State Soil Mechanics*. McGraw-Hill, England.

Smith, I.M. (1970). Incremental numerical solution of a simple deformation problem in soil mechanics. *Geotechnique*, 20(4), 357–372.

Yu, H.S. (1993). Cavity expansion in heavily OC clays under fully drained loading conditions. *Unpublished notes*. The University of Newcastle, NSW 2308, Australia.

Yu, H.S. (1994). State parameter from self-boring pressuremeter tests in sand. *Journal of Geotechnical Engineering*, ASCE, 120(12), 2118–2135.

Yu, H.S. (1998). CASM: a unified state parameter model for clay and sand. *International Journal for Numerical and Analytical Methods in Geomechanics*, 22, 621–653.

Yu, H.S. and Rowe, R.K. (1999). Plasticity solutions for soil behaviour around contracting cavities and tunnels. *International Journal for Numerical and Analytical Methods in Geomechanics*, 23, 1245–1279.

Zytynski, M., Randolph, M.F., Nova, R. and Wroth, C.P. (1978). On modelling the unloading-reloading behaviour of soils. *International Journal for Numerical and Analytical Methods in Geomechanics,* 2, 87–93.

CHAPTER 5

FURTHER ELASTOPLASTIC SOLUTIONS

5.1 INTRODUCTION

In Chapter 4, analytical solutions were presented for cavity expansion and contraction in soils modelled by critical state strain hardening/softening models. Although critical state models are widely used in soil mechanics, other types of plasticity models have also been used to describe the non-linear stress-strain behaviour of soil and rock. For this reason, this chapter describes some additional solutions for cavity problems in elastic-plastic strain hardening/softening soils and rocks.

5.2 CAVITY EXPANSION IN HARDENING/SOFTENING SOILS

5.2.1 Undrained expansion of a cylindrical cavity in strain hardening/softening clays

By neglecting elastic deformation and using a hyperbolic type of stress strain relations, Prevost and Hoeg (1975) presented an approximate small strain solution for undrained expansion of a cylindrical cavity in strain hardening/softening clays. This section briefly describes this solution procedure. It is noted that a non-linear elastic model can also be used to describe the observed initial non-linear stress strain curve in cavity expansion analysis (Bolton and Whittle, 1999).

(a) Modelling strain hardening/softening soil behaviour

The total stress variables used in this development are given by:

$$p = \frac{\sigma_r + \sigma_\theta + \sigma_z}{3} \tag{5.1}$$

$$q = \frac{1}{\sqrt{2}}[(\sigma_r - \sigma_\theta)^2 + (\sigma_\theta - \sigma_z)^2 + (\sigma_z - \sigma_r)^2]^{1/2} \tag{5.2}$$

where σ_r, σ_θ and σ_z are the total radial, tangential and axial stresses respectively for cylindrical cavity problems.

Corresponding to the stress variables, p and q, the following plastic strain increment invariants are introduced:

$$d\delta^p = d\varepsilon_r^p + d\varepsilon_\theta^p \tag{5.3}$$

$$d\gamma^p = \frac{2}{3}[(d\varepsilon_r^p)^2 - d\varepsilon_r^p d\varepsilon_\theta^p + (d\varepsilon_\theta^p)^2] \tag{5.4}$$

For a monotonously strain hardening soil, a hyperbolic relationship between the shear stress q and the plastic shear strain γ^p may be assumed:

$$q = F(\gamma^p) = \frac{\gamma^p}{D + \gamma^p} q_{ult} \tag{5.5}$$

where D is a material constant and q_{ult} is the ultimate shear stress.

It is observed in the laboratory and field testing that many soils behave in such a way that they first strain hardens and subsequently strain softens. To model this type of behaviour, the following relationship may be used:

$$q = F(\gamma^p) = A\frac{B(\gamma^p)^2 + \gamma^p}{1 + (\gamma^p)^2} \tag{5.6}$$

in which A and B are material constants. With equation (5.6), the shear strength reaches a peak when the shear strain is equal to:

$$\gamma^p{}_{peak} = B + \sqrt{1 + B^2} \tag{5.7}$$

Also the residual strength q_{res} at a large shear strain can be expressed in terms of the constants A and B as follows:

$$q_{res} = AB \tag{5.8}$$

With the above two equations, the material constants A and B may be determined from experimental stress strain curves.

(b) Solution procedure

The expansion of a cylindrical cavity is assumed to occur under plane strain and undrained loading conditions. Initially the stress state is isotropic with a stress p_0 acting everywhere in the soil. We are interested in the stress and displacement fields around the cavity when the internal pressure increases from its initial value p_0.

The problem can be considerably simplified by ignoring elastic deformation in the soil. With this assumption, there is no need to distinguish total strains from plastic strains. One consequence of using the strain hardening/softening models described by equations (5.5) and (5.6) is that plastic deformation will occur as long as the shear stress is increased. As a result, there is no elastic zone in the soil. In other words, once the internal pressure is increased, the entire soil mass becomes plastic.

For the cylindrical cavity, the total stresses must satisfy the following equilibrium equation:

$$r\frac{d\sigma_r}{dr} + \sigma_r - \sigma_\theta = 0 \tag{5.9}$$

The non-zero strain rates can then be expressed in terms of the radial displacement as follows:

$$\varepsilon_r = -\frac{\partial u}{\partial r} \tag{5.10}$$

$$\varepsilon_\theta = -\frac{u}{r} \tag{5.11}$$

The plastic flow rule, namely the incompressibility condition in the undrained case, suggests that:

$$\frac{\varepsilon_r}{\varepsilon_\theta} = -1 \tag{5.12}$$

which can be used to express the displacement u at radius r in terms of cavity wall displacement u_1 and cavity radius a:

$$u = \frac{a}{r}u_1 \tag{5.13}$$

The problem is therefore kinematically determined, and the non-zero strain components are:

$$\varepsilon_r = -\frac{\partial u}{\partial r} = \frac{a}{r^2}u_1 \tag{5.14}$$

$$\varepsilon_\theta = -\frac{u}{r} = -\frac{a}{r^2}u_1 \tag{5.15}$$

The shear strain can then be obtained as follows:

$$\gamma = \frac{2}{3}[(\varepsilon_r)^2 - \varepsilon_r\varepsilon_\theta + (\varepsilon_\theta)^2]^{1/2} = \frac{2}{\sqrt{3}}\frac{a}{r^2}u_1 \tag{5.16}$$

Using the plane strain condition in the z direction, it can be shown (Prevost and Hoeg, 1975) that the axial stress can be expressed in terms of the other two stresses:

$$\sigma_z = \frac{1}{2}(\sigma_r + \sigma_\theta) \tag{5.17}$$

With the above equation, the stress variables p and q reduce to:

$$p = \frac{1}{2}(\sigma_r + \sigma_\theta) \tag{5.18}$$

$$q = \frac{\sqrt{3}}{2}(\sigma_r - \sigma_\theta) \tag{5.19}$$

Strain hardening

For strain hardening soils, the stresses need to satisfy the equilibrium equation (5.9) and stress-strain relationship (5.5), both of which can be combined to give the following equation:

$$r\frac{d\sigma_r}{dr} = -\frac{2\gamma}{\sqrt{3}(D+\gamma)}q_{ult} \tag{5.20}$$

By integrating the above equation subject to the stress boundary condition at the outer boundary, the following is obtained:

$$\sigma_r = p_0 + \frac{q_{ult}}{\sqrt{3}}\ln(1+\frac{\gamma}{D}) \tag{5.21}$$

and from the stress strain relation (5.5), the hoop stress can be determined as follows:

$$\sigma_\theta = \sigma_r - \frac{2q_{ult}}{\sqrt{3}}\ln(\frac{\gamma}{D+\gamma}) \tag{5.22}$$

In particular, for the cavity wall we have $\sigma_r = \psi$ when $r = a$, which leads to the following cavity expansion relationship:

$$\psi = p_0 + \frac{q_{ult}}{\sqrt{3}}\ln(1+\frac{2}{\sqrt{3}D}\frac{u_1}{a}) \tag{5.23}$$

where D and q_{ult} are known soil constants.

Strain hardening and softening

For strain hardening and softening soils, the stress strain relation (5.6) is used. Following the usual procedure, the stress field is determined as follows:

$$\sigma_r = p_0 + \frac{A}{\sqrt{3}}[\frac{B}{2}\ln(1+\gamma^2) + \arctan(\gamma)] \tag{5.24}$$

$$\sigma_\theta = \sigma_r - \frac{2A}{\sqrt{3}}\frac{(\gamma + B\gamma^2)}{(1+\gamma^2)} \tag{5.25}$$

For the cavity wall $\sigma_r = \psi$ when $r = a$, the following cavity expansion relationship is obtained:

$$\psi = p_0 + \frac{A}{\sqrt{3}}\left\{\frac{B}{2}\ln[1 + (\frac{2}{\sqrt{3}}\frac{u_1}{a})^2] + \arctan(\frac{2}{\sqrt{3}}\frac{u_1}{a})\right\} \tag{5.26}$$

in which A and B are material constants.

5.2.2 Undrained cavity expansion from zero radius in clays

This section deals with the undrained expansion of cavities from zero initial radius in clays modelled by an assumed relationship between shear stress and shear strain.

The solution procedure presented was suggested by Ladanyi (1963). As it does not distinguish between elastic and plastic deformations, the procedure is therefore very simple and may be used to deal with any type of shear stress-strain relation. It is, however, noted that the shear strain expression used by Ladanyi (1963) was not entirely correct. For illustration, we will adopt similar stress-strain expressions as equations (5.5) and (5.6) in the following presentation.

(a) Shear stress-shear strain relationship

The shear stress and shear strain variables used in Ladanyi (1963) are defined by:

$$q = \sigma_r - \sigma_\theta \tag{5.27}$$

$$\gamma = \varepsilon_r - \varepsilon_\theta \tag{5.28}$$

It is assumed that the experimental stress-strain relation for undrained loading of a clay can be approximated by one of the following two equations:

$$q = F(\gamma) = \frac{\gamma}{D + \gamma} q_{ult} \tag{5.29}$$

$$q = F(\gamma) = A \frac{B(\gamma)^2 + \gamma}{1 + (\gamma)^2} \tag{5.30}$$

in which A, B and D are material constants.

(b) Shear strain distribution for cavity expansion from zero radius

It is well known that for cavity expansion from zero radius the stress and strain are function of a single variable r/a, where a is current cavity radius and r is the radius of a given particle.

For undrained cavity expansion, it can be shown (Collins and Yu, 1996; Davis *et al*, 1984) that the shear strain, as defined in equation (5.28), can be expressed by the following equation:

$$\gamma = -\ln[1 - (\tfrac{a}{r})^{1+k}] \tag{5.31}$$

where $k=1$ for cylindrical cavity and $k=2$ for spherical cavity. It is clear from the above equation that shear strain becomes infinity at the cavity wall and this is well known result for cavity expansion from zero initial radius. The shear strain expression used by Ladanyi (1963) gives $\pi/2$ for the cavity wall and therefore is not correct.

(c) Stress distribution

Once the shear strain is defined as a function of the normalised radius a/r, the stress-strain relation (5.29) or (5.30) can be assumed to determine the shear stress distribution around the cavity:

$$q = -\frac{\ln[1-(\frac{a}{r})^{1+k}]}{D-\ln[1-(\frac{a}{r})^{1+k}]} q_{ult} \tag{5.32}$$

$$q = A\frac{B\{\ln[1-(\frac{a}{r})^{1+k}]\}^2-\ln[1-(\frac{a}{r})^{1+k}]}{1+\{\ln[1-(\frac{a}{r})^{1+k}]\}^2} \tag{5.33}$$

for strain hardening or strain hardening/softening clays respectively.

The stress components in the soil must also satisfy the following equilibrium equation:

$$\frac{d\sigma_r}{dr} + k\frac{q}{r} = 0 \tag{5.34}$$

Substituting equation (5.32) or (5.33) into the equilibrium equation (5.34) leads to a differential equation which can be integrated numerically to give a solution for the distribution of radial stress σ_r. Once the radial stress is determined, the hoop stress can be obtained as $\sigma_\theta = \sigma_r - q$.

Closed form solutions

When the shear strain is expressed in the form of (5.31), the solution can only be obtained by numerical integration. To ensure that the solution can be obtained in closed form, we may rewrite the shear strain expression as follows:

$$\gamma = -\ln[1-(\tfrac{a}{r})^{1+k}] \doteq (\tfrac{a}{r})^{1+k} + \frac{(\frac{a}{r})^{2(1+k)}}{2} + \frac{(\frac{a}{r})^{3(1+k)}}{3} + \ldots \tag{5.35}$$

The above equation is valid because $0 \le \frac{a}{r} \le 1$.

For example, when the strain hardening soil behaviour is modelled, the equilibrium equation can be written in terms of the shear strain:

$$d\sigma_r = -kq_{ult}\frac{\gamma}{D+\gamma}\frac{dr}{r} \tag{5.36}$$

By using the expression (5.35) for shear strain, the above equilibrium equation may be integrated analytically.

5.3 CAVITY CONTRACTION IN BRITTLE/PLASTIC ROCK

The prediction of the wall movements of tunnels excavated in rock at great depth is an important problem in rock mechanics and mining engineering. It is well established that cavity contraction theory can be used to obtain reasonable solutions for this type of problems when circular tunnels are considered. For this reason, this section presents some fundamental analytical solutions for cavity contraction in rock materials.

5.3.1 Cavity unloading in brittle-plastic rock using the Mohr-Coulomb criterion

Cavity unloading solutions for perfectly plastic Mohr-Coulomb materials have already been presented in Chapter 3. Here we are considering a situation in which the rock strength drops suddenly to a residual value once yielding has occurred in the rock. This is necessary in order to correctly model the behaviour of jointed rock masses.

In this case, a cylindrical cavity (tunnel) is located in an infinitely large rock mass. Initially the rock is subjected to a hydrostatic stress so that all the stress components are p_0. The stress and displacement fields around the opening when the cavity pressure is gradually reduced to a lower value p are of interest. The analytical solutions presented here for brittle-plastic rock follow closely the solutions of Wilson (1980), Fritz (1984) and Reed (1986).

For most tunnel problems, it is reasonable to assume that the axial stress σ_z is intermediate stress between the radial and hoop stresses σ_r and σ_θ. For an unloading cavity, the hoop stress is the major principal stress. The linear Mohr-Coulomb yield function is assumed to indicate the initiation of yielding in the rock mass:

$$\sigma_\theta - \alpha\sigma_r = Y \tag{5.37}$$

where α and Y are related to friction angle and cohesion ϕ and C by:

$$\alpha = \frac{1 + \sin\phi}{1 - \sin\phi} \quad \text{and} \quad Y = \frac{2C\cos\phi}{1 - \sin\phi} \tag{5.38}$$

After yielding, the strength of rock is assumed to drop suddenly and to allow for such a post-yield behaviour we assume that the stresses after yielding are governed by the following residual failure function:

$$\sigma_\theta - \alpha'\sigma_r = Y' \tag{5.39}$$

where $\alpha' = (1 + \sin\phi')/(1 - \sin\phi')$ and $Y' = (2C'\cos\phi')/(1 - \sin\phi')$ are the residual strength parameters of the rock (see Figure 5.1).

(a) Stress analysis

Elastic response and initial yielding

When the internal cavity pressure reduces, the rock initially behaves elastically and the stresses are:

$$\sigma_r = p_0 - (p_0 - p)(\frac{a}{r})^2 \tag{5.40}$$

$$\sigma_\theta = p_0 + (p_0 - p)(\frac{a}{r})^2 \tag{5.41}$$

Yield will initiate at the cavity wall when the elastic stress fields satisfy the yield function (5.37) which occurs when the cavity pressure is equal to:

$$p = p_{1y} = \frac{2p_0 - Y}{\alpha + 1} \tag{5.42}$$

If the internal cavity pressure p drops below the value defined in equation (5.42), a plastic zone with radius c will form around the cavity. The rock in the zone outside the radius c will still behave elastically.

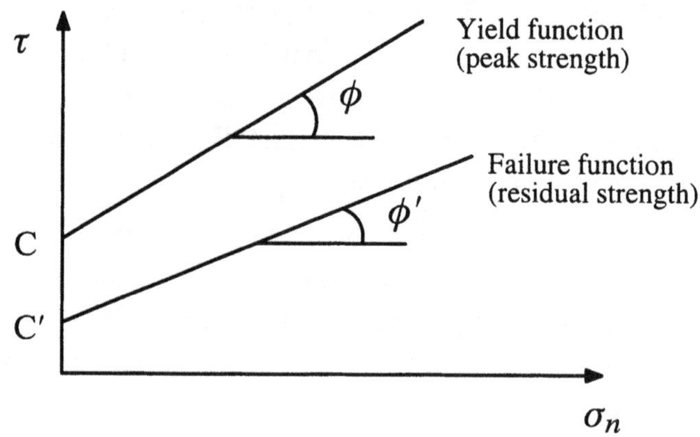

Figure 5.1: Rock strength parameters before and after yielding

Stresses in elastic zone $c \leq r \leq \infty$

After yielding has occurred, the stress fields in the outer elastic zone can be shown to be:

$$\sigma_r = p_0 - (p_0 - p_{1y})(\frac{c}{r})^2 \tag{5.43}$$

$$\sigma_\theta = p_0 + (p_0 - p_{1y})(\frac{c}{r})^2 \tag{5.44}$$

Stresses in plastic zone $a \leq r \leq c$

The stresses in the plastic region must satisfy the residual failure equation (5.39) and the equilibrium equation:

$$r\frac{d\sigma_r}{dr} = \sigma_\theta - \sigma_r \tag{5.45}$$

both of which can be combined to determine the stresses in the plastic zone. The derived expressions are:

$$\sigma_r = \frac{Y' + (\alpha'-1)p}{\alpha'-1}\left(\frac{r}{a}\right)^{\alpha'-1} - \frac{Y'}{\alpha'-1} \tag{5.46}$$

$$\sigma_\theta = \alpha'\frac{Y' + (\alpha'-1)p}{\alpha'-1}\left(\frac{r}{a}\right)^{\alpha'-1} - \frac{Y'}{\alpha'-1} \tag{5.47}$$

The continuity of the radial stress at the elastic-plastic interface $r=c$ required by equilibrium can be used to determine the radius of the plastic region:

$$\frac{c}{a} = \left\{\frac{Y' + (\alpha'-1)p_{1y}}{Y' + (\alpha'-1)p}\right\}^{\frac{1}{\alpha'-1}} \tag{5.48}$$

It is noted that the hoop stress is only continuous across the elastic-plastic boundary if the rock is perfectly plastic (i.e. $Y = Y'$).

The out-of-plane stress σ_z can be determined from the plane strain condition:

$$\sigma_z = \nu(\sigma_r + \sigma_\theta) + (1-2\nu)p_0 \tag{5.49}$$

(b) Displacement analysis

For cavity unloading problems, all radial displacements are towards the centre of the cavity, and therefore are negative in order to be consistent with the stress notation of compression positive. It is however more convenient for cavity unloading to deal with the radial convergence which is positive for inward radial displacements. For simplicity, the inward radial displacements, denoted by u, will be used in the following discussion.

Displacement in the outer elastic zone

The radial displacement in the outer elastic zone can be obtained as follows:

$$u = \frac{1+\nu}{E}(p_0 - p_{1y})\frac{c^2}{r} \tag{5.50}$$

In particular, for the elastic-plastic interface the displacement is:

$$u_c = u|_{r=c} = \frac{1+\nu}{E}(p_0 - p_{1y})c \tag{5.51}$$

Displacement in the plastic zone

To determine the displacement field in the plastic zone, a plastic flow rule needs to be used. As usual, a non-associated flow rule can be assumed to give the following relation:

$$d\varepsilon_r^p + \beta d\varepsilon_\theta^p = 0 \tag{5.52}$$

where $\beta = (1 + \sin\psi)/(1 - \sin\psi)$ and ψ is dilation angle. The above equation may be integrated to give the relationship between plastic strains:

$$\varepsilon_r^p + \beta\varepsilon_\theta^p = 0 \tag{5.53}$$

For small strain problems, the strains can be expressed in terms of inward radial displacement u as follows:

$$\varepsilon_r = \frac{du}{dr} \quad \text{and} \quad \varepsilon_\theta = \frac{u}{r} \tag{5.54}$$

After using (5.54) and putting the solution for elastic strains into equation (5.53), one obtains the following governing equation:

$$\frac{du}{dr} + \beta\frac{u}{r} = \frac{1+\nu}{E}[A(\frac{r}{c})^{\alpha'-1} + B] \tag{5.55}$$

in which

$$A = [(1 + \alpha'\beta)(1-\nu)-(\alpha' + \beta)\nu](p_{1y} + \frac{Y'}{\alpha'-1}) \tag{5.56}$$

$$B = -(1-2\nu)(1 + \beta)(p_0 + \frac{Y'}{\alpha'-1}) \tag{5.57}$$

The solution of equation (5.55), subject to the boundary condition $u = u_c$ at $r = c$, is shown to be:

$$u = \frac{1+\nu}{E} r[K_1(\frac{r}{c})^{\alpha'-1} + K_2(\frac{c}{r})^{\beta+1} + K_3] \tag{5.58}$$

where

$$K_1 = [\frac{1 + \alpha'\beta}{\alpha' + \beta}(1-\nu)-\nu](p_{1y} + \frac{Y'}{\alpha'-1}) \tag{5.59}$$

$$K_3 = -(1-2\nu)(p_0 + \frac{Y'}{\alpha'-1}) \tag{5.60}$$

$$K_2 = p_0-p_{1y}-K_1-K_3 \tag{5.61}$$

(c) Validity of $\sigma_r < \sigma_z < \sigma_\theta$

The solution obtained so far is valid only if the out-of-plane stress σ_z is the intermediate stress. This condition prevails until such a time when $\sigma_z = \sigma_\theta$ occurs at the cavity wall. Using the stress solutions just obtained, it can be shown (Reed, 1988) that such a condition is satisfied when the cavity wall pressure reduces to:

$$p = p_{2y} = \frac{1-2\nu}{\alpha'(1-\nu)-\nu}(p_0 + \frac{Y'}{\alpha'-1})-\frac{Y'}{\alpha'-1} \tag{5.62}$$

If the cavity pressure reduces further, an inner plastic zone will form in which the out-of-plane stress is equal to the hoop stress (i.e. $\sigma_z = \sigma_\theta$). The other possibility is that an inner plastic zone may develop so that the out-of-plane stress become the major principal stress (i.e. $\sigma_r \le \sigma_\theta \le \sigma_z$). This second case has been shown to be impossible (Reed, 1988)

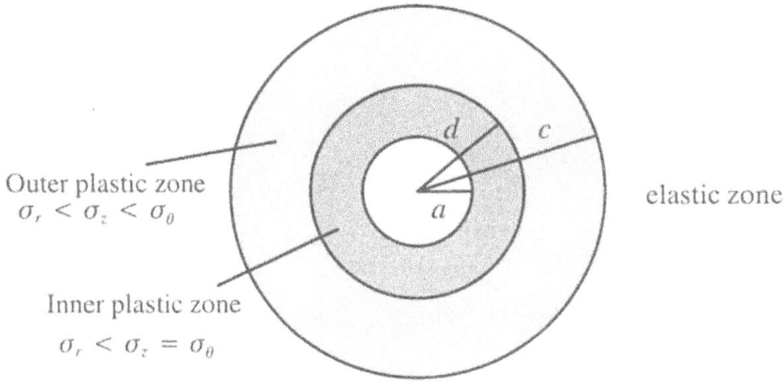

Figure 5.2: Two different plastic zones around cavity

The stress solutions (5.46)-(5.47) obtained previously will still apply in the inner plastic zone, but in this zone the out-of-plane stress should be equal to the hoop stress. By equating $\sigma_z = \sigma_\theta$ from (5.49) and (5.47), the boundary between the inner and outer plastic zones is at:

$$\frac{d}{a} = \left\{ \frac{Y' + (\alpha'-1)p_{2y}}{Y' + (\alpha'-1)p} \right\}^{\frac{1}{\alpha'-1}} \tag{5.63}$$

where d is the radius of the inner plastic zone, shown in Figure 5.2.

It is assumed that the displacement solution (5.58) obtained before is still valid in the outer plastic zone. The displacement field in the inner plastic zone is different from that in the outer plastic zone because the stress state in the inner plastic zone lies on the intersection of the two yield surfaces. In this case, the flow rule can be obtained by summing the contributions from the two corresponding plastic potentials.

It can be shown that such a plastic flow rule gives the following plastic relationship between plastic strain rates:

$$d\varepsilon_r^p + \beta d\varepsilon_\theta^p + \beta d\varepsilon_z^p = 0 \tag{5.64}$$

Using the condition (5.53) and $\varepsilon_z^p = 0$ as the initial values for the plastic strains, equation (5.64) can be integrated to give:

$$\varepsilon_r^p + \beta\varepsilon_\theta^p + \beta\varepsilon_z^p = 0 \tag{5.65}$$

The above equation together with $\varepsilon_z = 0$ and $\sigma_z = \sigma_\theta$ can be used to lead to the following differential equation for the displacement in the inner plastic zone:

$$\frac{du}{dr} + \beta\frac{u}{r} = \frac{1}{E}[M(\tfrac{r}{c})^{\alpha'-1} + N] \tag{5.66}$$

where

$$M = [1 + 2\alpha'\beta - 2(\alpha' + \beta + \alpha'\beta)\nu](p_{1y} + \frac{Y'}{\alpha'-1}) \tag{5.67}$$

$$N = -(1 + 2\beta)(1-2\nu)(p_0 + \frac{Y'}{\alpha'-1}) \tag{5.68}$$

The solution of equation (5.66) can be shown to be:

$$u = \frac{r}{E}[L_1(\tfrac{r}{c})^{\alpha'-1} + L_2(\tfrac{c}{r})^{\beta+1} + L_3] \tag{5.69}$$

in which

$$L_1 = \frac{1}{\alpha' + \beta}M \tag{5.70}$$

$$L_3 = \frac{1}{1 + \beta}N \tag{5.71}$$

and L_2 can be obtained with the continuity of the displacement u at $r = d$ from equation (5.58).

5.3.2 Cavity unloading in brittle-plastic rock using the Hoek-Brown criterion

In addition to the linear Mohr-Coulomb yield criterion, a number of researchers have also used non-linear yield criteria to analyse cavity unloading problems in rock. Notable examples have been presented by Hobbs (1966) who used a power law criterion, and Kennedy and Lindberg (1978) who used piece-wise linear approximations to non-linear Mohr envelopes. However, a more popular development is the cavity unloading solutions presented by Brown *et al.* (1983) who used the empirical Hoek-Brown non-linear criterion (Hoek and Brown, 1980) to describe the behaviour of rock masses. While the stress analysis described in this section is based on Brown *et al.* (1983), the displacement analysis presented here is new and represents a more rigorous solution than that of Brown *et al.* (1983).

(a) Hoek-Brown criterion

It is assumed that the yield of rock is governed by the Hoek and Brown criterion in the following form:

$$\sigma_1 = \sigma_3 + \sqrt{mY\sigma_3 + sY^2} \tag{5.72}$$

where σ_1 and σ_3 are the major and minor principal stresses; Y is the uniaxial compressive strength of the intact rock material; and m and s are constants depending on the nature of the rock mass and the extent to which it is broken prior to being subjected to the principal stresses σ_1 and σ_3.

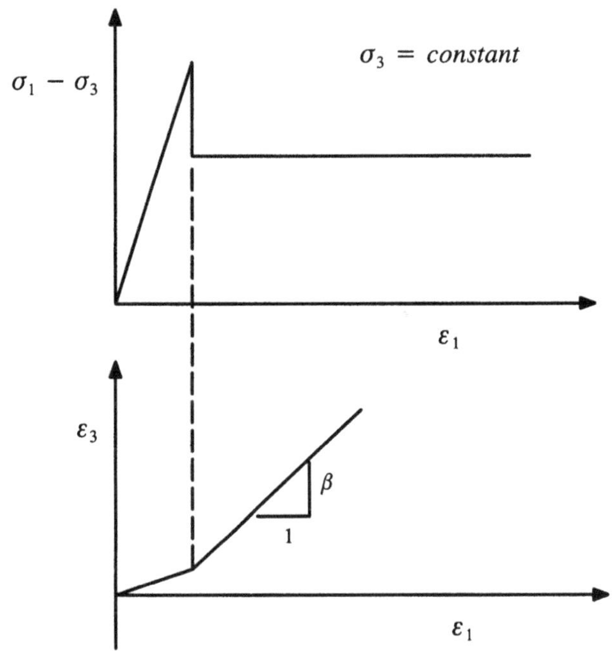

Figure 5.3: Idealised stress-strain behaviour of Hoek-Brown criterion

After yielding, the strength parameter m and s drop to residual values m' and s'. The unconfined compressive strength of the rock mass changes from its peak value $\sqrt{s}\,Y$ to its residual value $\sqrt{s'}\,Y$. The residual strength for the broken rock mass is therefore:

$$\sigma_1 = \sigma_3 + \sqrt{m'Y\sigma_3 + s'Y^2} \tag{5.73}$$

The Hoek-Brown criterion offers some advantages over other approaches in determining the overall strength of *in-situ* rock masses because it is based on one simple material property, Y, and rock mass quality data that may be systematically collected and evaluated during site investigation.

To ensure that closed form solutions can be obtained, it is necessary to further assume that after yield the strength of rock drops *suddenly* to its residual values. Hence the stress-strain relationship can be shown in Figure 5.3.

(b) Stress analysis

As in the case of Mohr-Coulomb criterion discussed in the previous section, the stresses in the outer elastic region are:

$$\sigma_r = p_0 - (p_0 - p_{1y})(\frac{c}{r})^2 \tag{5.74}$$

$$\sigma_\theta = p_0 + (p_0 - p_{1y})(\frac{c}{r})^2 \tag{5.75}$$

where c is the radius of elastic-plastic boundary and p_{1y} is the radial stress at the elastic-plastic boundary. For cavity unloading problems, the Hoek-Brown yield function takes the following form:

$$\sigma_\theta = \sigma_r + \sqrt{mY\sigma_r + sY^2} \tag{5.76}$$

The above yield criterion must be satisfied at the outside boundary of the elastic zone at $r=c$. Putting equations (5.74) and (5.75) with $r=c$ into equation (5.76) leads to the following solution for the radial stress p_{1y} at the elastic-plastic boundary:

$$p_{1y} = p_0 - MY \tag{5.77}$$

in which

$$M = \frac{1}{2}[(\frac{m}{4})^2 + m\frac{p_0}{Y} + s]^{1/2} - \frac{m}{8} \tag{5.78}$$

Within the inner plastic region (or broken zone), stresses need to satisfy equilibrium and the following residual strength criterion:

$$\sigma_\theta = \sigma_r + \sqrt{m'Y\sigma_r + s'Y^2} \tag{5.79}$$

Both the equilibrium equation and the above failure criterion can be combined to give the following solution for radial stress in the plastic zone:

$$\sigma_r = \frac{m'Y}{4}[\ln(\frac{r}{a})]^2 + \sqrt{m'Yp + s'Y^2}\ln(\frac{r}{a}) + p \tag{5.80}$$

where p denotes the cavity pressure. To simplify the above expression, let us introduce the following two quantities:

$$A = \sqrt{m'Yp + s'Y^2} \quad \text{and} \quad B = \frac{1}{4}m'Y \tag{5.81}$$

with which equation (5.80) can be rewritten as:

$$\sigma_r = p + A\ln(\tfrac{r}{a}) + B\ln^2(\tfrac{r}{a}) \tag{5.82}$$

From the residual failure function (5.79), the hoop stress in the plastic zone can be obtained as follows:

$$\sigma_\theta = p + A + (A + 2B)\ln(\tfrac{r}{a}) + B\ln^2(\tfrac{r}{a}) \tag{5.83}$$

Equations (5.82) and (5.83) fully determine the stress fields in the plastic zone. Now it remains to determine the radius of the elastic-plastic boundary c. This can be solved by using the continuity of the radial stress at the elastic-plastic boundary:

$$\tfrac{c}{a} = \exp\left\{N - \frac{2}{m'Y}\sqrt{m'Yp + s'Y^2}\right\} \tag{5.84}$$

in which

$$N = \frac{2}{m'Y}\sqrt{m'Yp_0 + s'Y^2 - m'Y^2M} \tag{5.85}$$

It is noted that a plastic zone will exist only if the cavity pressure reduces to the critical value $p \leq p_{1y}$.

(c) Displacement analysis

Displacement in the outer elastic zone

The displacements in the outer elastic zone can be derived as follows:

$$u = \frac{1+\nu}{E}(p_0 - p_{1y})\frac{c^2}{r} \tag{5.86}$$

In particular, at the elastic-plastic interface the displacement is:

$$u_c = u|_{r=c} = \frac{1+\nu}{E}(p_0 - p_{1y})c \tag{5.87}$$

Displacement in the plastic zone

To determine the displacement field in the plastic zone, a plastic flow rule is needed. If a non-associated flow rule similar to that of the Mohr-Coulomb criterion can be used (see Figure 5.3), then we have:

$$d\varepsilon_r^{\ p} + \beta d\varepsilon_\theta^{\ p} = 0 \tag{5.88}$$

where $\beta = (1 + \sin\psi)/(1 - \sin\psi)$ and ψ is dilation angle of the material. The above equation may be integrated to give the relationship between strains:

$$\varepsilon_r^{\ p} + \beta\varepsilon_\theta^{\ p} = 0 \tag{5.89}$$

which can be further written as:

$$\varepsilon_r + \beta\varepsilon_\theta = \varepsilon_r^e + \beta\varepsilon_\theta^e \tag{5.90}$$

Elastic strains are related to stress changes by elasticity theory which gives

$$\varepsilon_r^e = \frac{1+\nu}{E}[(1-\nu)\sigma_r - \nu\sigma_\theta - (1-2\nu)p_0] \tag{5.91}$$

$$\varepsilon_\theta^e = \frac{1+\nu}{E}[(1-\nu)\sigma_\theta - \nu\sigma_r - (1-2\nu)p_0] \tag{5.92}$$

For small deformation the strains can be expressed in terms of inward radial displacement u as follows:

$$\varepsilon_r = \frac{du}{dr} \quad \text{and} \quad \varepsilon_\theta = \frac{u}{r} \tag{5.93}$$

Substituting equations (5.91)-(5.93) into equation (5.90) gives:

$$\frac{du}{dr} + \beta\frac{u}{r} = g(r) \tag{5.94}$$

where

$$g(r) = D_1 + D_2\ln(\frac{r}{a}) + D_3\ln^2(\frac{r}{a}) \tag{5.95}$$

and

$$D_1 = \frac{1+\nu}{E}\{(1+\beta)(1-2\nu)(p-p_0) + (\beta-\nu\beta-\nu)A\} \tag{5.96}$$

$$D_2 = \frac{1+\nu}{E}\{(1-\nu)(A+\beta A+2\beta B) - \nu(\beta A + A + 2B)\} \tag{5.97}$$

$$D_3 = \frac{1+\nu}{E}(1+\beta)(1-2\nu)B \tag{5.98}$$

The general solution of equation (5.94) can be shown to be:

$$u = n_1\frac{r}{1+\beta} + n_2[\frac{r\ln r}{1+\beta} - \frac{r}{(1+\beta)^2}]$$
$$+ D_3[\frac{r\ln^2 r}{1+\beta} - \frac{2r\ln r}{(1+\beta)^2} + \frac{2r}{(1+\beta)^3}] + C_0 r^{-\beta} \tag{5.99}$$

in which

$$n_1 = D_1 - D_2\ln a + D_3\ln^2 a \tag{5.100}$$

$$n_2 = D_2 - 2D_3\ln a \tag{5.101}$$

and C_0 is an integration constant that can be determined using the known displacement at the elastic-plastic boundary (i.e. equation (5.87)):

$$C_0 = \frac{1+\nu}{E}(p_0 - p_{1y})c^{1+\beta} - n_1\frac{c^{1+\beta}}{1+\beta}$$

$$-n_2[\frac{\ln c}{1+\beta} - \frac{1}{(1+\beta)^2}]c^{1+\beta}$$

$$-D_3[\frac{\ln^2 c}{1+\beta} - \frac{2\ln c}{(1+\beta)^2} + \frac{2}{(1+\beta)^3}]c^{1+\beta} \quad (5.102)$$

By combining equations (5.102) and (5.99), the displacement field in the plastic zone is fully defined. The cavity wall displacement can then be obtained as a special case by setting $r=a$ in equation (5.99).

5.4 SOLUTIONS FOR PIECE-WISE MOHR-COULOMB CRITERION

In Chapter 3, cavity expansion solutions for the linear Mohr-Coulomb criterion were presented. It is well known that the real behaviour of soil or rock may be better modelled by a curved Mohr-Coulomb criterion. In other words, the value of internal friction angle is not a constant but rather a function of normal stress. Research has been done using a non-linear Mohr-Coulomb criterion to analyse cavity expansion or contraction problems. For example, Baligh (1976) used a numerical method to extend Vesic's solution to include the effect of dependence of friction angle on normal stress. On the other hand, Kennedy and Lindberg (1978) derived a small strain analytical solution by using a piece-wise Mohr-Coulomb yield surface. The solution procedure presented in this section follows Kennedy and Lindberg (1978) and is an extension of that of Florence and Schwer (1978) for the linear Mohr-Coulomb criterion. As in the previous section, the inward radial displacement is taken as positive and denoted by u.

(a) Piece-wise approximation of a non-linear yield surface

A general non-linear yield surface can be represented to any degree of accuracy by a series of straight line segments ($f_1, f_2, ... f_{n+1}$) with different slopes ($\tan\omega_1, \tan\omega_2, ... \tan\omega_{n+1}$). Shown in Figure 5.4 are two such segments.

Along each segment, it is possible to define a yield function. For example, the yield functions can be given by:

$$f_1 = 0 \quad \text{for} \quad \sigma_3 < p_1 \quad (5.103)$$

$$f_2 = 0 \quad \text{for} \quad p_1 < \sigma_3 < p_2 \quad (5.104)$$

$$f_3 = 0 \quad \text{for} \quad p_2 < \sigma_3 < p_3 \quad (5.105)$$

$$f_j = 0 \quad \text{for} \quad p_{j-1} < \sigma_3 < p_j \quad (5.106)$$

The yield function for the jth segment can be expressed by the following equation:

$$f_j = \sigma_1 - a_j\sigma_3 - Y_j = 0 \tag{5.107}$$

where

$$a_j = 1 + \tan\omega_j = \frac{1+\sin\phi_j}{1-\sin\phi_j} \tag{5.108}$$

$$Y_j = \frac{2C_j\cos\phi_j}{1-\sin\phi_j} \tag{5.109}$$

in which f_j, C_j, Y_j and ϕ_j are shown in Figure 5.4.

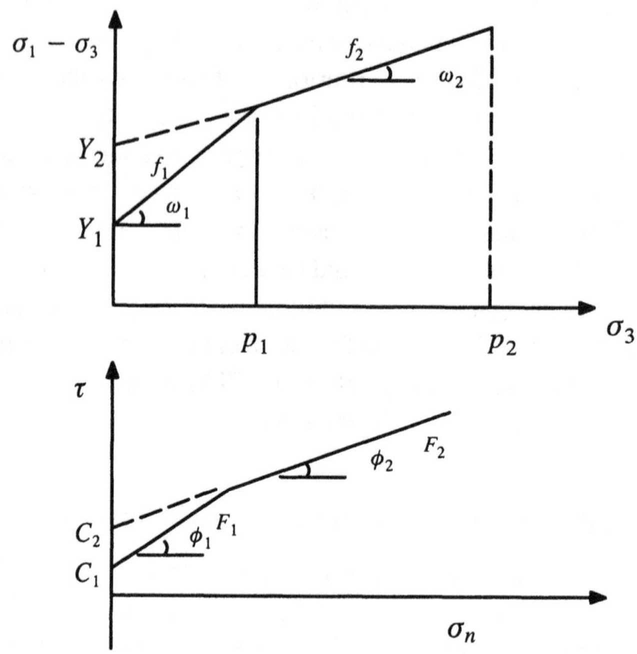

Figure 5.4: Piece-wise approximation of non-linear Mohr-Coulomb criterion

A piece-wise linearization of the non-linear yield function will now be used to analyse the elastic-plastic closure of a cylindrical cavity. It is assumed that the cavity has a radius of a and initially the material is stress free. An internal pressure p_0 and an external pressure p are gradually applied at the cavity wall and at an infinite boundary respectively. It is also assumed that the internal pressure is less than the free-field (or external) pressure and both are applied in such a way that the internal pressure reaches its maximum value before yielding of the material occurs. As the external pressure p increases, yielding will take place first on the inner boundary of the cavity. As p increases further, the yielded region will move out to some radius c until the pressure has reached its peak. As a result, in the region defined by $a \leq r \leq c$, the material will behave plastically while a completely elastic zone will exist for the region $c < r < \infty$.

(b) Governing equations

Although the elastic and plastic zones are analysed separately, it is convenient to first consider the governing equations that are valid for both regions. These governing equations are given below:

Equilibrium equation

$$\frac{d\sigma_r}{dr} + \frac{\sigma_r - \sigma_\theta}{r} = 0 \tag{5.110}$$

Boundary conditions

$$\sigma_r = p_0 \quad at \quad r = a \tag{5.111}$$

$$\sigma_r = p \quad at \quad r = \infty \tag{5.112}$$

Strain-displacement relations

$$\varepsilon_r = \frac{du}{dr} \quad \text{and} \quad \varepsilon_\theta = \frac{u}{r} \tag{5.113}$$

Compatibility equation

$$\frac{d\varepsilon_\theta}{dr} + \frac{\varepsilon_\theta - \varepsilon_r}{r} = 0 \tag{5.114}$$

Stress-strain relations

If the strains and displacement are measured from the stress-free state, the plane strain stress-strain relations may be written as follows:

$$\varepsilon_r - \varepsilon_r^p = \varepsilon_r^e = \frac{1}{E'}(\sigma_r - \nu'\sigma_\theta) \tag{5.115}$$

$$\varepsilon_\theta - \varepsilon_\theta^p = \varepsilon_\theta^e = \frac{1}{E'}(\sigma_\theta - \nu'\sigma_r) \tag{5.116}$$

where

$$E' = \frac{E}{1-\nu^2} \quad \text{and} \quad \nu' = \frac{\nu}{1-\nu} \tag{5.117}$$

(c) Division of plastic zone

From the solution for the linear Mohr-Coulomb yield function, it is known that the major principal stress is the hoop stress and the minor principal stress is the radial stress which is known to be a monotonically increasing function of r. This follows that close to the cavity wall where the radial stress is between p_0 and p_1, $f_1 = 0$ will be the yield function. This yield function will hold until $\sigma_r = p_1$ at $r = c_1$. For $p_1 \leq \sigma_r \leq p_2$, $f_2 = 0$ will be the yield function and so on. This process continues until the elastic-plastic boundary is reached at $r = c_m$, where $m \leq n + 1$ and n+1 is the number of straight segments in the yield function, Figure 5.5. As a result, the plastic zone is divided into a number of circular rings, where different yield functions apply as the yield function.

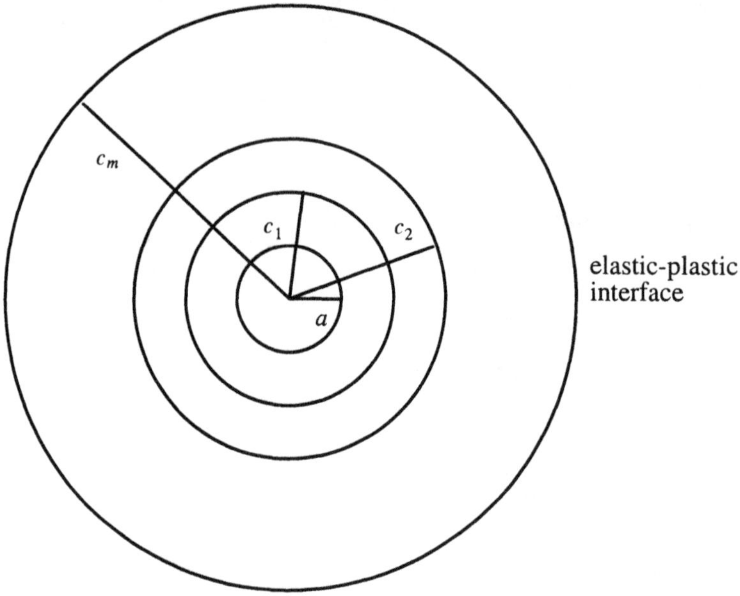

Figure 5.5: Division of the plastic region into m circular rings

(d) Stress analysis

Stress calculation in plastic zone

As mentioned above, the plastic region is divided into m zones in which different straight line segments of the yield function apply as the yield condition. As a result, each zone has to be analysed individually.

For zone 1 which is closest to the cavity wall, the yield function is given by:

$$f = f_1 = \sigma_\theta - \alpha_1 \sigma_r - Y_1 = 0 \tag{5.118}$$

Substituting the above yield function into the equilibrium equation (5.110) gives the following equation in terms of radial stress:

$$r\frac{d\sigma_r}{dr} + (1-\alpha_1)\sigma_r = Y_1 \tag{5.119}$$

which can be solved for σ_r as follows:

$$\sigma_r = \frac{Y_1}{1-\alpha_1} + Ar^{\alpha_1-1} \tag{5.120}$$

The integration constant A can be determined by using the boundary condition at the cavity wall:

$$\sigma_r|_{r=a} = p_0 \tag{5.121}$$

as

$$A = (\frac{Y_1}{\alpha_1-1} + p_0)a^{1-\alpha_1} \tag{5.122}$$

Hence, the stresses in zone 1 can be expressed as follows:

$$\sigma_r = \frac{Y_1}{1-\alpha_1} + (\frac{Y_1}{\alpha_1-1} + p_0)(\frac{r}{a})^{\alpha_1-1} \tag{5.123}$$

$$\sigma_\theta = \frac{Y_1}{1-\alpha_1} + \alpha_1(\frac{Y_1}{\alpha_1-1} + p_0)(\frac{r}{a})^{\alpha_1-1} \tag{5.124}$$

By using a similar procedure to above, expressions for stresses in other zones can also be obtained. For example for zone j, the stresses are:

$$\sigma_r = \frac{Y_j}{1-\alpha_j} + (\frac{Y_j}{\alpha_j-1} + p_{j-1})(\frac{r}{a})^{\alpha_j-1} \tag{5.125}$$

$$\sigma_\theta = \frac{Y_j}{1-\alpha_j} + \alpha_j(\frac{Y_j}{\alpha_j-1} + p_{j-1})(\frac{r}{a})^{\alpha_j-1} \tag{5.126}$$

The values for the radii of c_1, c_2,... and c_m are determined by employing the requirement of continuity of radial stress from zone $j-1$ to zone j. The results are shown to be:

$$c_1 = \left\{\frac{Y_1 + p_1(\alpha_1-1)}{Y_1 + p_0(\alpha_1-1)}\right\}^{1/(\alpha_1-1)} a \tag{5.127}$$

$$c_2 = \left\{\frac{Y_2 + p_2(\alpha_2-1)}{Y_2 + p_1(\alpha_2-1)}\right\}^{1/(\alpha_2-1)} c_1 \tag{5.128}$$

$$c_3 = \left\{ \frac{Y_3 + p_3(\alpha_3 - 1)}{Y_3 + p_2(\alpha_3 - 1)} \right\}^{1/(\alpha_3 - 1)} c_2 \tag{5.129}$$

$$\ldots\ldots \tag{5.130}$$

$$c_m = \left\{ \frac{Y_m + p_m(\alpha_m - 1)}{Y_m + p_{m-1}(\alpha_m - 1)} \right\}^{1/(\alpha_m - 1)} c_{m-1} \tag{5.131}$$

Note that p_m is the radial stress at the elastic-plastic boundary which can be determined by considering stress fields in the outer elastic zone.

Stress calculation in elastic zone

It is well known that the stresses in the outer elastic zone can be written as follows:

$$\sigma_r = p - (p - p_m)\left(\frac{c_m}{r}\right)^2 \tag{5.132}$$

$$\sigma_\theta = p + (p - p_m)\left(\frac{c_m}{r}\right)^2 \tag{5.133}$$

By using the condition that at the elastic-plastic boundary $r = c_m$ the stresses should satisfy the yield function for zone m:

$$f = f_m = \sigma_\theta - \alpha_m \sigma_r - Y_m = 0 \tag{5.134}$$

we obtain the expression for the radial stress at the elastic-plastic interface:

$$p_m = \frac{2p - Y_m}{1 + \alpha_m} \tag{5.135}$$

(e) Displacement analysis

This subsection describes how the strains and displacement can be determined in the elastic and plastic zones. Expressions of strains in the elastic zone can be obtained simply by combining equations (5.115)-(5.116) and (5.132)-(5.133).

In the plastic region, however, a plastic flow rule is needed to determine the strains and displacement. Although we can easily use a general non-associated Mohr-Coulomb flow rule in the calculation, for simplicity we will only present results obtained with both the associated (i.e. plastic potential is identical to the yield function) and fully non-associated (i.e. no volume change) flow rules.

Associated flow rule

Let us first consider zone m, defined by $c_{m-1} \leq r \leq c_m$, that is closest to the outer elastic region. The associated flow rule gives the following relationship between the plastic strains:

$$\varepsilon_r^p + \alpha_m \varepsilon_\theta^p = 0 \tag{5.136}$$

which can also be written as:

$$\varepsilon_r = \varepsilon_r^e - \alpha_m \varepsilon_\theta^p \tag{5.137}$$

$$\varepsilon_\theta = \varepsilon_\theta^e + \varepsilon_\theta^p \tag{5.138}$$

where the hoop plastic strain ε_θ^p needs to be determined. Substituting equations (5.137)-(5.138) into the compatibility condition (5.114) results in the following equation in terms of ε_θ^p:

$$r\frac{d\varepsilon_\theta^p}{dr} + (1 + \alpha_m)\varepsilon_\theta^p = g_m(r) \tag{5.139}$$

where

$$g_m(r) = -\frac{\alpha_m^2-1}{E'}[\frac{Y_m}{\alpha_m-1} + p_{m-1}](\frac{r}{c_{m-1}})^{\alpha_m-1} \tag{5.140}$$

The equation (5.139) can be solved to give the following general solution:

$$\varepsilon_\theta^p = -\frac{\alpha_m^2-1}{2\alpha_m E'}[\frac{Y_m}{\alpha_m-1} + p_{m-1}](\frac{r}{c_{m-1}})^{\alpha_m-1} + Br^{-1-\alpha_m} \tag{5.141}$$

where the integration constant B is determined using the boundary condition at the elastic-plastic boundary:

$$\varepsilon_\theta^p|_{r=c_m} = 0 \tag{5.142}$$

The derived expression for B is:

$$B = \frac{\alpha_m^2-1}{2\alpha_m E'}[\frac{Y_m}{\alpha_m-1} + p_{m-1}](\frac{c_m}{c_{m-1}})^{\alpha_m-1}(c_m)^{\alpha_m-1} \tag{5.143}$$

As a result, the strains and displacement in zone m may be written as follows:

$$\varepsilon_r = \frac{1}{E'}\left\{\frac{1-\nu'}{1-\alpha_m}Y_m + [1-\nu'\alpha_m + \frac{1}{2}(\alpha_m^2-1)]\right.$$

$$\left.(\frac{Y_m}{\alpha_m-1} + p_{m-1})(\frac{r}{c_{m-1}})^{\alpha_m-1}\right\}$$

$$-\frac{\alpha_m^2-1}{2\alpha_m E'}[\frac{Y_m}{\alpha_m-1} + p_{m-1}](\frac{c_m}{c_{m-1}})^{\alpha_m-1}(\frac{c_m}{r})^{\alpha_m+1} \tag{5.144}$$

$$\varepsilon_\theta = \frac{1}{E'}\left\{\frac{1-\nu'}{1-\alpha_m}Y_m - [\frac{\alpha_m^2-1}{2\alpha_m} - \alpha_m + \nu']\right.$$

$$(\frac{Y_m}{\alpha_m-1} + p_{m-1})(\frac{r}{c_{m-1}})^{\alpha_m-1}\Big\}$$

$$+ \frac{\alpha_m{}^2-1}{2\alpha_m E'}[\frac{Y_m}{\alpha_m-1} + p_{m-1}](\frac{c_m}{c_{m-1}})^{\alpha_m-1}(\frac{c_m}{r})^{\alpha_m+1} \tag{5.145}$$

$$u = r\varepsilon_\theta \tag{5.146}$$

The strains and displacement in the other zones can be obtained by following a similar procedure. The continuity of displacement is used to determine the integration constants. In particular, it is found that the displacement at the cavity wall $r=a$ is:

$$\frac{u_a}{a} = \frac{1}{E'}[\frac{1-\nu'}{1-\alpha_1}Y_1 + (\alpha_1-\nu')(\frac{Y_1}{\alpha_1-1} + p_0)]$$

$$+ \frac{\alpha_1{}^2-1}{2E'\alpha_1}(\frac{Y_1}{\alpha_1-1} + p_0)[(\frac{c_1}{a})^{2\alpha_1}-1]$$

$$+ \frac{\alpha_2{}^2-1}{2E'\alpha_2}(\frac{Y_2}{\alpha_2-1} + p_1)[(\frac{c_2}{c_1})^{2\alpha_2}-1](\frac{c_1}{a})^{\alpha_1+1}$$

$$+ \frac{\alpha_3{}^2-1}{2E'\alpha_3}(\frac{Y_3}{\alpha_3-1} + p_2)[(\frac{c_3}{c_2})^{2\alpha_3}-1](\frac{c_2}{c_1})^{\alpha_2+1}(\frac{c_1}{a})^{\alpha_1+1}$$

$$+ \ldots\ldots\ldots$$

$$+ \frac{\alpha_m{}^2-1}{2E'\alpha_m}(\frac{Y_m}{\alpha_m-1} + p_{m-1})[(\frac{c_m}{c_{m-1}})^{2\alpha_m}-1]$$

$$\times (\frac{c_{m-1}}{c_{m-2}})^{\alpha_{m-1}+1}(\frac{c_{m-2}}{c_{m-3}})^{\alpha_{m-2}+1}\ldots(\frac{c_2}{c_1})^{\alpha_2+1}(\frac{c_1}{a})^{\alpha_1+1} \tag{5.147}$$

Non-associated flow rule

Consider zone m, defined by $c_{m-1} \leq r \leq c_m$, that is closest to the outer elastic region. The non-associated flow rule gives the following relationship between the plastic strains:

$$\varepsilon_r{}^p + \varepsilon_\theta{}^p = 0 \tag{5.148}$$

When this equation is used instead of the associated plastic flow rule (5.136), the cavity displacement is shown to be:

$$\frac{u_a}{a} = \frac{1}{E'}[\frac{1-\nu'}{1-\alpha_1}Y_1 + (\alpha_1-\nu')(\frac{Y_1}{\alpha_1-1} + p_0)]$$

$$+ \frac{1}{E'}[Y_1 + (\alpha_1-1)p_0][(\frac{c_1}{a})^{\alpha_1+1}-1]$$

$$+ \frac{1}{E'}[Y_2 + (\alpha_2-1)p_1][(\frac{c_2}{c_1})^{\alpha_1+1}-1](\frac{c_1}{a})^2$$

$$+ \frac{1}{E'}[Y_3 + (\alpha_3-1)p_2][(\frac{c_3}{c_2})^{\alpha_2+1}-1](\frac{c_2}{a})^2$$

$$+ \ \ldots\ldots$$

$$+ \frac{1}{E'}[Y_m + (\alpha_m-1)p_{m-1}][(\frac{c_m}{c_{m-1}})^{\alpha_m+1}-1](\frac{c_{m-1}}{a})^2 \quad (5.149)$$

5.5 INVERSE CAVITY EXPANSION PROBLEMS

Cavity expansion problems may be divided into two main groups: (i) for a given soil stress-strain relationship, determine cavity expansion curves and stress and displacement fields around the cavity; (ii) for a given plastic flow rule and cavity expansion curve, determine the actual soil stress-strain relationship as well as stress and displacement fields. All the solutions presented so far in this book belong to the first group. This section describes solution procedures for solving the second type of problems that may be termed the inverse cavity expansion problems.

5.5.1 Cavity expansion in undrained clay

The inverse problems for cavity expansion in undrained clay appear to have been first studied at the same time by Palmer (1972), Baguelin *et al.* (1972) and Ladanyi (1972). Presented here is a small strain solution.

(a) Kinematics of small strain undrained cavity expansion

The non-zero strain rates are related to the outward radial displacement by:

$$\varepsilon_r = -\frac{\partial u}{\partial r} \quad (5.150)$$

$$\varepsilon_\theta = -\frac{u}{r} \quad (5.151)$$

The incompressibility condition dictates:

$$\varepsilon_r + \varepsilon_\theta = 0 \quad (5.152)$$

which can be used to express the displacement u at radius r in terms of cavity wall displacement u_1 and cavity radius a:

$$u = \frac{a}{r}u_1 \quad (5.153)$$

The problem is therefore kinematically determined, and the non-zero strain components are:

$$\varepsilon_r = -\frac{\partial u}{\partial r} = \frac{a}{r^2}u_1 \quad (5.154)$$

$$\varepsilon_\theta = -\frac{u}{r} = -\frac{a}{r^2}u_1 \quad (5.155)$$

We now define a strain ε which is the negative of the hoop strain, namely:

$$\varepsilon = \frac{u}{r} = \frac{a}{r^2}u_1 \quad (5.156)$$

which can be used to give:

$$-\frac{d\varepsilon}{2\varepsilon} = \frac{dr}{r} \quad (5.157)$$

Note that at the cavity wall ε is equal to the cavity strain $\varepsilon_c = u_1/a$.

(b) Equilibrium equation

For the cylindrical cavity, the total stresses have to satisfy the following equilibrium equation:

$$r\frac{d\sigma_r}{dr} + \sigma_r - \sigma_\theta = 0 \quad (5.158)$$

which can be rewritten as:

$$d\sigma_r = -(\sigma_r - \sigma_\theta)\frac{dr}{r} \quad (5.159)$$

It is now assumed that the stress-strain relationship of the soil can be described by the following equation:

$$\sigma_r - \sigma_\theta = 2s_u(\varepsilon) \quad (5.160)$$

Note the undrained shear strength $s_u(\varepsilon)$ in equation (5.160) is not a constant value but rather an unknown function of the strain ε.

Substituting equation (5.160) into equation (5.159) gives:

$$d\sigma_r = -2s_u(\varepsilon)\frac{dr}{r} \quad (5.161)$$

With equation (5.157), the above equation can be rewritten as:

$$d\sigma_r = \frac{s_u(\varepsilon)}{\varepsilon}d\varepsilon \quad (5.162)$$

Integrating the above equation from $r = \infty\,(\varepsilon = 0)$ to $r = a(\varepsilon = \varepsilon_c)$ gives:

$$\psi = p_0 + \int_0^{\varepsilon_c} \frac{s_u(\varepsilon)}{\varepsilon} d\varepsilon \tag{5.163}$$

where ψ is the cavity pressure. This equation can be used to give the following function for the undrained shear strength:

$$s_u(\varepsilon_c) = \varepsilon_c \frac{d\psi}{d\varepsilon_c} \tag{5.164}$$

in which the derivative $d\psi/d\varepsilon_c$ is readily obtained if the cavity expansion curve $\psi - \varepsilon_c$ is known (for example, from pressuremeter tests).

5.5.2 Cavity expansion in cohesionless sand

The procedure for undrained inverse cavity expansion analysis presented above may also be applied to solve cavity expansion problems in sand under fully drained conditions. Due to dilation, the analysis for sand is not as simple as for undrained clay. In particular, no closed form solutions can be obtained and instead a finite difference method needs to be used to obtain the stress-strain relationship from the given cavity expansion curve. The analysis presented here has been derived by Manassero (1989).

(a) Kinematics of small strain drained cavity expansion

For dilatant sands, the relationship between strains may be assumed to be related by a unknown function *f*:

$$\varepsilon_r = f(\varepsilon_\theta) \tag{5.165}$$

with a condition that $\varepsilon_r = f = 0$ when $\varepsilon_\theta = 0$. The function f must be determined.

Equation (5.165) can be used to give:

$$\frac{d\varepsilon_r}{d\varepsilon_\theta} = f' \tag{5.166}$$

where f' denotes the derivative of f with respect to ε_θ.

The strain compatibility condition is:

$$\frac{d\varepsilon_\theta}{dr} = \frac{\varepsilon_r - \varepsilon_r}{r} \tag{5.167}$$

and using equation (5.165), the above equation becomes:

$$\frac{dr}{r} = \frac{d\varepsilon_\theta}{f - \varepsilon_\theta} \tag{5.168}$$

(b) Stress-dilatancy relation and equilibrium equation

By ignoring elastic deformation, Rowe's stress-dilatancy relationship can be written as follows:

$$\frac{\sigma_r}{\sigma_\theta} = -K\frac{d\varepsilon_\theta}{d\varepsilon_r} \tag{5.169}$$

where $K = (1 + \sin\phi_{cv})/(1 - \sin\phi_{cv})$ and ϕ_{cv} is the critical state friction angle.

For the cylindrical cavity, the total stresses have to satisfy the following equilibrium equation:

$$r\frac{d\sigma_r}{dr} + \sigma_r - \sigma_\theta = 0 \tag{5.170}$$

which can be rewritten as:

$$d\sigma_r = -(\sigma_r - \sigma_\theta)\frac{dr}{r} \tag{5.171}$$

By using equations (5.168) and (5.169), this equation can be written as:

$$\frac{1}{\sigma_r} \times \frac{d\sigma_r}{d\varepsilon_\theta} = -\frac{1 + \frac{f'}{K}}{f - \varepsilon_\theta} \tag{5.172}$$

Unfortunately the above equation can not be integrated analytically. However, when applying it at the cavity wall, the finite difference method can be used to solve for a numerical function f and therefore the relationship between ε_r and ε_θ. This is possible because at the cavity wall both σ_r and the derivative $d\sigma_r/d\varepsilon_\theta$ are given, for example, by measured pressuremeter curves.

Once the function f is determined, all other variables (i.e. stresses and displacement) can be readily determined using the equations presented.

5.6 SUMMARY

1. Apart from Tresca, Von Mises, Mohr-Coulomb, and critical state plasticity models, other types of elastoplastic models have also been used to analyse cavity expansion problems. This chapter deals with some additional elastoplastic cavity expansion solutions that are relevant to geotechnical engineering.

2. By neglecting elastic deformation and using a hyperbolic type of stress-strain behaviour, Prevost and Hoeg (1975) derived an approximate small strain solution for undrained expansion of a cylindrical cavity in strain hardening/softening clays. For a monotonously hardening, undrained soil, the shear stress-strain rela-

tionship is defined by the hyperbolic equation (5.5). With this assumption, the cavity expansion curve is defined by equation (5.23). For many soils, however, the behaviour is such that the soil first strain hardens and subsequently strain softens. This behaviour can be described by a simple shear stress-strain relationship (5.6). Using equation (5.6), the cavity expansion curve is shown to be defined by equation (5.26). The application of these solutions to pressuremeter analysis will be discussed in Chapter 8.

3. Ladanyi (1963) showed that by assuming a total shear stress-strain relationship (e.g., equation (5.29) or (5.30)), it is possible to derive a large strain solution for the special case of undrained cavity expansion from zero radius. Undrained cavity expansion is a relatively simple problem because the kinematic relations (i.e. shear strain distribution around the cavity) can be obtained purely from the incompressibility condition without considering stresses. Collins and Yu (1996) showed that the large shear strain generated by undrained cavity expansion from zero radius is defined by equation (5.31). It is noted that the shear strain expression used by Ladanyi (1963) was not entirely correct. A major limitation of this approach is that it is not clear how the solution procedure can be used to analyse cavity unloading–reloading problems.

4. The prediction of the wall movements of tunnels excavated in rock at great depth is an important problem in underground engineering. It is well established that cavity contraction theory can be used to obtain reasonable solutions for circular tunnels. If the rock can be modelled by a brittle-plastic Mohr-Coulomb theory, the rigorous small strain solution for displacement around a circular tunnel is given by equation (5.58). On the other hand, if a brittle-plastic Mohr-Coulomb theory is used, the rigorous displacement solution is defined by equation (5.99). The solution presented in this book is more rigorous than the solution derived by Brown *et al.* (1983). A more detailed discussion of these brittle-plastic solutions and their application to underground excavations will be given in Chapter 10.

5. In Chapter 3, cavity expansion solutions for a linear Mohr-Coulomb yield surface were presented. It is well known, however, that the real behaviour of soil or rock can be better modelled by a curved Mohr-Coulomb criterion. In other words, the value of internal friction angle is not a constant but a function of normal stress. Research has been conducted in using a non-linear Mohr-Coulomb criterion to analyse cavity expansion or contraction problems. For example, Kennedy and Lindberg (1978) derived a small strain analytical solution by using piece-wise Mohr-Coulomb yield surface. Assuming an associated flow rule, the solution of Kennedy and Lindberg (1978) is defined by equation (5.147). Its counterpart for

a fully non-associated flow rule is given by equation (5.149). Kennedy and Lindberg (1978) showed that these solutions are applicable to tunnel problems.

6. Cavity expansion problems can be divided into two main groups: (i) for a given soil stress-strain relationship, determine cavity expansion curves and stress and displacement fields around the cavity; (ii) for a given plastic flow rule and cavity expansion curve, determine the actual soil stress-strain relationship as well as stress and displacement fields. The second type of problem is termed the inverse cavity expansion problems. The inverse problems for cavity expansion are of particular interest to researchers working on pressuremeter analysis. The problem in undrained clay was first studied by Palmer (1972), Baguelin *et al.* (1972) and Ladanyi (1972). The small strain solution is defined by equations (5.163) and (5.164). Equation (5.164) has been used as a theoretical basis for determining soil stress-strain behaviour from measured pressuremeter curves. Because of dilation, the analysis for sand is not as simple as for undrained clay. As shown by Manassero (1989), closed form solutions cannot be obtained, and instead a finite difference method must be used to determine the stress-strain relationship from a known cavity expansion curve.

REFERENCES

Baguelin, F., Jezequel, J.F., Lemee, E. and Mehause, A. (1972). Expansion of cylindrical probes in cohesive soils. *Journal of the Soil Mechanics and Foundations Division*, ASCE, 98(SM11), 1129–1142.

Baligh, M.M. (1976). Cavity expansion in sands with curved envelopes. *Journal of the Geotechnical Engineering Division*, ASCE, 102(GT11), 1131–1147.

Bolton, M.D. and Whittle, R.W. (1999). A non-linear elastic perfectly plastic analysis for plane strain undrained expansion tests. *Geotechnique,* 49(1), 133–141.

Brown, E.T., Bray, J.W., Ladanyi, B. and Hoek, E. (1983). Ground response curves for rock tunnels. *Journal of Geotechnical Engineering*, ASCE, 109(1), 15–39.

Collins, I.F. and Yu, H.S. (1996). Undrained expansions of cavities in critical state soils. *International Journal for Numerical and Analytical Methods in Geomechanics*, 20(7), 489–516.

Davis, R.O., Scott, R.F. and Mullenger, G. (1984). Rapid expansion of a cylindrical cavity in a rate type soil. *International Journal for Numerical and Analytical Methods in Geomechanics*, 8, 3–23.

Florence, A.L. and Schwer, L.E. (1978). Axisymmetric compression of a Mohr-Coulomb medium around a circular hole. *International Journal for Numerical and Analytical Methods in Geomechanics*, 2, 367–379.

Fritz, P. (1984). An analytical solution for axisymmetric tunnel problems in elasto-visco-plastic media. *International Journal for Numerical and Analytical Methods in Geomechanics,* 8, 325–342.

Hobbs, D.W. (1966). A study of the behaviour of broken rock under triaxial compression and its application to mine roadways. *International Journal for Rock Mechanics and Mining Sciences,* 3, 11–43.

Hoek, E. and Brown, E.T. (1980). *Underground Excavations in Rock.* The Institution of Mining and Metallurgy, London, England.

Kennedy, T.C. and Lindberg, H.E. (1978). Tunnel closure for non-linear Mohr-Coulomb functions. *Journal of the Engineering Mechanics Division*, ASCE, 104(EM6), 1313–1326.

Ladanyi, B. (1963). Expansion of a cavity in a saturated clay medium. *Journal of the Soil Mechanics and Foundations Division*, ASCE, 89(SM4), 127–161.

Ladanyi, B. (1972). *In-situ* determination of undrained stress-strain behaviour of sensitive clays with the pressuremeter. *Canadian Geotechnical Journal*, 9(3), 313–319.

Manassero, M. (1989). Stress-strain relationships from drained self-boring pressuremeter tests in sands. *Geotechnique,* 39, 293–307.

Palmer, A.C. (1972). Undrained plane strain expansion of a cylindrical cavity in clay: a simple interpretation of the pressuremeter test. *Geotechnique,* 22(3), 451–457.

Prevost, J.H. and Hoeg, K. (1975). Analysis of pressuremeter in strain softening soil. *Journal of the Geotechnical Engineering Division*, ASCE, 101(GT8), 717–732.

Reed, M.B. (1986). Stresses and displacement around a cylindrical cavity in soft rock. *IMA Journal of Applied Mathematics,* 36, 223–245.

Reed, M.B. (1988). Influence of out of plane stress on a plane strain problem in rock mechanics. *International Journal for Numerical and Analytical Methods in Geomechanics,* 12, 173–181.

Wilson, A.H. (1980). A method of estimating the closure and strength of lining required in drivages surrounded by a yield zone. *International Journal for Rock Mechanics and Mining Sciences,* 17, 349–355.

CHAPTER 6

TIME-DEPENDENT SOLUTIONS

6.1 INTRODUCTION

This chapter presents some time-dependent solutions for cavity expansion in rock and soil. In rock mechanics, cavity solutions derived from constitutive models based on visco-elasticity and visco-plasticity have been applied to tunnel problems. As far as soil mechanics is concerned, studies have been carried out on the pore pressure changes around a driven pile and an excavated tunnel in soils. This chapter will concentrate on the solutions with applications to these two types of geotechnical problems.

6.2 VISCO-ELASTIC SOLUTIONS

A detailed account of stress analysis using viscoelastic stress strain laws can be found in Lee (1965), Flugge (1967), and Jaeger and Cook (1976). This section will, however, only give a brief summary of its application to cavity problems by following closely on the description of Jaeger and Cook (1976).

6.2.1 Visco-elastic models and method of stress analysis

(a) Visco-elastic stress-strain laws

The Maxwell model

The so-called Maxwell material consists of a spring μ and a dash-pot η arranged in series, Figure 6.1(a). If the stress applied at the material is $\sigma(t)$, it can be easily shown that the stress-strain relation of the Maxwell material is:

$$\frac{1}{\mu}\dot{\sigma}(t) + \frac{1}{\eta}\sigma(t) = \dot{\varepsilon}(t) \tag{6.1}$$

where μ and η are constants related to the spring and the dash-pot.

If the system is unstrained at $t = 0$ and a constant pressure p_0 is applied suddenly at $t = 0^+$, equation (6.1) can be integrated to yield:

$$\varepsilon(t) = \frac{p_0}{\mu} + \frac{p_0}{\eta}t \tag{6.2}$$

where the first term represents the instantaneous elastic deformation and the second term denotes the steady state creep deformation.

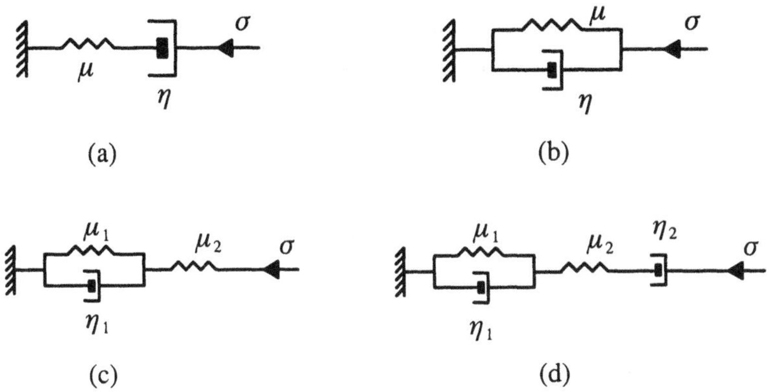

Figure 6.1: Visco-elastic models: (a) The Maxwell model, (b) The Kelvin model, (c) The generalised Kelvin model, (d) The Burgers model

The Kelvin model

If a spring μ and a dash-pot η are put together in parallel, then the system is known as the Kelvin substance, Figure 6.1(b). The stress-strain relation for the Kelvin material is:

$$\sigma(t) = \mu\varepsilon(t) + \eta\dot{\varepsilon}(t) \tag{6.3}$$

If the system is unstrained at $t = 0$ and a constant pressure p_0 is applied suddenly at $t = 0^+$, equation (6.3) can again be integrated to give the expression for strain:

$$\varepsilon(t) = \frac{p_0}{\mu}\left\{1-\exp(-\frac{\mu t}{\eta})\right\} \tag{6.4}$$

This model gives zero strain at $t = 0^+$ when the pressure p_0 is applied and this does not represent the observed behaviour.

The generalized Kelvin model

To overcome the zero initial strain of the Kelvin model, a second spring μ_2 can be added to the Kelvin system (μ_1, η_1), Figure 6.1(c). The second spring and the Kelvin material are put together in series. This is known as the generalized Kelvin model. The stress-strain relation of the generalized Kelvin material is:

$$\eta_1\dot{\sigma}(t) + (\mu_1 + \mu_2)\sigma = \mu_2\eta_1\dot{\varepsilon}(t) + \mu_1\mu_2\varepsilon(t) \tag{6.5}$$

For an initially unstrained system, the strain due to a sudden application of a constant pressure p_0 at $t = 0^+$ can be derived by integrating equation (6.5):

$$\varepsilon(t) = \frac{p_0}{\mu_2} + \frac{p_0}{\mu_1}\left\{1 - \exp(-\frac{\mu_1 t}{\eta_1})\right\} \tag{6.6}$$

The above equation suggests that the strain starts off with an initial value p_0/μ_2 and then increases to approach a constant limit value $(1/\mu_2 + 1/\mu_1)p_0$.

The Burgers model

The Burgers material consists of a Kelvin element (μ_1, η_1) and a Maxwell element (μ_2, η_2) arranged in series, Figure 6.1(d). The stress-strain relation for the Burgers model is:

$$\frac{\eta_1}{\mu_2}\ddot{\sigma} + (1 + \frac{\mu_1}{\mu_2} + \frac{\eta_1}{\eta_2})\dot{\sigma} + \frac{\mu_1}{\eta_2}\sigma = \eta_1\ddot{\varepsilon} + \mu_1\dot{\varepsilon} \tag{6.7}$$

Upon integration, the strain of an initially unstrained system caused by a sudden application of a constant pressure p_0 can be derived as follows:

$$\varepsilon(t) = \frac{p_0}{\mu_2} + \frac{p_0}{\mu_1}\left\{1 - \exp(-\frac{\mu_1 t}{\eta_1})\right\} + \frac{p_0}{\eta_2}t \tag{6.8}$$

It is clear from equation (6.8) that the Burgers model may be satisfactorily used to describe materials that show an instantaneous strain, transient creep and steady state creep under constant loading conditions.

(b) Method of stress analysis in visco-elastic materials

One of the most useful methods for deriving solutions for visco-elastic materials is to use the Laplace transformation that enables many elastic solutions to be converted to solutions for visco-elastic problems (Lee, 1955; Flugge, 1967; Jaeger and Cook, 1976). This section gives a brief description of how this method can be applied to obtain solutions for cavity expansion problems in visco-elastic materials.

Elastic stress-strain relations in three dimensions

In elasticity theory, it is often convenient to decompose stresses and strains into mean and deviatoric counterparts. In this way, the elastic stress-strain relations are:

$$s_{ij} = 2Ge_{ij} \quad \text{and} \quad s = 3Ke \tag{6.9}$$

where $s_{ij} = \sigma_{ij} - s$ and $s = \sigma_{ii}/3$ are the deviatoric and mean stresses respectively; $e_{ij} = \varepsilon_{ij} - e$ and $e = e_{ii}/3$ are the corresponding deviatoric and mean normal strains; and G and K are shear and bulk moduli of the material.

Visco-elastic stress-strain relations in three dimensions

As shown in the previous section, a most general form of stress-strain relation for visco-elastic materials can be written in the form of:

$$P_0\sigma(t) + P_1\dot{\sigma}(t) + P_2\ddot{\sigma}(t) + \ldots = Q_0\varepsilon(t) + Q_1\dot{\varepsilon}(t) + Q_2\ddot{\varepsilon}(t) + \ldots \quad (6.10)$$

or

$$\sum_0^m P_k \frac{d^k\sigma(t)}{dt^k} = \sum_0^n Q_k \frac{d^k\varepsilon(t)}{dt^k} \quad (6.11)$$

When equation (6.11) is subject to the Laplace transformation, we obtain the following equation between the Laplace transforms $\bar{\sigma}(s)$ and $\bar{\varepsilon}(s)$ of stress and strain:

$$P(s)\bar{\sigma}(s) = Q(s)\bar{\varepsilon}(s) \quad (6.12)$$

where

$$P(s) = \sum_0^m P_k s^k \quad \text{and} \quad Q(s) = \sum_0^n Q_k s^k \quad (6.13)$$

and

$$\bar{\sigma}(s) = \int_0^\infty e^{-st}\sigma(t)dt \quad ; \quad \bar{\varepsilon}(s) = \int_0^\infty e^{-st}\varepsilon(t)dt \quad (6.14)$$

in which s is a real positive number sufficiently large to make the integrals converge. It is noted from equation (6.12) that by using the Laplace transformation the time is temporarily removed from the viscoelastic stress-strain laws.

Applying equation (6.12) to both devatoric and mean stresses and strains leads to the following general three-dimensional visco-elastic stress-strain relations:

$$P'(s)\bar{s}_{ij}(s) = Q'(s)\bar{e}_{ij}(s) \quad (6.15)$$

$$P''(s)\bar{s}(s) = Q''(s)\bar{e}(s) \quad (6.16)$$

These visco-elastic stress-strain laws become identical to their elastic counterparts (6.9) if we make the following substitutions:

$$2G \rightarrow \frac{Q'(s)}{P'(s)} \quad \text{and} \quad 3K \rightarrow \frac{Q''(s)}{P''(s)} \quad (6.17)$$

or in terms of Young's modulus E:

$$E \rightarrow \frac{3Q'(s)Q''(s)}{2P'(s)Q''(s) + Q'(s)P''(s)} \quad (6.18)$$

and Poisson's ratio ν

$$\nu \rightarrow \frac{P'(s)Q''(s) - Q'(s)P''(s)}{P'(s)Q''(s) + Q'(s)P''(s)} \tag{6.19}$$

If the material is incompressible, $\nu = 0.5$ and $K \rightarrow \infty$, the following equation becomes valid:

$$E \rightarrow \frac{3Q'(s)}{2P'(s)} \quad ; \quad G \rightarrow \frac{Q'(s)}{2P'(s)} \tag{6.20}$$

Correspondence principle

The correspondence principle that has been widely used to derive solutions for visco-elastic materials can be described simply as follows. If the solution of an elastic problem is known, the Laplace transform of the solutions to the corresponding visco-elastic problem may be found by replacing the elastic constants G and K according to (6.17) and the actual loads by their Laplace transforms.

6.2.2 Solutions for two simple cavity problems

To illustrate how the correspondence principle can be used to derive viscoelastic solution for cavity expansion problems, this section presents the viscoelastic solutions for two simple unlined tunnel problems. A more detailed review of the time-dependent response of rock around tunnels can be found in Ladanyi (1993).

(a) A stress-free cavity subject to an isotropic stress at infinity

Associated elastic solution

A stress-free cylindrical cavity is subject to an isotropic pressure p_0 at infinity. It is assumed that the pressure p_0 is applied suddenly at time $t = 0^+$. The elastic solutions of stress and displacement for this problem are presented in Chapter 1. For simplicity, we assume that the material is incompressible. For this case, the stresses and displacement around the cavity are:

$$\sigma_r = p_0[1 - (\tfrac{a}{r})^2] \tag{6.21}$$

$$\sigma_\theta = p_0[1 + (\tfrac{a}{r})^2] \tag{6.22}$$

$$u = \frac{3p_0}{2E}(\frac{a^2}{r}) \tag{6.23}$$

where a is the radius of cavity.

Visco-elastic solution

Since the expressions for the elastic stresses (6.21)-(6.22) are independent of any elastic constants, they are therefore also the solutions for visco-elastic materials.

The elastic displacement (6.23) is however a function of Young's modulus. As a result, the visco-elastic displacement must be determined by the correspondence principle.

For change in shape with no volume change ($\nu = 0.5$), the simple visco-elastic stress-strain laws presented previously can be used to give the following transformations:

$$E \rightarrow \frac{3Q'(s)}{2P'(s)} = \frac{3s}{2(s/\mu_1 + 1/\eta_1)} \quad \text{for } \textit{Maxwell model} \tag{6.24}$$

$$E \rightarrow \frac{3Q'(s)}{2P'(s)} = \frac{3}{2}(\eta_2 s + \mu_2) \quad \text{for } \textit{Kelvin model} \tag{6.25}$$

$$E \rightarrow \frac{3Q'(s)}{2P'(s)} = \frac{3(\mu_2\eta_1 s + \mu_1\mu_2)}{2(\eta_1 s + \mu_1 + \mu_2)} \quad \text{for } \textit{generalised Kelvin model} \tag{6.26}$$

$$E \rightarrow \frac{3Q'(s)}{2P'(s)} = \frac{3(\eta_1 s^2 + \mu_1 s)}{2[\frac{\eta_1}{\mu_2}s + (1 + \frac{\mu_1}{\mu_2} + \frac{\eta_1}{\eta_2})s + \frac{\mu_1}{\eta_2}]} \quad \text{for } \textit{Burgers model} \tag{6.27}$$

The Laplace transform of the elastic displacement expression (6.23) can be written as follows by replacing p_0 by p_0/s and E by $\overline{E}(s) = 3Q'(s)/2P'(s)$:

$$\overline{u} = \frac{3p_0/s}{2\overline{E}(s)}(\frac{a^2}{r}) \tag{6.28}$$

The inverse transformation of equation (6.28) combined with (6.24) gives the following visco-elastic displacement for Maxwell materials:

$$u(t) = (\frac{p_0}{\mu_1} + \frac{p_0}{\eta_1}t) \times \frac{b^2}{r} \tag{6.29}$$

The inverse transformation of equation (6.28) combined with (6.25) gives the following visco-elastic displacement for Kelvin materials:

$$u(t) = [\frac{p_0}{\mu_2} - \frac{p_0}{\mu_2}\exp(-\frac{\mu_2}{\eta_2}t)] \times \frac{b^2}{r} \tag{6.30}$$

The inverse transformation of equation (6.28) combined with (6.27) gives the following visco-elastic displacement for Burgers materials:

$$u(t) = \left\{\frac{p_0}{\mu_2} + \frac{p_0}{\eta_2}t + \frac{p_0}{\mu_1}[1 - \exp(-\frac{\mu_1}{\eta_1}t)]\right\} \times \frac{b^2}{r} \tag{6.31}$$

(b) A stress-free cavity subject to biaxial stresses at infinity

Associated elastic solution

A stress-free cylindrical cavity is subject to biaxial stresses of p_{h0} and p_{v0} at infinity. It is assumed that the pressures p_{h0} and p_{v0} are applied suddenly at time $t = 0^+$. The elastic solutions of stress and displacement for this problem are presented in Chapter 1. Because the elastic stresses do not depend on elastic constants and only the displacement need to be considered for viscoelastic materials. The elastic expression for displacement is:

$$u = -\frac{1-\nu^2}{E}\left\{p_m(r + \frac{a^2}{r}) + p_d(r - \frac{a^4}{r^3} + \frac{4a^2}{r})\cos 2\theta\right\}$$

$$+ \frac{\nu(1+\nu)}{E}\left\{p_m(r - \frac{a^2}{r}) - p_d(r - \frac{a^4}{r^3})\cos 2\theta\right\} \tag{6.32}$$

where

$$p_m = \frac{p_{h0} + p_{v0}}{2} \quad \text{and} \quad p_d = \frac{p_{h0} - p_{v0}}{2} \tag{6.33}$$

Visco-elastic solution

For change in shape with no volume change ($\nu = 0.5$), solutions for viscoelastic displacement can be obtained by following the correspondence principle procedure used previously for any viscoelastic models. For example, the solution for Burgers model is (Goodman, 1989):

$$u(t) = -[A + B(\frac{1}{2} - \frac{a^2}{4r^2})][\frac{1}{\mu_2} + \frac{1}{\mu_1} - \frac{1}{\mu_1}\exp(-\frac{\mu_1}{\eta_1}t) + \frac{t}{\eta_2}] \tag{6.34}$$

where

$$A = \frac{p_m}{2} \times \frac{a^2}{r} \tag{6.35}$$

$$B = (2p_d \cos 2\theta) \times \frac{a^2}{r} \tag{6.36}$$

6.3 ELASTIC-VISCOPLASTIC SOLUTIONS

Analytical solutions for cavity expansion problems with time-dependent material properties are relatively rare. Examples of this type of solution has been given by Salamon (1974), Nonaka (1981) and Fritz (1984). This section gives a brief description of the analytical solution developed by Fritz (1984).

In Fritz's time-dependent analysis, the elastic-brittle-plastic solution presented in Section 5.3.1 of Chapter 5 has been used as the limiting solution at time $t \to \infty$. The elastic-viscoplastic model used by Fritz (1984) consists of a spring, a dashpot and a sliding element. The dashpot and the sliding element are arranged in parallel. Only cylindrical cavity is considered in the analysis.

6.3.1 Elastic-viscoplastic stress-strain relations

At the first instant of arbitrary loading, the material is assumed to behave elastically. The elastic stresses and displacement around a cylindrical cavity that is subjected to the internal and external pressures of p and p_0 are:

$$\sigma_r^e = p_0-(p_0-p)(\tfrac{a}{r})^2 \tag{6.37}$$

$$\sigma_\theta^e = p_0 + (p_0-p)(\tfrac{a}{r})^2 \tag{6.38}$$

$$u^e = \frac{1+\nu}{E}(p_0-p)\frac{a^2}{r} \tag{6.39}$$

With time, the dashpot relaxes in zones in which the yield function is satisfied. To make it possible to derive an analytical solution for the time-dependent problem, the plastic strain rates may be defined in a similar way to Hooke's law:

$$\dot{\varepsilon}_r^p = \frac{1+\nu_p}{\eta}[(1-\nu_p)(\sigma_r(t)-\sigma_r(\infty))-\nu_p(\sigma_\theta(t)-\sigma_\theta(\infty))] \tag{6.40}$$

$$\dot{\varepsilon}_\theta^p = \frac{1+\nu_p}{\eta}[(1-\nu_p)(\sigma_\theta(t)-\sigma_\theta(\infty))-\nu_p(\sigma_r(t)-\sigma_r(\infty))] \tag{6.41}$$

where η is the viscosity of the dashpot; ν_p is a measure of the plastic volume dilatancy which is assumed to be constant. $\sigma_r(\infty)$ and $\sigma_\theta(\infty)$ are the radial and tangential stresses obtained in Section 5.3.1 using a time-independent plasticity model. The above equations are obtained by using the condition that the plastic strain rate is zero at time $t \to \infty$.

Using the following abbreviations:

$$\bar{\sigma}_r = \sigma_r(t)-\sigma_r(\infty) \tag{6.42}$$

$$\bar{\sigma}_\theta = \sigma_\theta(t)-\sigma_\theta(\infty) \tag{6.43}$$

the total strain rates in the plastic zone can be obtained by summing the elastic parts and the visco-plastic parts:

$$\dot{\varepsilon}_r = \frac{1+\nu}{E}\frac{d}{dt}[(1-\nu)\bar{\sigma}_r-\nu\bar{\sigma}_\theta] + \frac{1+\nu_p}{\eta}[(1-\nu_p)\bar{\sigma}_r-\nu_p\bar{\sigma}_\theta] \tag{6.44}$$

$$\dot{\varepsilon}_\theta = \frac{1+\nu}{E}\frac{d}{dt}[(1-\nu)\bar{\sigma}_\theta-\nu\bar{\sigma}_r] + \frac{1+\nu_p}{\eta}[(1-\nu_p)\bar{\sigma}_\theta-\nu_p\bar{\sigma}_r] \tag{6.45}$$

6.3.2 Stresses and displacement in the initial plastic zone

The displacement compatibility condition can be shown to be:

$$r\frac{d\dot{\varepsilon}_\theta}{dr} = \dot{\varepsilon}_r - \dot{\varepsilon}_\theta \tag{6.46}$$

The equilibrium equation for stresses both at time t and at time $t \to \infty$ can be combined to give:

$$r\frac{d\bar{\sigma}_r}{dr} = \bar{\sigma}_\theta - \bar{\sigma}_r \tag{6.47}$$

By using equations (6.44)-(6.47) and the following boundary conditions:

$$\sigma_r(t) = \sigma_r^e \quad at \quad t = 0 \tag{6.48}$$

$$\sigma_r(t) = p \quad at \quad r = a \tag{6.49}$$

the following stress solutions may be obtained:

$$\sigma_r(t) = \sigma_r^e e^{-mt} + \sigma_r(\infty)(1-e^{-mt}) + [1-(\tfrac{a}{r})^2]f(t) \tag{6.50}$$

$$\sigma_\theta(t) = \sigma_\theta^e e^{-mt} + \sigma_\theta(\infty)(1-e^{-mt}) + [1 + (\tfrac{a}{r})^2]f(t) \tag{6.51}$$

where $f(t)$ is a function that still needs to be determined and

$$m = \frac{(1-\nu_p^2)E}{(1-\nu^2)\eta} \tag{6.52}$$

It should be noted that the above stress expressions are only valid in the initial (at time $t = 0^+$) plastic zone $a \leq r \leq c_0$ only. The radius of the initial plastic zone, c_0, may be derived by substituting the elastic stress solutions (6.37)-(6.38) in the peak yield condition:

$$\sigma_\theta(t) - \alpha\sigma_r(t) = Y \tag{6.53}$$

as follows:

$$\frac{c_0}{a} = \sqrt{\frac{(\alpha + 1)(p_0 - p)}{Y + (\alpha - 1)p_0}} \tag{6.54}$$

With the boundary condition

$$u(t) = u^e \quad at \quad t = 0 \tag{6.55}$$

equations (6.40) and (6.41) can be used to yield the following solution for displacement in the initial plastic zone $a \leq r \leq c_0$:

$$u(t) = \frac{1+\nu}{E}[1-2\nu + (\tfrac{a}{r})^2]rf(t)$$

$$+ \frac{1+\nu_p}{\eta}[1-2\nu_p + (\tfrac{a}{r})^2]r\int_0^t f(t)dt$$

$$+\frac{1+\nu}{E}[(1-\nu)\sigma_\theta^e-\nu\sigma_r^e]re^{-mt}$$

$$+\frac{1+\nu}{E}[(1-\nu)\sigma_\theta(\infty)-\nu\sigma_r(\infty)]r(1-e^{-mt})$$

$$+\frac{1+\nu_p}{\eta}[(1-\nu_p)(\sigma_\theta^e-\sigma_\theta(\infty))-\nu_p(\sigma_r^e-\sigma_r(\infty))]r\frac{1-e^{-mt}}{a}$$

$$-\frac{(1+\nu)(1-2\nu)}{E}rp_0 \tag{6.56}$$

6.3.3 Stresses and displacement in the time-dependent plastic zone

It may be further assumed that the size of the initial plastic zone will increase with time. If the radius of the plastic zone is $c(t)$ at time t, it remains to determine the stresses and displacement in the region defined by $c_0 \leq r \leq c(t)$. The stresses and displacement in this region may be obtained in a similar way as the elastic-plastic solutions presented in Section 5.3.1. If the radial stress at $r = c_0$ is denoted as $\sigma_r^{c_0}$, the plastic radius $c(t)$ at time t is:

$$\frac{c(t)}{c_0} = \left\{\frac{Y' + (\alpha'-1)p_{1y}}{Y' + (\alpha'-1)\sigma_r^{c_0}}\right\}^{\frac{1}{\alpha'-1}} \tag{6.57}$$

The stresses and displacement in this region can be shown to be as follows (see Section 5.3.1 for details):

$$\sigma_r(t) = \frac{Y' + (\alpha'-1)\sigma_r^{c_0}}{\alpha'-1}\left(\frac{r}{c_0}\right)^{\alpha'-1} - \frac{Y'}{\alpha'-1} \tag{6.58}$$

$$\sigma_\theta(t) = \alpha'\frac{Y' + (\alpha'-1)\sigma_r^{c_0}}{\alpha'-1}\left(\frac{r}{c_0}\right)^{\alpha'-1} - \frac{Y'}{\alpha'-1} \tag{6.59}$$

$$u(t) = \frac{1+\nu}{E}r\left\{K_1\left(\frac{r}{c(t)}\right)^{\alpha'-1} + K_2\left(\frac{c(t)}{r}\right)^{\beta+1} + K_3\right\} \tag{6.60}$$

where

$$p_{1y} = \frac{2p_0-Y}{\alpha+1} \tag{6.61}$$

$$K_1 = \left[\frac{1+\alpha'\beta}{\alpha'+\beta}(1-\nu)-\nu\right]\left(p_{1y} + \frac{Y'}{\alpha'-1}\right) \tag{6.62}$$

$$K_3 = -(1-2\nu)\left(p_0 + \frac{Y'}{\alpha'-1}\right) \tag{6.63}$$

$$K_2 = p_0 - p_{1y} - K_1 - K_3 \tag{6.64}$$

The unknown function $f(t)$ can be determined by using the displacement continuity at the radius $r = c_0$ which gives a differential equation for $f(t)$:

$$Q_1(f(t)) + Q_2 \int f(t)dt + Q_3(t) = Q_4 \tag{6.65}$$

where

$$Q_1 = \frac{1+\nu}{E}[1-2\nu + (\frac{a}{c_0})^2]f(t)$$

$$-\frac{1+\nu}{E}[(1-\nu)\frac{1+\alpha'\beta}{\alpha'+\beta} - \nu][1-(\frac{a}{c_0})^2]f(t)$$

$$-\frac{1+\nu}{E}K_2\left\{\frac{Y' + (\alpha'-1)p_{1y}}{Y' + (\alpha'-1)\sigma_r^{c_0}}\right\}^{\frac{\beta+1}{\alpha'-1}} \tag{6.66}$$

$$Q_2 = \frac{1+\nu_p}{\eta}[1-2\nu_p + (\frac{a}{c_0})^2] \tag{6.67}$$

$$Q_3 = e^{-mt}\frac{1-\nu^2}{E}[\sigma_\theta^e|_{r=c_0} - \frac{1+\alpha'\beta}{\alpha'+\beta}\sigma_r^e|_{r=c_0}]$$

$$+ (1-e^{-mt})\frac{1-\nu^2}{E}[\sigma_\theta(\infty)|_{r=c_0} - \frac{1+\alpha'\beta}{\alpha'+\beta}\sigma_r(\infty)|_{r=c_0}]$$

$$+ \frac{(1-e^{-mt})(1+\nu_p)}{m\eta}[(1-\nu_p)(\sigma_\theta^e|_{r=c_0} - \sigma_\theta(\infty)|_{r=c_0})]$$

$$-\frac{(1-e^{-mt})(1+\nu_p)}{m\eta}\nu_p[\sigma_r^e|_{r=c_0} - \sigma_r(\infty)|_{r=c_0}] \tag{6.68}$$

$$Q_4 = \frac{1-\nu^2}{E}K_4 \tag{6.69}$$

$$K_4 = \frac{1}{1-\nu}(\frac{\beta}{\beta+1} - \nu)[2p_0 - (\beta+1)p_{1y} - Y']$$

$$+ (\frac{1+\alpha'\beta}{\beta+\alpha'} - 1) \times \frac{Y'}{\alpha'-1} \tag{6.70}$$

The boundary condition that must be used to solve equation (6.65) is:

$$f(t) = \int f(t)dt = 0 \quad when \quad t = 0 \tag{6.71}$$

The constant ν_p is found by equating the time-dependent cavity displacement at time $t \to \infty$ according to equation (6.56) to the time-independent cavity displacement derived in Section 5.3.1. The result is:

$$\nu_p = \frac{-B + \sqrt{B^2 - 4AC}}{2A} \tag{6.72}$$

in which

$$A = 2[\sigma_r^e|_{r=a} - \sigma_r(\infty)|_{r=a} - (\sigma_r^e|_{r=c_0} - \sigma_r(\infty)|_{r=c_0}) + P(a) - P(c_0) \tag{6.73}$$

$$\begin{aligned} B = &-[1 + (\tfrac{a}{c_0})^2](\sigma_r^e|_{r=a} - \sigma_r(\infty)|_{r=a}) \\ &+ 2[\sigma_r^e|_{r=c_0} - \sigma_r(\infty)|_{r=c_0}] \\ &-[3 + (\tfrac{a}{c_0})^2][P(a) - K_4] + 4[P(c_0) - K_4] \end{aligned} \tag{6.74}$$

$$C = [1 + (\tfrac{a}{c_0})^2][P(a) - K_4] - 2[P(c_0) - K_4] \tag{6.75}$$

$$P(r) = \sigma_\theta^e - \frac{1 + \alpha'\beta}{\beta + \alpha'}\sigma_r(\infty) - \frac{K_2}{1-\nu}\left\{\frac{Y' + (\alpha'-1)p_{1y}}{Y' + (\alpha'-1)\sigma_r(\infty)}\right\}^{\frac{\beta+1}{\alpha'-1}} \tag{6.76}$$

6.4 CONSOLIDATION SOLUTIONS

6.4.1 Consolidation of soil around an expanding cavity

To predict the variation of bearing capacity of a driven pile with time, Randolph and Wroth (1979) developed an analytical solution for radial consolidation of soil around a cylindrical cavity. The initial pore pressures are assumed to be those predicted by using a total stress cavity expansion analysis with the Tresca yield criterion. During the consolidation stage, the soil skeleton is assumed to be purely elastic to ensure that analytical treatment is possible. The solution of Randolph and Wroth (1979) is summarized below.

(a) Governing equations for radial consolidation

The basic equations that govern the consolidation process are briefly given now. To distinguish the pore pressure from the radial displacement of soil particles, we will use U and u to denote pore pressure and radial displacement respectively.

Stress-strain relations of soil skeleton

The elastic plane strain stress-strain relations are in terms of effective stresses:

$$\varepsilon_r = -\frac{\partial u}{\partial r} = \frac{1}{2G}[(1-\nu)d\sigma_r'-\nu d\sigma_\theta'] \tag{6.77}$$

$$\varepsilon_\theta = -\frac{u}{r} = \frac{1}{2G}[-\nu d\sigma_r' + (1-\nu)d\sigma_\theta'] \tag{6.78}$$

where the strain in the axial direction is zero. These relations can also be expressed in terms of effective stress changes:

$$d\sigma_r' = -\frac{2G}{1-2\nu}[(1-\nu)\frac{\partial u}{\partial r} + \nu\frac{u}{r}] \tag{6.79}$$

$$d\sigma_\theta' = -\frac{2G}{1-2\nu}[\nu\frac{\partial u}{\partial r} + (1-\nu)\frac{u}{r}] \tag{6.80}$$

where G and v are shear modulus and Poisson's ratio respectively.

Equilibrium equation

The equilibrium equation for the cylindrical cavity is in terms of total stresses:

$$\frac{\partial(rd\sigma_r)}{\partial r} - d\sigma_\theta = 0 \tag{6.81}$$

The total stress increments are:

$$d\sigma_r = d\sigma_r' + dU = d\sigma_r' + U-U_0 \tag{6.82}$$

$$d\sigma_\theta = d\sigma_\theta' + dU = d\sigma_\theta' + U-U_0 \tag{6.83}$$

where U and U_0 are current and initial excess pore pressures. With equations (6.82) and (6.83), equation (6.81) can be rewritten as follows:

$$\frac{\partial U}{\partial r} = \frac{\partial U_0}{\partial r} - \frac{\partial(d\sigma_r')}{\partial r} + \frac{d\sigma_\theta' - d\sigma_r'}{r} \tag{6.84}$$

Putting the stress-strain relations (6.79) and (6.80) into the above equation leads to:

$$\frac{\partial U}{\partial r} = \frac{\partial U_0}{\partial r} + G'\frac{\partial}{\partial r}\left\{\frac{1}{r}\frac{\partial}{\partial r}(ru)\right\} \tag{6.85}$$

in which

$$G' = \frac{2G(1-\nu)}{1-2\nu} \tag{6.86}$$

Continuity of volumetric strain rate and flow of pore water

Using the Darcy law, the velocity of the pore water relative to the soil particles can be expressed in terms of the pore pressure as follows:

$$v = -k\frac{\partial(U/\gamma_w)}{\partial r} = -\frac{k}{\gamma_w}\frac{\partial U}{\partial r} \tag{6.87}$$

where k is the permeability of the soil and γ_w is the unit weight of water.

Assuming incompressible water and soil particles, the continuity condition insists that the rate of volumetric strain should be related to the flow of pore water into and out of any element by:

$$\frac{\partial}{\partial t}(\varepsilon_r + \varepsilon_\theta) = \frac{1}{r}\frac{\partial(rv)}{\partial r} \tag{6.88}$$

Equations (6.87) and (6.88) can be combined to give the following equation:

$$\frac{k}{\gamma_w}\left\{\frac{1}{r}\frac{\partial}{\partial r}\left(r\frac{\partial U}{\partial r}\right)\right\} = \frac{\partial}{\partial t}\left(\frac{1}{r}\frac{\partial(ru)}{\partial r}\right) \tag{6.89}$$

Integration of the above equation leads to:

$$\frac{k}{\gamma_w}\frac{\partial U}{\partial r} = \frac{\partial u}{\partial t} + \frac{f(t)}{r} \tag{6.90}$$

where $f(t)$ is a constant of integration that can be determined with boundary conditions.

We now have two equations in terms of pore pressure U and displacement u defined by (6.85) and (6.90). These two equations can be combined to eliminate the displacement u to give:

$$\frac{\partial U}{\partial t} = c\left\{\frac{\partial^2 U}{\partial r^2} + \frac{1}{r}\frac{\partial U}{\partial r}\right\} + g(t) \tag{6.91}$$

in which $g(t)$ is another constant of integration and c is the coefficient of consolidation defined by:

$$c = \frac{k}{\gamma_w}G' = \frac{k}{\gamma_w} \times \frac{2G(1-\nu)}{1-2\nu} \tag{6.92}$$

Alternatively, equations (6.85) and (6.90) can also be used to eliminate the pore pressure to give the following equation in terms of displacement:

$$\frac{\partial u}{\partial t} = \frac{k}{\gamma_w} \times \frac{\partial U_0}{\partial r} + c\left\{\frac{\partial}{\partial r}\left[\frac{1}{r}\frac{\partial(ru)}{\partial r}\right]\right\} - \frac{f(t)}{r} \tag{6.93}$$

The governing equation (6.91) was first derived by Soderberg (1962).

(b) Solution procedure

Boundary conditions

If the cavity wall is assumed to be impermeable, the boundary conditions for excess pore pressures are:

$$U = U_0 \quad at\ t = 0 \quad \text{for } r \geq a \tag{6.94}$$

$$U \to 0 \quad as\ t \to \infty \quad \text{for } r \geq a \tag{6.95}$$

$$\frac{\partial U}{\partial r} = 0 \quad at\ r = a \quad \text{for } t > 0 \tag{6.96}$$

$$U \to 0 \quad as\ r \to \infty \quad \text{for } t \geq 0 \tag{6.97}$$

The condition (6.97) implies that the integration constant that appears in equation (6.91) must be zero: $g(t)=0$.

The boundary conditions for displacement of soil particles are:

$$u = 0 \quad at\ t = 0 \quad \text{for } \ r \geq a \tag{6.98}$$

$$u = 0 \quad at\ r = a \quad \text{for } \ t \geq 0 \tag{6.99}$$

$$u \to 0 \quad at\ r \to \infty \quad \text{for } \ t \geq 0 \tag{6.100}$$

The conditions (6.96) and (6.99) dictate that the integration constant $f(t)$ appearing in equation (6.93) must be zero: $f(t)=0$.

Initial excess pore pressure distribution

If we consider consolidation of the soil around a created cavity, the initial excess pore pressure distribution may be obtained from the change in mean total stress. Assuming the Tresca yield criterion, the initial excess pore pressure in the plastic zone can be shown to be:

$$U_0 = 2s_u \ln(\tfrac{R}{r}) \quad \text{for } \ r \leq R \tag{6.101}$$

where the radius of the plastic zone is given by $R = \sqrt{G/s_u}\,a$. Outside the plastic region, the initial excess pore pressures are zero everywhere, Figure 6.2.

Consolidation solutions

As shown in Randolph and Wroth (1979), the solution of equation (6.91) may be obtained by separating the variables (Carslaw and Jaegar, 1959) to give, for a separation constant of $-\alpha^2$:

$$U = \sum_{n=1}^{\infty} B_n \exp(-\alpha_n^2 t) B_0(\lambda_n r) \quad \text{for } \ a \leq r \leq r^* \tag{6.102}$$

$$U = 0 \qquad \text{for } \quad r > r^* \tag{6.103}$$

where r^* is a large radius introduced to replace infinity in the boundary conditions to enable analytical treatment possible. Typically it is satisfactory to use values of r^* being five to ten times of R. $B_i(\lambda r)$ is a cylinder function of the ith order.

The coefficient B_n is given by

$$B_n = \frac{4s_u}{\lambda_n^2} \times \frac{B_0(\lambda_n a) - B_0(\lambda_n R)}{[r^* B_1(\lambda_n r^*)]^2 - [aB_0(\lambda_n a)]^2} \tag{6.104}$$

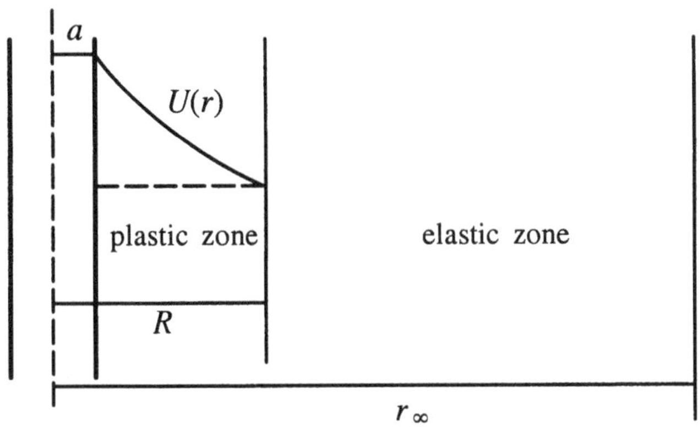

Figure 6.2: Radial consolidation around a cavity

6.4.2 Consolidation of soil around a contracting cavity

In a related study, Carter (1988) presented a semi-analytical solution for consolidation of soil around a contracting cavity. This solution has applications to the problem of a vertical borehole or a bored pile.

Laplace transforms of the governing equation

The governing equation for the excess pore pressure generated by a cavity unloading is same as that caused by a loading cavity. In other words, the governing equation is defined by equation (6.91):

$$\frac{\partial U}{\partial t} = c\left\{\frac{\partial^2 U}{\partial r^2} + \frac{1}{r}\frac{\partial U}{\partial r}\right\} \tag{6.105}$$

where U is the excess pore pressure at time t and radius r.

Taking Laplace transforms of equation (6.105) leads to:

$$\frac{\partial^2 \overline{U}}{\partial r^2} + \frac{1}{r}\frac{\partial \overline{U}}{\partial r} - q^2\overline{U} = -\frac{U_0}{c} \tag{6.106}$$

in which

$$\overline{U} = \int_0^{\infty} e^{-st} U dt \tag{6.107}$$

$$q = \sqrt{\frac{s}{c}} \tag{6.108}$$

and U_0 is the initial excess pore water pressure at time $t=0$, immediately after the borehole is created.

The solution of equation (6.106) can be expressed in terms of modified Bessel functions I_0 and K_0 by the following equation:

$$\overline{U} = A_1 K_0(qr) + A_2 I_0(qr) + P \tag{6.109}$$

where P is a particular integral of equation (6.106), that depends on the initial excess pore pressure distribution U_0. The coefficients A_1 and A_2 can be determined from the boundary condition of the problem.

Initial excess pore pressure distribution

The *in-situ* stress condition in the soil prior to the drilling of the vertical hole is assumed to be defined by the vertical effective stress σ_v', the horizontal effective stress $k_0\sigma_v'$, and the hydrostatic pore water pressure U_i. k_0 is the coefficient of earth pressure at rest.

When the total radial stress acting across the cylindrical boundary of the borehole is reduced, the soil will respond elastically at first, until the yield criterion is satisfied at the cavity wall. If the total stress at the cavity wall is further reduced, a plastic zone will form around the cavity wall, within which negative excess pore pressures (suctions) will be generated. Outside this plastic region, soil unloads elastically with no change in mean total stress. As a result, no excess pore pressures are generated in the elastic zone.

Let us assume that the total cavity pressure at any stage during the unloading process is defined by $\lambda|U_i|$ $(0 \le \lambda \le 1)$. With the Tresca yield criterion, the radius of the plastic zone can be shown to be as follows:

$$R = a \exp\left\{\frac{\sigma_R - \lambda|U_i|}{2s_u}\right\} \tag{6.110}$$

where s_u is the undrained shear strength of the soil and

$$\sigma_R = k_0\sigma_v' + U_i - s_u \tag{6.111}$$

The excess pore pressure distribution in the plastic zone $a \leq r \leq R$ at time $t=0$ is given by the following equation:

$$U_0(r) = 2s_u \ln(\tfrac{r}{a}) - \sigma_R + \lambda |U_i| \tag{6.112}$$

Boundary conditions and consolidation solution

In the elastic region $r > R$, the initial excess pore pressures are zero everywhere. With the boundary condition that the pore pressure must be zero when $r \to \infty$, the Laplace transforms of the pore water pressure in the elastic region should take the following form:

$$\overline{U} = A_3 K_0(qr), \qquad \text{for} \quad r > R \tag{6.113}$$

The coefficients A_1, A_2 and A_3 can be determined from the continuity and boundary conditions. First of all, both the pressure U and its derivative $\partial U/\partial r$ must be continuous across the elastic-plastic interface $r=R$. Secondly, at the cavity wall $r=a$, a hydraulic condition needs to be given.

If, for example, the excess pore pressure at $r=a$ is maintained at a value of $U(a) = U_c$, for $t > 0$, then the constants A_1, A_2 and A_3 are given by the following equation:

$$\begin{bmatrix} K_0(qa) & I_0(qa) & 0 \\ K_0(qR) & I_0(qR) & -K_0(qR) \\ -K_1(qR) & I_1(qR) & K_1(qR) \end{bmatrix} \begin{Bmatrix} A_1 \\ A_2 \\ A_3 \end{Bmatrix} = \begin{Bmatrix} -P(a) + \dfrac{U_c}{s} \\ 0 \\ -\dfrac{\phi(R)}{q} \end{Bmatrix} \tag{6.114}$$

where $\phi(r)$ and $P(r)$ are defined by :

$$\phi(r) = \frac{\partial U}{\partial r} \tag{6.115}$$

$$P(r) = \frac{U_0(r)}{s} \tag{6.116}$$

For the general case of a supported borehole with a permeable cavity wall, the excess pore pressure at the cavity wall is given by:

$$U_c = \lambda |U_i| - U_i \tag{6.117}$$

On the other hand, if the hydraulic condition at the cavity wall is impermeable, then the constants A_1, A_2 and A_3 are given by the following equation:

$$\begin{bmatrix} -K_1(qa) & I_1(qa) & 0 \\ K_0(qR) & I_0(qR) & -K_0(qR) \\ -K_1(qR) & I_1(qR) & K_1(qR) \end{bmatrix} \begin{Bmatrix} A_1 \\ A_2 \\ A_3 \end{Bmatrix} = \begin{Bmatrix} -\dfrac{\phi(a)}{q} \\ 0 \\ -\dfrac{\phi(R)}{q} \end{Bmatrix} \tag{6.118}$$

Once the Laplace transforms of the excess pore pressure are determined, the excess pore pressures can then be obtained by the inversion of the Laplace transforms through a numerical integration scheme (Talbot, 1979).

6.5 SUMMARY

1. Relatively few studies have been carried out to derive time-dependent cavity expansion solutions in soils and rocks. Based on the assumptions of visco-elasticity and visco-plasticity, cavity solutions have been developed and applied to tunnel problems in rock. In soil mechanics, time-dependent consolidation solutions are derived to predict the pore pressure changes around a driven pile and an excavated tunnel in soils.

2. A most useful method for deriving solutions for visco-elastic materials is the Laplace transformation that enables many elastic solutions to be converted to solutions for visco-elastic problems (Lee, 1955). The so-called correspondence principle can be described simply as follows. If the solutions of an elastic problem is known, the Laplace transform of the solutions to the corresponding visco-elastic problem may be found by replacing the elastic constants G and K according to (6.17) and the actual loads by their Laplace transforms.

3. Time-dependent analytical solutions for cavity expansion problems using elastic-viscoplastic models are rare. For illustration, a brief description of the analytical solution developed by Fritz (1984) is given. In Fritz's time-dependent analysis, the elastic-brittle-plastic solution presented in Section 5.3.1 of Chapter 5 has been used as the limiting solution at time $t \rightarrow \infty$. The elastic-viscoplastic model used by Fritz (1984) consists of a spring, a dashpot and a sliding element. The dashpot and the sliding element are arranged in parallel. Although very complex, this solution was shown by Fritz to be relevant in predicting the time-dependent tunnel behaviour.

4. In soil mechanics, time-dependent cavity expansion solutions are mainly related to consolidation analysis which has applications to modelling soil behaviour around driven piles and underground excavations. To predict the variation of bearing capacity of a driven pile with time, Randolph and Wroth (1979) developed an analytical solution for radial consolidation of soil around a cylindrical cavity. This solution will be used in Chapter 9 to provide a satisfactory explanation for the observed increase in capacity of piles with time. In a related study, Carter (1988) presented a semi-analytical solution for consolidation of soil around a contracting cavity which is relevant to the problem of a bored pile.

REFERENCES

Carslaw, H.S. and Jaeger, J.C. (1959). *Conduction of Heat in Solids*. Clarendon Press, Oxford.

Carter, J.P. (1988). A semi-analytical solution for swelling around a borehole. *International Journal for Numerical and Analytical Methods in Geomechanics*, 12, 197–212.

Carter, J.P and Booker, J.R. (1982). Elastic consolidation around a deep circular tunnel. *International Journal of Solids and Structures*, 18(12), 1059–1074.

Flugge, W. (1967). *Viscoelasticity*. Blaisdell Publishing Company, Massachusetts.

Fritz, P. (1984). An analytical solution for axisymmetric tunnel problems in elasto-visco-plastic media. *International Journal for Numerical and Analytical Methods in Geomechanics*, 8, 325–342.

Gnirk, P.F. and Johnson, R.E. (1974). The deformation behaviour of a circular mine shaft situated in a visco-elastic medium under hydrostatic stress. *Proceedings of the 6th Symposium on Rock Mechanics*, 231–259.

Goodman, R.E. (1989). *Introduction to Rock Mechanics*. Chapman and Hall.

Jaeger, J.C. and Cook (1976). *Fundamentals of Rock Mechanics*. Chapman and Hall.

Ladanyi, B. (1993). Time-dependent response of rock around tunnels. In: *Comprehensive Rock Engineering* (Editor: J.A. Hudson), Pergamon Press, Oxford, Vol. 2, 77–112.

Lee, E.H. (1955). Stress analysis in viscoelastic bodies. *Quarterly Applied Mathematics*, 13 (2), 183–190.

Nonaka, T. (1981). A time-independent analysis for the final state of an elasto-visco-plastic medium with internal cavities. *International Journal of Solids and Structures*, 17, 961–967.

Randolph, M.F. and Wroth, C.P. (1979). An analytical solution for the consolidation around a driven pile. *International Journal for Numerical and Analytical Methods in Geomechanics*, 3, 217–229.

Salamon, M.D.G. (1974). Rock mechanics of underground excavations. *Proceedings of the 3rd Congress of International Society of Rock Mechanics*, 1(B), 994–1000.

Scott, R.F. (1990). Radial consolidation of a phase-change soil. *Geotechnique*, 40(2), 211–221.

Soderberg, L.O. (1962). Consolidation theory applied to time effects. *Geotechnique*, 12(3), 217–225.

Talbot, A. (1979). The accurate numerical inversion of Laplace transforms. *Journal of the Institute of Mathematical Applications*, 23, 97–120.

CHAPTER 7

FINITE ELEMENT SOLUTIONS

7.1 INTRODUCTION

This book is mainly concerned with analytical and semi-analytical solutions for cavity expansion problems. In practice, however, it is inevitable that numerical methods must be used and this is particularly true when more sophisticated soil models are employed. For this reason, this chapter presents the basic formulations for finite element analysis of cavity expansion problems. To validate the finite element formulations, the analytical solutions presented in previous chapters need to be used for comparison.

7.2 UNCOUPLED DRAINED AND UNDRAINED ANALYSIS

For many practical problems, the analysis of soil behaviour can be adequately reduced to either a fully drained or an undrained problem. In this case, the uncoupled finite element formulation can be used in the calculation. In particular, the fully undrained problem can be analysed using a total stress formulation with simple plasticity models such as the Tresca and Von Mises models. On the other hand, frictional soils under fully drained conditions may be modelled by ignoring the excess pore water pressures. Finite element work using this type of uncoupled formulations for fully drained and undrained cavity expansion problems has been carried out by Carter and Yeung (1985), Reed (1986), Yeung and Carter (1989), Yu (1990), Yu and Houlsby (1990), Yu (1994*a*,1996), and Shuttle and Jefferies (1998) among others.

7.2.1 Finite element formulation

The finite element formulation presented in this section was developed by Yu (1990) and Yu and Houlsby (1990) to accurately model soil behaviour.

To carry out the analysis for both cylindrical and spherical cavities at the same time, the symbol $k=1$ for cylindrical and $k=2$ for spherical is used. The strain rate vector $\dot{\varepsilon}$ can be expressed in terms of the radial displacement (or velocity) $\dot{u}$:

$$\dot{\varepsilon} = \boldsymbol{L}\dot{u} \tag{7.1}$$

where

$$\dot{\varepsilon} = [\dot{\varepsilon}_r, \dot{\varepsilon}_z, \dot{\varepsilon}_\theta]^T \tag{7.2}$$

$$\boldsymbol{L} = [\frac{\partial}{\partial r}, \frac{(k-1)}{r}, \frac{1}{r}]^T \tag{7.3}$$

The stress rate vector containing the radial, axial and tangential stresses is defined as follows:

$$\dot{\boldsymbol{\sigma}} = [\dot{\sigma}_r, \dot{\sigma}_z, \dot{\sigma}_\theta]^T \tag{7.4}$$

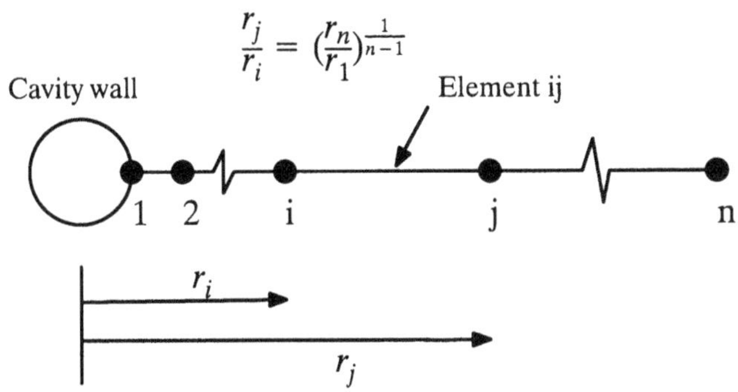

Figure 7.1: Cavity expansion and finite element mesh

In the study of Yu (1990) and Yu and Houlsby (1990), a two-noded one dimensional element was used so that the velocity field around a cavity is represented by values at the connecting nodes, Figure 7.1. For an element connected by nodes i and j, we have the following node velocity vector:

$$\dot{\boldsymbol{u}} = [\dot{u}_i, \dot{u}_j]^T \tag{7.5}$$

The velocity value at any point within the element ij can be approximated by the following equation through a shape function matrix:

$$\dot{u} = \boldsymbol{N}\dot{\boldsymbol{u}} \tag{7.6}$$

where the shape function matrix is related to the shape function associated with node number in the form:

$$\boldsymbol{N} = [N_i, N_j] \tag{7.7}$$

Combining equations (7.1) and (7.6) results in:

$$\dot{\boldsymbol{\varepsilon}} = \boldsymbol{L}\boldsymbol{N}\dot{\boldsymbol{u}} = \boldsymbol{B}\dot{\boldsymbol{u}} \tag{7.8}$$

where the matrix linking the node velocity vector and strain vector is given by:

$$B = \begin{bmatrix} \frac{\partial N_i}{\partial r} & \frac{\partial N_j}{\partial r} \\ \frac{N_i(k-1)}{r} & \frac{N_j(k-1)}{r} \\ \frac{N_i}{r} & \frac{N_j}{r} \end{bmatrix} \tag{7.9}$$

If we ignore the effects of element distortion, the rate form of the virtual work principle may be written as follows:

$$\dot{p} = K\dot{u} \tag{7.10}$$

where $\dot{\mathbf{p}}$ is the incremental node force vector and the stiffness matrix $\mathbf{K}$ is defined by:

$$K = \pi \int B^T D^{ep} B \, (2r)^k dr \tag{7.11}$$

and $\mathbf{D}^{ep}$ is the elastic-plastic matrix which links stress rate to strain rate for a given material model. A general expression for $\mathbf{D}^{ep}$ in an elastic-strain hardening plastic model may be shown to be:

$$D^{ep} = D^e - \frac{D^e b a^T D^e}{H + a^T D^e b} \tag{7.12}$$

where H is a hardening parameter and its value is zero for perfectly plastic models; $\mathbf{D}^e$ is the elastic stress-strain matrix; and **a** and **b** are derivatives of the yield function f=0 and the plastic potential g=0 with respect to stresses:

$$a = \frac{\partial f}{\partial \sigma} \quad \text{and} \quad b = \frac{\partial g}{\partial \sigma} \tag{7.13}$$

In order to model an infinite medium using an outer infinite spring element, the interface between the last annular element and the spring element should be selected such that the spring element always remains elastic. Hence, the radial stiffness K_{ss}, which relates the radial force acting at the interface to the radial displacement at the interface, can be computed from the cavity expansion theory in an infinite elastic medium and may be shown to be:

$$K_{ss} = 4k^2 \pi G (r_n)^{k-1} \tag{7.14}$$

where G denotes shear modulus and r_n is the radius of the interface between the last annular element and the infinite spring element.

To evaluate the shape function matrix **N**, some assumptions about the displacement interpolation are required. Following the conventional approach, in which a linear displacement interpolation along an element is assumed, the two-noded element ij has the following shape function matrix:

$$N = \left[\frac{r_j - r}{r_j - r_i}, \frac{r - r_i}{r_j - r_i}\right] \tag{7.15}$$

The above shape function matrix is derived using the conventional linear displacement interpolation, namely:

$$\dot{u} = C_0 + C_1 r \tag{7.16}$$

In Yu (1990) and Yu and Houlsby (1990), the following novel non-linear displacement interpolation is assumed within each element:

$$\dot{u} = \frac{C_0}{r^k} + C_1 r \tag{7.17}$$

where C_0 and C_1 are constants for a given element.

Based on the new displacement interpolation (7.17), the shape function matrix can be derived as follows:

$$N = \left[\frac{(r_j^{k+1} - r^{k+1})r_i^k}{(r_j^{k+1} - r_i^{k+1})r^k}, \frac{(r^{k+1} - r_i^{k+1})r_j^k}{(r_j^{k+1} - r_i^{k+1})r^k}\right] \tag{7.18}$$

The rationale behind the new displacement interpolation (7.17) is as follows. As the material becomes almost incompressible, the displacements around the cavity in an infinite medium become of the form $\dot{u} = C/r^k$ where C is a constant. The conventional linear displacement expansion (7.16) is unable to capture this mode exactly, whereas the new expansion (7.17) does. Furthermore the new expansion captures exactly the elasticity solution for general cavity expansion in compressible or incompressible materials.

7.2.2 Plasticity models for soils

For the purpose of illustration, this section briefly describes the forms of elastic-plastic stress-strain relationship for several widely used soil models in cavity expansion analysis.

(a) Mohr-Coulomb and Tresca plasticity models

The Mohr-Coulomb and Tresca models are widely used in finite element analysis to represent the behaviour of frictional and cohesive soils respectively. Nevertheless, these models contain discontinuous yield surface with edges at which the yield function is not differentiable. These singularities deserve special treatment as they are particularly important for cavity expansion problems. Several approaches exist for dealing with these singularities. One of the classical approaches was developed Nayak and Zienkienicz (1972) and consists of using only one yield function in combination with a rounding off procedure for points at which two planes of the yield function meet (the so-called corner point). Sloan and Booker (1984) adopted a modified surface to round off the corners so that a smooth yield surface may be obtained. Although these approaches have proven to be effective to some extent, they are mathematically inconvenient and physically somewhat artificial.

In Yu (1990, 1994*b*), a closed form solution was derived for the Mohr-Coulomb and Tresca elastic-plastic stiffness matrices. In deriving this closed form stiffness matrices solution, the yield surface singularities were dealt with by using the assumption that the total plastic strain rate is the sum of the individual contributions of the two flow rules (Koiter, 1960). This assumption may be simply expressed as:

$$\dot{\varepsilon}^p = \lambda_1 \boldsymbol{b}_1 + \lambda_2 \boldsymbol{b}_2 \tag{7.19}$$

where

$$\boldsymbol{b}_1 = \frac{\partial g_1}{\partial \sigma} \quad \text{and} \quad \boldsymbol{b}_2 = \frac{\partial g_2}{\partial \sigma} \tag{7.20}$$

and λ_1 and λ_2 are scalar multipliers, and g_1 and g_2 are the plastic potentials corresponding to the two yield surfaces.

Assume that the corner is defined by two yield functions $f_1 = 0$ and $f_2 = 0$, then on plastic yielding, the stress state will stay on the yield surface, thus

$$\boldsymbol{a}_1 \dot{\sigma} = 0 \tag{7.21}$$

$$\boldsymbol{a}_2 \dot{\sigma} = \boldsymbol{0} \tag{7.22}$$

where

$$\boldsymbol{a}_1 = \frac{\partial f_1}{\partial \sigma} \quad \text{and} \quad \boldsymbol{a}_2 = \frac{\partial f_2}{\partial \sigma} \tag{7.23}$$

The elastic strain rate is related to the stress rate through Hooke's law:

$$\dot{\sigma} = \boldsymbol{D}^e \dot{\varepsilon}^e = \boldsymbol{D}^e (\dot{\varepsilon} - \dot{\varepsilon}^p) \tag{7.24}$$

where $\mathbf{D}^e$ is the elastic stress-strain matrix and $\dot{\boldsymbol{\varepsilon}}^e$ and $\dot{\boldsymbol{\varepsilon}}^p$ are elastic and plastic strain rates respectively.

By combining equations (7.19)–(7.24), we have the following two equations for λ_1 and λ_2:

$$\begin{bmatrix} a_1 D^e b_1 & a_1 D^e b_2 \\ a_2 D^e b_1 & a_2 D^e b_2 \end{bmatrix} \begin{bmatrix} \lambda_1 \\ \lambda_2 \end{bmatrix} = \begin{bmatrix} a_1 D^e \dot{\varepsilon} \\ a_2 D^e \dot{\varepsilon} \end{bmatrix} \tag{7.25}$$

The above equations can be used to solve for λ_1 and λ_2 as follows:

$$\lambda_1 = h_1 \dot{\varepsilon} \tag{7.26}$$

$$\lambda_2 = h_2 \dot{\varepsilon} \tag{7.27}$$

where h_1 and h_2 are simple functions of $\boldsymbol{a_1, a_2, b_1, b_2}$ **and** $\boldsymbol{D^e}$.

With equations (7.26) and (7.27), the relationship between stress rate and strain rate (7.24) can be further simplified as follows:

$$\dot{\sigma} = (D^e - D^p)\dot{\varepsilon} = D^{ep}\dot{\varepsilon} \tag{7.28}$$

where the elastic-plastic stress-strain matrix $\mathbf{D}^{ep}$ is defined by:

$$D^{ep} = D^e - D^e b_1 h_1 - D^e b_2 h_2 \tag{7.29}$$

Equation (7.29) is the general expression of elastic-plastic stiffness for stress points that lie on the intersection of two yield surfaces. Although the solution (7.29) seems somewhat complicated, for the Mohr-Coulomb and Tresca models, the final results are in fact very simple.

Stress states on a single yield surface

For illustration, let's take tensile stresses as positive and the three principal stresses are in the order $\sigma_1 < \sigma_2 < \sigma_3$. If the stress state lies on a single yield surface, then the Mohr-Coulomb yield function is defined as:

$$f = \sigma_3 - \sigma_1 + (\sigma_3 + \sigma_1)\sin\phi - 2c\cos\phi = 0 \tag{7.30}$$

where c and ϕ are soil cohesion and friction angle respectively. A plastic potential is generally obtained by using a dilation angle ψ:

$$g = \sigma_3 - \sigma_1 + (\sigma_3 + \sigma_1)\sin\psi = constant \tag{7.31}$$

With the above assumptions, the elastic-plastic stiffness matrix defined by equation (7.29) can be shown to be (Yu, 1994*b*):

$$D^{ep} = C_2 \begin{bmatrix} (K + \frac{G}{3})(1+s)(1+n) & (K - \frac{2G}{3})(1+s) & (K + \frac{G}{3})(1+s)(1-n) \\ (K - \frac{2G}{3})(1+n) & K(1+3sn) + \frac{4G}{3} & (K - \frac{2G}{3})(1-n) \\ (K + \frac{G}{3})(1-s)(1+n) & (K - \frac{2G}{3})(1-s) & (K + \frac{G}{3})(1-s)(1-n) \end{bmatrix} \tag{7.32}$$

where $s = \sin\phi$ and $n = \sin\psi$; K and G are bulk and shear moduli; and

$$C_2 = \frac{G}{G + (K + \frac{G}{3})sn} \tag{7.33}$$

Stress states on the intersection of two yield surfaces

If the stress state lies on the intersection of any two yield functions, for example, those defined by:

$$f_1 = \sigma_3 - \sigma_1 + (\sigma_3 + \sigma_1)\sin\phi - 2c\cos\phi = 0 \tag{7.34}$$

$$f_2 = \sigma_2 - \sigma_1 + (\sigma_2 + \sigma_1)\sin\phi - 2c\cos\phi = 0 \tag{7.35}$$

the corresponding plastic potentials may be written as:

$$g_1 = \sigma_3 - \sigma_1 + (\sigma_1 + \sigma_3)\sin\psi = constant \tag{7.36}$$

$$g_2 = \sigma_2 - \sigma_1 + (\sigma_2 + \sigma_1)\sin\psi = constant \tag{7.37}$$

With the above assumptions, the elastic-plastic stiffness matrix defined by equation (7.29) can be shown to be (Yu, 1994*b*):

$$\boldsymbol{D}^{ep} = C_3 \begin{bmatrix} (1+s)(1+n) & (1-n)(1+s) & (1+s)(1-n) \\ (1-s)(1+n) & (1-n)(1-s) & (1-s)(1-n) \\ (1-s)(1+n) & (1-n)(1-s) & (1-s)(1-n) \end{bmatrix} \tag{7.38}$$

where

$$C_3 = \frac{9GK}{12Ksn + G(3-s)(3-n)} \tag{7.39}$$

Similar matrices can be obtained for other combinations of stress magnitudes. The elastic-plastic stiffness matrices for the Tresca model can be obtained from the solution for the Mohr-Coulomb model simply by setting $\phi = 0$ and $\psi = 0$.

(b) von Mises plasticity model

Like the Tresca yield criterion, the Von Mises plasticity model is also used in finite element calculations to model the undrained behaviour of cohesive soils. Following the usual procedure, the plastic part of the elastic-plastic stiffness matrix can be shown to be:

$$\boldsymbol{D}^{p} = -\frac{3G}{4s_u} \begin{bmatrix} (\sigma_r - p)^2 & (\sigma_z - p)(\sigma_r - p) & (\sigma_\theta - p)(\sigma_r - p) \\ (\sigma_r - p)(\sigma_z - p) & (\sigma_z - p)^2 & (\sigma_z - p)(\sigma_\theta - p) \\ (\sigma_r - p)(\sigma_\theta - p) & (\sigma_\theta - p)(\sigma_z - p) & (\sigma_\theta - p)^2 \end{bmatrix} \tag{7.40}$$

where s_u is the undrained shear strength and p denotes the mean pressure.

(c) A plasticity model with shear strain softening

In the previous plasticity formulation for the Mohr-Coulomb criterion, the strength parameters c and ϕ are assumed to be constant during the entire loading process. In reality, however, the strength parameters may vary with displacement for many soils. In particular, for strain weakening soils the cohesion and friction angle decrease with plastic shear strain. To account for this observed softening behaviour, Carter and Yeung (1985) combined the Mohr-Coulomb yield function with a very simple strain softening law that is shown in Figure 7.2, where τ and γ are shear stress and shear strain; v is volumetric strain; γ^p is plastic shear strain; and the subscripts p and r denote peak and residual strength parameters.

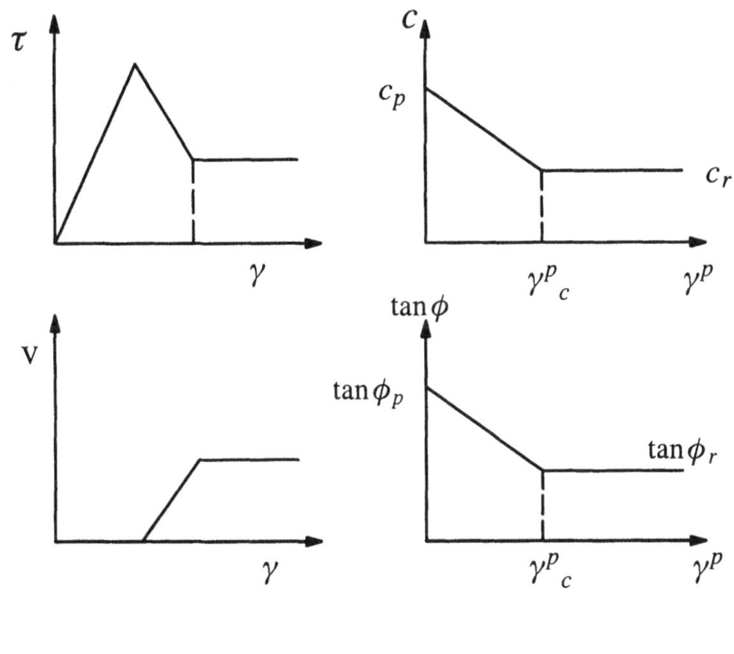

Figure 7.2: Idealised stress-strain behaviour with strain softening (adopted from Carter and Yeung, 1985)

Mathematically, the softening behaviour presented in Figure 7.2 can be expressed as follows:

$$c = \begin{cases} c_p - (c_p - c_r)\dfrac{\gamma^p}{\gamma^p{}_c} & 0 \le \gamma^p \le \gamma^p{}_c \\ c_r & \gamma^p > \gamma^p{}_c \end{cases} \tag{7.41}$$

$$\tan\phi = \begin{cases} \tan\phi_p - (\tan\phi_p - \tan\phi_r)\dfrac{\gamma^p}{\gamma^p{}_c} & 0 \le \gamma^p \le \gamma^p{}_c \\ \tan\phi_r & \gamma^p > \gamma^p{}_c \end{cases} \tag{7.42}$$

$$\psi = \begin{cases} \psi_0 & 0 \le \gamma^p \le \gamma^p{}_c \\ 0 & \gamma^p > \gamma^p{}_c \end{cases} \tag{7.43}$$

where $\gamma^p{}_c$ is the critical plastic strain before strain softening occurs and ψ_0 is the initial soil dilation angle.

With the assumptions made in (7.41)-(7.43), the hardening parameter H in the elastic-plastic stiffness matrix (7.12) can be derived. The details of such a derivation are given in Carter and Yeung (1985).

(d) A state parameter model for sand with strain hardening/softening

The state parameter model presented in this section is an elastic-plastic strain hardening (or softening) model which has been used by Yu (1994*a*, 1996) to analyse pressuremeter tests in sand.

General stress-strain relationship

The basic assumption of the state parameter sand model is the existence of a critical state at which the sand deforms without any plastic volume change so that the dilation angle is zero. As argued before, the material behaviour prior to the achievement of the critical state is controlled by the state parameter which is defined by Been and Jefferies (1985):

$$\xi = e + \lambda \ln(\frac{p}{p_1}) - \Gamma \tag{7.44}$$

where e is void ratio, p denotes the mean effective stress, and p_1 is the reference mean pressure which is often taken as unity in soil mechanics. It is noted that the state parameter ξ is zero at the critical state, positive at the loose side and negative at the dense side.

It is noted that the void ratio e can be eliminated from the equation (7.44) using the following expression:

$$\frac{dv}{v} = d\varepsilon_v \tag{7.45}$$

where $v = (1 + e)$ is known as specific volume and ε_v denotes volumetric strain. The above equation can be integrated to give

$$e = (1 + e_0)\exp(\varepsilon^e{}_v + \varepsilon^p{}_v) - 1 \tag{7.46}$$

Suppose now that the plastic behaviour of the materials can be modelled by a yield function:

$$f(\boldsymbol{\sigma}, \xi) = 0 \tag{7.47}$$

and a corresponding plastic potential is defined by:

$$g(\boldsymbol{\sigma}, \xi) = 0 \tag{7.48}$$

where $\boldsymbol{\sigma}$ is the effective stress vector and $\varepsilon^p{}_v$ denotes the plastic volumetric strain. As the state parameter ξ depends on the mean effective stress, it is necessary to be clear whether ξ or $\varepsilon^p{}_v$ is to be kept constant when differentiating both the yield function and plastic potential with respect to the components of effective stress. For conciseness we will denote derivatives of f with respective to stress evaluated at constant ξ by $f_{,ij}$, and at constant $\varepsilon^p{}_v$ by $\hat{f}_{,ij}$ and similarly for the plastic potential.

Since the plastic flow rule describes the mechanical properties of a material element in a given state, Collins (1990) suggested that the strain rates are given by:

$$\dot{\varepsilon}^p{}_{ij} = \Pi \, g_{,ij} \tag{7.49}$$

and the plastic multiplier Π could be easily eliminated from the plastic flow rule and the consistency condition for continued plastic deformation. This condition is obtained by differentiating the yield function with respective to time but in doing so it is necessary to regard f as a function of σ_{ij} and $\varepsilon^p{}_v$, so

$$\hat{f}_{,ij}\dot{\sigma}_{ij} + f_{,\varepsilon^p{}_v}\dot{\varepsilon}^p{}_v = 0 \tag{7.50}$$

where $\dot{\varepsilon}^p{}_v$ denotes the plastic volumetric strain rate. Combining equations (7.49) and (7.50) leads to:

$$\Pi = -\frac{\hat{f}_{,ij} \, \dot{\sigma}_{ij}}{f_{,\xi} \, \xi_{,\varepsilon^p{}_v} \, g_{,kk}} \tag{7.51}$$

The resulting expression for Π can then be substituted into the plastic flow rule (7.49) to give the plastic strain rates as follows:

$$\dot{\varepsilon}^p{}_{ij} = \frac{1}{H} g_{,kk} \, \hat{f}_{,ij} \, \dot{\sigma}_{ij} \tag{7.52}$$

where H is the hardening modulus defined as:

$$H = -f_{,\xi} \, \xi_{,\varepsilon^p{}_v} \, g_{,kk} \tag{7.53}$$

The plastic strain rates are then fully determined for each given stress rate when the yield function and plastic potential are specified. On the other hand, the elastic strain rates can be obtained from a simple linear elastic law. A complete stress-strain relationship is therefore obtained:

$$\dot{\sigma}_{ij} = [D_{ijkl} - \frac{D_{ijkl}\,\hat{F}_{,kl}\,G_{,ij}\,D_{ijkl}}{H + G_{,ij}\,D_{ijkl}\,\hat{F}_{,kl}}]\,\dot{\varepsilon}_{kl} \tag{7.54}$$

where D_{ijkl} is the elastic stress-strain matrix. For the one dimensional cavity expansion problems considered here it can be expressed in terms of shear modulus G and bulk modulus K.

Yield function and plastic potential

The choice of the constitutive functions for plastic potential and yield function is an important step in constitutive modelling, since strain rates essentially depend on their derivatives with respect to stresses.

In the study of Yu (1994*a*, 1996), the yield criterion proposed by Matsuoka (1976) has been adopted to describe sand behaviour. The Matsuoka yield function is normally written in the following elegant form in terms of three stress invariants:

$$f = \frac{I_1 I_2}{I_3} - (9 + 8\tan^2\phi_m) = 0 \tag{7.55}$$

where I_1, I_2 and I_3 are the first, second and third invariants of the effective stress tensor respectively, and ϕ_m denotes the mobilized friction angle under triaxial loading conditions.

To derive the plastic strain rates, Yu (1990) proposed the use of the following plastic potential

$$g = \frac{I^*_1 I^*_2}{I^*_3} - (9 + 8\tan^2\psi_m) = 0 \tag{7.56}$$

where ψ_m denotes the mobilized dilation angle under triaxial loading conditions and I_1^*, I_2^* and I^*_3 are the first, second and third invariants of a modified stress tensor which is defined by

$$\sigma^*_{ij} = \sigma_{ij} + l\delta_{ij} \tag{7.57}$$

in which δ_{ij} is the Kronecker delta. This plastic potential function is of a similar form to the Matsuoka yield function, but the friction angle is replaced by the dilation angle, and the apex of the surface is moved from the origin to the point in principal stress space with the coordinates ($-l$, $-l$, $-l$). The parameter l is calculated on the basis that the plastic potential and the yield function must coincide at the current stress state. As shown by Yu (1990), this condition can be successfully used to derive a cubic equation in l from which the required root may be selected. The plastic potential (7.56) can only be used for initially dense sand where dilation angle is positive. In order to model the behaviour of an initially loose sand whose dilation angle becomes negative, a plastic potential that takes the same form as the Mohr-Coulomb yield function, but uses dilation angle instead of friction angle can be adopted.

As discussed by Collins *et al.* (1992), it is reasonable to use the following exponential type of empirical relationship to fit experimental data:

$$\phi_m - \phi_{cv} = A[\exp(-\xi) - 1] \tag{7.58}$$

where ϕ_{cv} is the angle of internal friction at the critical state, and A is a curve fitting parameter ranging from 0.6-0.95 depending on the type of sands (the angles are measured in radians). According to the above relationship, the mechanical properties (i.e. yield function) of a sand prior to the achievement of the critical state are therefore dependent on both the void ratio and the mean effective stress through a single composite state parameter. The critical state friction angle ϕ_{cv} may be determined by conventional laboratory tests and normally ranges from 30–33 degrees.

To link the plastic potential and state parameter, a stress-dilatancy equation must be used which defines the relationship between the angles of friction and dilation. Perhaps the most successful stress-dilatancy model is that developed by Rowe (1962), which has been further simplified by Bolton (1986) as follows:

$$\psi_m = 1.25(\phi_m - \phi_{cv}) = 1.25A[\exp(-\xi) - 1] \tag{7.59}$$

The above assumption implies a non-associated flow rule as the angles of friction and dilation are not the same. Using the stress-dilatancy equation (7.59), the plastic potential (7.56) is therefore controlled by the state parameter.

7.2.3 Finite element program

A computer program for performing uncoupled drained and undrained analysis of cylindrical and spherical cavity expansion problems has been developed by the author (Yu, 1990; Yu and Houlsby, 1990; Yu, 1994*a*; Yu; 1996). The computer code is called CAVEXP and was first written at Oxford in 1989 when the author was a research student. However, the program has since been considerably modified and revised to include several strain hardening/softening soil models. It has been used by a number of researchers for analysing pressuremeter and CPT tests in clay and sand. It has also been used to model compaction grouting for soil improvement (Boulanger and Yu, 1997).

For convenience, the author will make the program CAVEXP available for use by readers of the book; it may not be transferred to others without the author's written permission. In any case, no responsibility will be accepted for any loss or damage resulting from the use of this program.

7.3 COUPLED CONSOLIDATION ANALYSIS

In a saturated cohesive soil, it is very useful to determine the generation and the dissipation of pore water pressure during and after cavity expansion. Such information can, for example, be used to estimate the increase in strength of the soil with time after a pile is driven into the ground. The solution for the dissipation of excess pore water pressure around a cavity can also be used to estimate the horizontal consolidation coefficient from pressuremeter holding tests.

Carter (1978) developed a finite element program for the coupled consolidation analysis of cylindrical cavity expansion in a saturated two-phase cohesive soil to meet these needs. In consolidation analysis, an effective stress soil model is assumed to describe the behaviour of soil skeleton. However, the equation of soil equilibrium is formulated in terms of total stresses. The effective and total stresses are related by Terzaghi's effective stress principle.

7.3.1 Finite element formulation

In a coupled consolidation analysis, the finite element formulation is based on the following assumptions:

Effective stress principle

The response of the soil skeleton to applied stress is coupled to the movement of pore fluid through the saturated soil. It is assumed that the total stress is equal to the effective stress plus the pore water pressure, namely:

$$\boldsymbol{\sigma} = \boldsymbol{\sigma}' + U_t \, \boldsymbol{m} \tag{7.60}$$

where $\boldsymbol{\sigma}$ and $\boldsymbol{\sigma}'$ are total and effective stresses respectively; U_t is total pore water pressure and $\boldsymbol{m} = (1, 1, 1)^T$.

The stress-strain relationship for soil skeleton

The stress-strain relationship for soil skeleton is governed by the rate of effective stress and strain rate. If the soil skeleton is modelled by plasticity theory, then the relationship can be written as follows:

$$\dot{\boldsymbol{\sigma}}' = (\boldsymbol{D}^e - \boldsymbol{D}^p)\dot{\boldsymbol{\varepsilon}} = \boldsymbol{D}^{ep}\dot{\boldsymbol{\varepsilon}} \tag{7.61}$$

where $\boldsymbol{D}^{ep}$ is the elastic-plastic stiffness matrix which depends on the particular soil model used in the analysis.

Darcy's law

The water flow through the soil is governing by Darcy's law, which can be expressed in terms of the pore water velocity relative to the soil skeleton:

$$n(v_f - v_s) = -\frac{k}{\gamma_w}\frac{\partial U}{\partial r} \tag{7.62}$$

where n is the soil porosity; v_f and v_s are the outward radial velocities of the pore water and the skeleton respectively; k and γ_w are the soil permeability coefficient and the unit weight of pore water respectively; and U is excess pore water pressure.

Volume continuity

Compared with the two-phase soil, the pore water and material forming the soil skeleton may be assumed to be incompressible. With this assumption, any change in volume of an element of soil is entirely due to the expulsion or absorption of water from the element, namely:

$$\dot{\varepsilon}_v = \frac{1}{r}\frac{\partial}{\partial r}[rn(v_f - v_s)] \tag{7.63}$$

where $\dot{\varepsilon}_v$ is the rate of volumetric strain of the soil element.

Combining with the virtual work equation, the above assumptions can be used to form the basic finite element equations for consolidation analysis of cavity expansion problems. As shown in Carter (1978), the resulting equations are:

$$\boldsymbol{K}\Delta\boldsymbol{u} - \boldsymbol{L}^T\Delta\boldsymbol{U} = \boldsymbol{f} \tag{7.64}$$

$$-\boldsymbol{L}\frac{\partial \boldsymbol{u}}{\partial t} - \boldsymbol{\Phi}\boldsymbol{U} = \boldsymbol{0} \tag{7.65}$$

where the vectors $\boldsymbol{u}$ and $\boldsymbol{U}$ are nodal values of the displacement and excess pore water pressure. The incremental quantities $\Delta\boldsymbol{u}$ and $\Delta\boldsymbol{U}$ are defined by:

$$\Delta\boldsymbol{u} = \boldsymbol{u}(r,t) - \boldsymbol{u}(r_o, t_o) \tag{7.66}$$

$$\Delta\boldsymbol{U} = \boldsymbol{U}(r,t) - \boldsymbol{U}(r_o, t_o) \tag{7.67}$$

and the general forms of the stiffness matrices contained in equations (7.64) and (7.65) are:

$$\boldsymbol{K} = \int \boldsymbol{B}^T\, \boldsymbol{D}^{ep}\, \boldsymbol{B}\, dV \tag{7.68}$$

$$\boldsymbol{L}^T = \int \boldsymbol{N}^T\, \boldsymbol{A}\, dV \tag{7.69}$$

$$\boldsymbol{\Phi} = \int \frac{k}{\gamma_w} \boldsymbol{E}^T \boldsymbol{E} \, dV \tag{7.70}$$

$$\boldsymbol{f} = -\int \boldsymbol{B}^T \sigma(r_o, t_o) \, dV + \int \boldsymbol{A}^T \boldsymbol{T} \, dS \tag{7.71}$$

where $\boldsymbol{T}$ is a traction vector applied on the boundary surface S.

Equations (7.64) and (7.65) form a set of differential-integral equations which may be integrated over the finite time interval (t_o, t) to obtain the following approximated equation:

$$\begin{bmatrix} \overline{\boldsymbol{K}} & -\overline{\boldsymbol{L}}^T \\ -\overline{\boldsymbol{L}} & -\beta \Delta t \overline{\boldsymbol{\Phi}} \end{bmatrix} \begin{pmatrix} \Delta \boldsymbol{u} \\ \Delta \boldsymbol{U} \end{pmatrix} = \begin{pmatrix} \bar{\boldsymbol{f}} \\ \overline{\boldsymbol{\Phi}} \boldsymbol{u}(r_0, t_0) \Delta t \end{pmatrix} \tag{7.72}$$

The quantities with a superior bar are calculated using an average configuration over the time interval. With equation (7.72), the solution for both displacement and excess pore pressure at time $t_o + \Delta t$ can be obtained as long as the solution at time t_0 is known. To ensure the stability of this solution process, the condition $\beta \geq 0.5$ must be imposed (Booker and Small, 1975).

Carter (1978) used a three-noded finite element so that within each element both the excess pore water pressure and displacement were assumed to be quadratic functions of the radial coordinate r. If an element is defined by radii (r_i, r_j, r_k), the element matrices can be shown to be:

$$\boldsymbol{M} = \begin{bmatrix} 1 & r_i & r_i^2 \\ 1 & r_j & r_j^2 \\ 1 & r_k & r_k^2 \end{bmatrix} \tag{7.73}$$

$$\boldsymbol{A}^T = (1, r, r^2)\boldsymbol{M}^{-1} \tag{7.74}$$

$$\boldsymbol{B} = (-1)\begin{bmatrix} 0 & 1 & 2r \\ 0 & 0 & 0 \\ 1/r & 1 & r \end{bmatrix} \boldsymbol{M}^{-1} \tag{7.75}$$

$$\boldsymbol{N}^T = (-1)(1/r, 2, 3r)\boldsymbol{M}^{-1} \tag{7.76}$$

$$\boldsymbol{E}^T = (0, 1, 2r)\boldsymbol{M}^{-1} \tag{7.77}$$

7.3.2 The modified Cam clay model

Stress definitions and yield function

The effective mean and shear stresses p', q are used in the modified Cam clay model (Roscoe and Burland, 1968). For the cavity expansion problems considered, they are given by:

$$p' = \frac{1}{3}(\sigma_r' + \sigma_z' + \sigma_\theta') \tag{7.78}$$

$$q = \sqrt{\frac{1}{2}[(\sigma_r'-\sigma_\theta')^2 + (\sigma_\theta'-\sigma_z')^2 + (\sigma_z'-\sigma_r')^2]} \tag{7.79}$$

In the modified Cam clay model, the yield is assumed to occur as long as the stresses satisfy the following function:

$$q^2-M^2[p'(p_c'-p')] = 0 \tag{7.80}$$

where p_c' is the hardening parameter which defines the intersection of the current yield locus and the p' axis in principal effective stress space. M is the slope of the critical state line (CSL) in the q–p' plot.

Elastic-plastic stiffness

Using an associated plastic flow rule, the modified Cam clay model gives the following stiffness matrix to relate the effective stress rate and strain rate:

$$\boldsymbol{D}^{ep} = \boldsymbol{D}^e - \frac{\boldsymbol{D}^e\boldsymbol{a}\boldsymbol{a}^T\boldsymbol{D}^e}{H + \boldsymbol{a}^T\boldsymbol{D}^e\boldsymbol{a}} \tag{7.81}$$

where the hardening modulus H is given by:

$$H = -\alpha(1,1,1)\boldsymbol{a} \tag{7.82}$$

and α is shown to be:

$$\alpha = p'p_c'\frac{1+e}{\lambda-\kappa} \tag{7.83}$$

where e is the void ratio and λ, κ are slopes of the critical state line (CSL) and unloading-reloading curve in the e–lnp' plot.

7.3.3 Finite element program

A finite element computer program for consolidation analysis of cylindrical cavity expansion problems was written by Carter (1978). The computer code, called CAMFE, was published in a soil mechanics research report at the University of Cambridge. It has been used by researchers for analysing pressuremeter tests in clay.

7.4 SUMMARY

1. There are two types of finite element formulation: uncoupled and coupled analysis. Many practical problems can be adequately formulated as either a fully

drained or an undrained problem. In both of these cases, an uncoupled finite element formulation can be used in the calculation. In particular, a fully undrained problem can be analysed using a total stress formulation with simple plasticity models such as the Tresca and Von Mises models. On the other hand, frictional soils under fully drained conditions can be modelled by ignoring the excess pore water pressures.

2. A computer program has been developed by the author (Yu, 1990; Yu and Houlsby, 1990; Yu, 1994*a*; Yu; 1996) for performing uncoupled drained and undrained analysis of cylindrical and spherical cavity expansion problems. The computer code is called CAVEXP and incorporates several well known plasticity models. CAVEXP has been used by several researchers to analyse pressuremeter and CPT tests in clay and sand. Some of these studies will be described in Chapter 8. It has also been used to model compaction grouting for soil improvement.

3. When considering the generation and dissipation of pore water pressure during and after cavity expansion, a coupled finite element consolidation formulation is required. Such information on pore water pressures can, for example, be used to estimate the increase in strength of the soil with time after a pile is driven into the ground. The solution for the dissipation of excess pore water pressure around a cavity can also be used to estimate the horizontal consolidation coefficient from pressuremeter holding tests. Carter (1978) developed a finite element program for performing coupled consolidation analysis of cylindrical cavity expansion in a saturated two-phase cohesive soil. The modified Cam clay model was used. Carter's finite element computer code, CAMFE, was published as a Soil Mechanics Research Report at the University of Cambridge. CAMFE has been used by a number of researchers for analysing pressuremeter tests in clay.

REFERENCES

Been, K and Jefferies, M.G. (1985). A state parameter for sands. *Geotechnique,* 35(2), 99–112.

Bolton, M. D. (1986). The strength and dilatancy of sands. *Geotechnique*, 36(1), 65–78.

Booker, J.R. and Small, J.C. (1975). An investigation of the stability of numerical solutions of Biot's equations of consolidation. *International Journal of Solids and Structures,* 11, 907–917.

Boulanger, R.W. and Yu, H.S. (1997). Theoretical aspects of compaction grouting in sands. *Civil Engineering Research Report No.* 149.07.97, The University of Newcastle, NSW 2308, Australia.

Carter, J.P. (1978). CAMFE: A computer program for the analysis of a cylindrical cavity expansion in soil. *Report CUED/C – Soils TR52, Department of Engineering, University of Cambridge.*

Carter, J.P and Yeung, S.K. (1985). Analysis of cylindrical cavity expansion in strain weakening material. *Computers and Geotechnics,* 1, 161–180.

Carter, J.P, Randolph, M.F. and Wroth, C.P. (1979). Stress and pore pressure changes in clay during and after the expansion of a cylindrical cavity. *International Journal for Numerical and Analytical Methods in Geomechanics,* 3, 305–322.

Collins, I.F. (1990). On the mechanics of state parameter models for sands. *Proceedings of the 7th International Conference on Computer Methods and Advances in Geomechanics*, Cairns, Australia, 1, 593–598.

Collins, I.F., Pender, M.J. and Wan, Y. (1992). Cavity expansion in sands under drained loading conditions. *International Journal for Numerical and Analytical Methods in Geomechanics,*16(1), 3–23.

Koiter, W.T. (1960). General theorems for elastic-plastic solids. In: *Progress in Solid Mechanics* (editors: I.N. Sneddon and R. Hill), 165–221.

Matsuoka, H. (1976). On significance of the spatial mobilized plane. *Soils and Foundations*, 16(1), 91–100.

Nayak, G.C. and Zienkiewicz, O.C. (1972). Elasto-plastic stress analysis: a generalization for various constitutive relations including strain softening. *International Journal for Numerical Methods in Engineering*, 5, 113–135.

Randolph, M.F. and Wroth, C.P. (1979). An analytical solution for the consolidation around a driven pile. *International Journal for Numerical and Analytical Methods in Geomechanics*, 3, 217–229.

Reed, M.B. (1986). Stresses and displacement around a cylindrical cavity in soft rock. *IMA Journal of Applied Mathematics*, 36, 223–245.

Roscoe, K.H. and Burland, J.B. (1968). On the generalised behaviour of wet clay. In: *Engineering Plasticity* (editors: J. Heyman and F.A. Leckie), Cambridge University Press, 535–609.

Rowe, P. W. (1962). The stress–dilatancy relation for static equilibrium of an assembly of particles in contact. *Proceedings of Royal Society,* 267, 500–527.

Shuttle, D. and Jefferies, M. (1998). Dimensionless and unbiased CPT interpretation in sand. *International Journal for Numerical and Analytical Methods in Geomechanics,* 22, 351–391.

Sloan, S.W. and Booker, J.R. (1984). Removal of singularities in Tresca and Mohr-Coulomb yield functions. *Communications in Applied Numerical Methods*, 2, 173–179.

Yeung, S.K. and Carter, J.P (1989). An assessment of the bearing capacity of calcareous and silica sands. *International Journal for Numerical and Analytical Methods in Geomechanics,* 13, 19–36.

Yu, H.S. (1990). *Cavity Expansion Theory and Its Application to the Analysis of Pressuremeters*, DPhil Thesis, University of Oxford, England.

Yu, H.S. (1994*a*). State parameter from self-boring pressuremeter tests in sand. *Journal of Geotechnical Engineering,* ASCE, 120(12), 2118–2135.

Yu, H.S. (1994*b*). A closed form solution of stiffness matrix for Tresca and Mohr-Coulomb plasticity models. *Computers and Structures*, 53(3), 755–757.

Yu, H.S. (1996). Interpretation of pressuremeter unloading tests in sands. *Geotechnique*, 46(1), 17–31.

Yu, H.S. and Houlsby, G.T. (1990). A new finite element formulation for one dimensional analysis of elastic plastic materials. *Computers and Geotechnics,* 9, 241–25

CHAPTER 8

IN-SITU SOIL TESTING

8.1 INTRODUCTION

In-situ soil testing is an essential part of geotechnical engineering. While many devices have been developed over the years for measuring soil properties *in-situ*, the pressuremeter and the cone penetrometer (CPT) are arguably the two most widely used *in-situ* soil testing devices. Unlike many laboratory tests, pressuremeter and CPT tests are indirect tests in that the measured data needs to be interpreted to give fundamental soil properties. Due to similar mechanical action generated by cavity expansion and cone penetration and pressuremeter expansion, cavity expansion theory has been used with considerable success in the interpretation of these two types of *in-situ* soil tests (e.g. Wroth, 1984, Clarke, 1995, Yu and Mitchell, 1998; Lunne *et al.*, 1997). The purpose of this chapter is to summarise some key applications of cavity expansion solutions to the interpretation of pressuremeter and CPT tests in both clay and sand.

8.1.1 The principle of pressuremeter testing

A pressuremeter is defined as a cylindrical probe that has an expandable flexible membrane designed to apply a uniform pressure to the wall of a borehole (Clarke, 1995). When a pressuremeter test is carried out in soils, the measured pressure displacement curve can be used to back-calculate the mechanical properties of soils (see Figure 8.1).

The main advantages of the pressuremeter test over other *in-situ* tests are (i) the boundary conditions imposed by the pressuremeter test are relatively well defined; (ii) the pressuremeter test can be used to measure both deformation and strength parameters at the same time; (iii) the self-boring pressuremeter potentially offers the closest approach to undisturbed soil testing of any *in-situ* test due to its ability to tunnel into the ground with minimal soil disturbance prior to a test being carried out (Mair and Wood, 1987). The basis of the test is the expansion of a long cylindrical membrane installed in the ground, and therefore cavity expansion theory can be used to analyse the pressuremeter test.

8.1.2 Types of pressuremeter

Depending on how the pressuremeter is installed into the ground, the pressuremeters can be classified into three groups. They are pre-bored pressuremeters, self-boring pressuremeter, and pushing-in pressuremeters. The Menard pressuremeter is the most well known example of the pre-bored pressuremeters in which the device is lowered into a pre-formed hole. Camkometer and PAF are two examples of the self-boring pressuremeter where the pressuremeter bores its own way into the ground by a self-boring technique. The so-called cone or full-displacement pressuremeter can be grouped into the category of push-in pressuremeter where the device is pushed into the ground with a cone attached at the front of the pressuremeter.

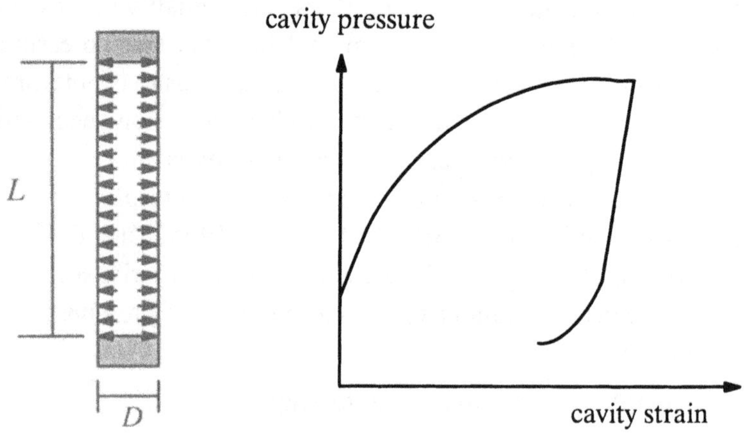

Figure 8.1: Definition of a pressuremeter

8.1.3 Cone penetrometer testing

At the present time, the cone penetrometer test (CPT) is arguably the most widely used *in-situ* test device in the world. In cone penetrometer testing, a cone on the end of a series of rods is pushed into the ground at a constant rate and continuous or intermittent measurements are made of the resistance to the penetration of the cone. Measurements are also made of either the combined resistance to penetration of the cone and outer surface of the rods or the resistance of a surface sleeve. The values of measured cone tip resistance and sleeve friction can be used to estimate soil types. In addition, the cone tip resistance is also directly correlated to strength properties of soils.

With a recent form of cone penetrometer (i.e. piezocones), measurement can also be made of pore water pressure around the cone and shaft. This additional information may be used to correlate with soil stress history and consolidation coefficients (Lunne *et al.*, 1997).

8.2 SELF-BORING PRESSUREMETER TESTS IN CLAY

Over the last three decades, the self-boring pressuremeter has been established as one of the best *in-situ* tests for geotechnical investigation. The results of pressuremeter tests in soils are now widely used in many parts of the world to derive fundamental soil properties.

Almost all the theoretical interpretation methods developed for pressuremeters are based on the fundamental assumption that pressuremeter tests can be simulated as the expansion or/and contraction of an infinitely long, cylindrical cavity in soils. This assumption allows the analytical correlation to be developed between cavity expansion curves and soil properties. This theoretical correlation can then be used to backfigure the actual soil properties from measured pressuremeter curves.

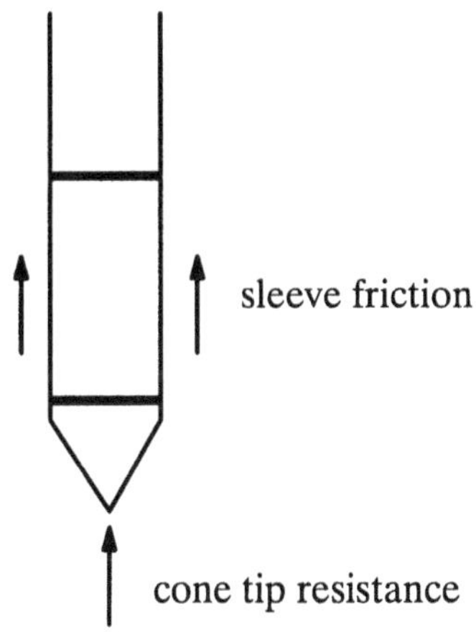

Figure 8.2: Definition of a cone penetrometer

The main soil properties that can be deduced from pressuremeter tests in clay include shear modulus, total horizontal *in-situ* stress, undrained shear strength, and the coefficient of horizontal consolidation. This section briefly describes how cavity expansion solutions are used to develop procedures for deriving these fundamental soil properties.

In the following description, all methods are based on the cylindrical cavity expansion assumption. However, they differ as several plasticity models are used to represent the stress-strain behaviour of soils. With one or two exceptions, most interpretation methods developed for clays are based on total stress formulations and therefore total rather than effective stresses are used in the analysis.

8.2.1 Shear modulus

As pointed out by Wroth (1982), one of the major uses of self-boring pressuremeters is to measure soil stiffness. If the pressuremeter results are presented in terms of cavity pressure ψ against cavity strain ε_c, the cylindrical cavity expansion solution suggests that the shear modulus of soil is equal to half the slope of an unloading-reloading loop, Figure 8.3.

If soil is assumed to behave as a linear elastic-perfectly plastic material, the shear modulus obtained from an unloading-reloading loop is not dependent on the magnitude of strain or pressure. In other words, the measured modulus does not depend on the depth and location of the loop. In reality, however, most soils do not behave exactly as a linear elastic plastic material, and as a result the pressuremeter modulus is a function of both pressure and unloading-reloading loop strain levels.

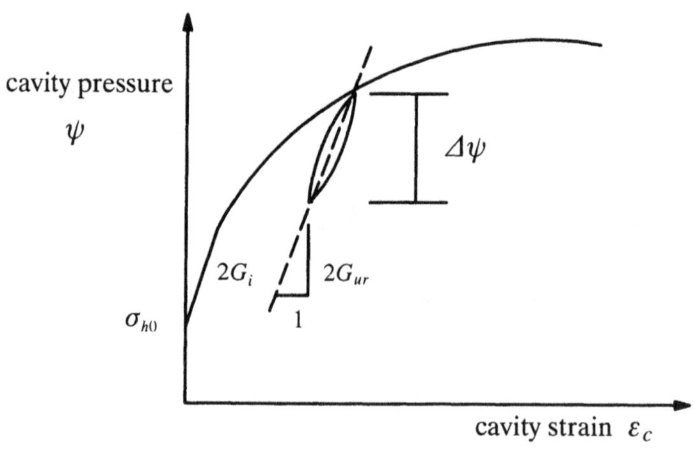

Figure 8.3: Shear modulus from pressuremeter curves

While conducting unloading-reloading loops, it is important to ensure that the loop remains in an elastic region. For an elastic-perfectly plastic Tresca soil, cavity expansion theory can be used to show that for an entirely elastic unloading, the maximum cavity pressure reduction must be less than the following value (Wroth, 1982):

$$(\Delta\psi)_{\max} = 2s_u \tag{8.1}$$

where s_u is the undrained shear strength of the soil.

Remarks

Due to the soil disturbance caused by the installation of a pressuremeter, the shear modulus obtained from the initial pressuremeter curve will be lower than that derived from an unloading-reloading loop in the plastic phase. As a result, the use of initial pressuremeter moduli may often lead to a conservative design.

8.2.2 *In-situ* total horizontal stress

One of the main reasons for the initial development of self-boring pressuremeters was to enable direct measurements of total horizontal stress to be made. If the self-boring pressuremeter is installed in the ground in such a way that there is little or no disturbance caused to the surrounding soil, the initial pressuremeter pressure corresponding to zero cavity strain should theoretically be equal to the *in-situ* total horizontal stress. This approach for determining the total horizontal stress is often termed the 'lift-off' method.

Apart from the 'lift-off' method, several other empirically based, curve-fitting methods have also been used to estimate the *in-situ* horizontal stress. However, most of these curve-fitting methods are developed mainly for application to push-in pressuremeters whereas the 'lift-off' approach is clearly not applicable due to the significant initial disturbance.

Remarks

The 'lift-off' approach for determining the *in-situ* total horizontal stress is based on the assumption that the pressuremeter probe is drilled into the ground without changing the magnitude of the *in-situ* horizontal stress. It therefore follows that prior to lift off there should be no movement of the inside pressuremeter membrane. After lift off (i.e. when the applied pressuremeter pressure is equal to or exceeds the *in-situ* horizontal stress), the membrane should start to expand. In practice, however, extreme care must be taken in order to obtain a reliable result. This is because most self-boring pressuremeter tests cause some degree of soil disturbance which tends to have a significant effect on the initial portion of the pressuremeter curves. As detailed in Clarke (1995), both the drilling techniques used and the system compliance of the probe have a considerable effect on the shapes of the initial portion of self-boring pressuremeter tests.

8.2.3 Undrained shear strength

Self-boring pressuremeters are frequently used to determine the undrained shear strength of clays. Most interpretation methods can be divided into two categories. First, a complete stress-strain relation is assumed for the soil based on which the theoretical pressuremeter curves can be obtained using the cylindrical cavity assumption, either analytically or numerically. By matching some major parts of the theoretical pressuremeter curves with those of a real pressuremeter test curve, the undrained shear strength of soils can be estimated. The interpretation methods of

Gibson and Anderson (1961), Jefferies (1988) and Yu and Collins (1998) are examples of this approach.

In the second approach, only a plastic flow rule is assumed for the clay (i.e. assuming the pressuremeter test is carried out under undrained conditions so that the soil is incompressible). As shown in Section 5.5, using the inverse cavity expansion solutions, the incompressible flow rule combined with a real pressuremeter curve can be used to derive the actual stress-strain relationship for the clay. The interpretation methods suggested by Palmer (1972), Baguelin *et al.* (1972) and Ladanyi (1972) belong to this second approach.

Shear strength derived from a total stress loading analysis

Gibson and Anderson (1961) were among the first to use cavity expansion theory to develop interpretation methods for deriving soil properties from pressuremeter test results. In the analysis of Gibson and Anderson (1961), the soil is assumed to behave as an elastic-perfectly plastic Tresca material. The pressuremeter test is idealised as the expansion of an infinitely long cylindrical cavity in soils under undrained conditions. For simplicity, a total stress analysis is used in the Gibson and Anderson method.

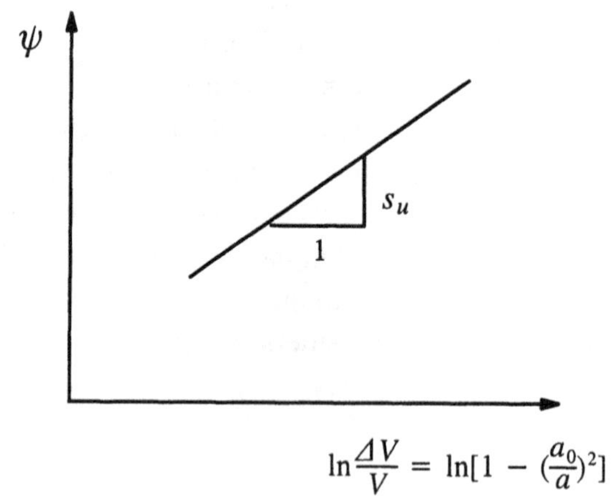

Figure 8.4: Graphical method using the analysis of Gibson and Anderson (1961)

By assuming the Tresca soil model, the analytical solution for the cavity expansion curve in the plastic stage is derived in Section 3.2.2 as follows:

$$\psi = \sigma_{h0} + s_u[1 + \ln(\frac{G}{s_u})] + s_u \ln\frac{\Delta V}{V} \tag{8.2}$$

where $\Delta V/V = (a^2 - a_0^2)/a^2$ is the volumetric strain; a and a_0 are current and initial cavity radii; ψ and σ_{h0} are total pressuremeter pressure and total *in-situ* horizontal stress; G and s_u are shear modulus and undrained shear strength of the soil.

The theoretical pressuremeter curve, as defined by equation (8.2), indicates that if the pressuremeter results are plotted in terms of the cavity pressure against the logarithm of the volumetric strain, the slope of the plastic portion (which should be a straight line) is equal to the undrained shear strength of the soil s_u. This interpretation method is graphically shown in Figure 8.4.

Shear strength derived from a total stress unloading analysis

To analyse pressuremeter unloading results, Jefferies (1988) and Houlsby and Withers (1988) independently extended the Gibson and Anderson method to include cavity unloading solutions. Jefferies (1988) derived the unloading solution for application to self-boring pressuremeter tests and therefore some small strain assumptions were used to simplify the mathematics. On the other hand, Houlsby and Withers (1988) used the cavity contraction solution to develop an interpretation method for cone pressuremeters where a large strain assumption is necessary.

As this section is concerned with the interpretation of self-boring pressuremeter tests, the cavity unloading solution derived by Jefferies (1988) is used to correlate the pressuremeter unloading curve with the undrained shear strength. Assuming a pressuremeter has been expanded in a clay to a plastic phase with a total cavity pressure of ψ_{max}, the pressuremeter is then unloaded slowly. Initially, the pressuremeter unloading curve is linear elastic which is followed by a nonlinear plastic unloading response. Using a small strain assumption, the following cavity contraction relationship may be derived:

$$\psi = \psi_{max} - 2s_u[1 + \ln(\frac{G}{2s_u})] - 2s_u \ln[\frac{a_{max}}{a} - \frac{a}{a_{max}}] \qquad (8.3)$$

where a_{max} is the cavity radius at the end of loading stage, and a is cavity radius at any stage of pressuremeter unloading; G and s_u are shear modulus and undrained shear strength of the soil.

The theoretical pressuremeter unloading curve, defined by (8.3), suggests that if the pressuremeter unloading results are presented as the cavity pressure ψ versus $-\ln[a_{max}/a - a/a_{max}]$, the slope of the plastic unloading portion (which is a straight line) is equal to twice the undrained shear strength of the soil $2s_u$. This simple interpretation method, that has not yet been widely used in practice, is graphically shown in Figure 8.5.

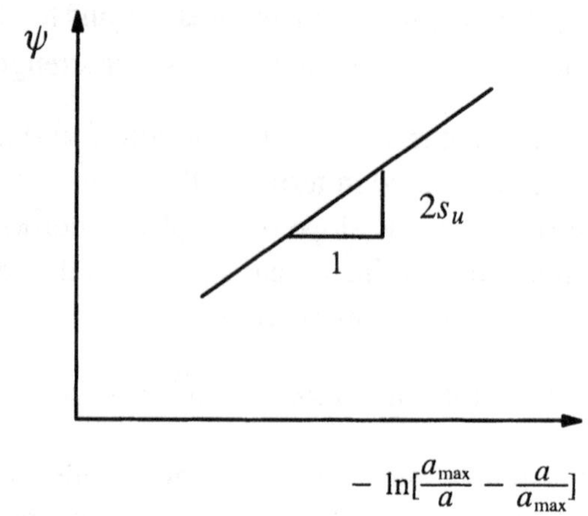

Figure 8.5: Graphical method using the unloading analysis of Jefferies (1988)

Shear strength derived from an effective stress loading analysis

The self-boring pressuremeter test in clay is usually interpreted using undrained cavity expansion theory in terms of total stress. This is reasonably accurate for normally consolidated or lightly overconsolidated clays where the shear resistance of the soil does not change significantly during the undrained pressuremeter test. For heavily overconsolidated soils, however, the shear resistance may vary considerably with deformation history and this cannot be accounted for by total stress analysis with a perfectly plastic soil model.

This section presents an effective stress analysis of the self-boring pressuremeter test in undrained clays that was originally developed by Yu and Collins (1998). A critical state soil model was used in the analysis so that the dependence of soil strength with effective stresses is properly taken into account. Following the Gibson and Anderson procedure of deriving undrained shear strength from the self-boring pressuremeter test, a theoretical correlation between the ratio of pressuremeter strength to triaxial undrained shear strength and OCR of the soil was established. This correlation can then be used to properly deduce the actual undrained shear strength of the soil from the apparent undrained shear strength derived from self-boring pressuremeter tests in overconsolidated clays.

Until recently, the analysis of cavity expansion problems in clay has been restricted to the *total stress* analysis (Hill, 1950; Gibson and Anderson, 1961; Palmer, 1972; Houlsby and Withers, 1988). In fact, in almost all the analyses of the undrained pressuremeter tests, the behaviour of soils has been modelled by an elastic-perfectly plastic model. In the analysis of undrained deformations, it is common to work with total stress. However, this is no longer appropriate in models in which the strength of the soil is a variable, since the strength is a function of the effective stress rather than the total stress. In particular, unlike the effective stress approach, the total stress analysis does not take account of the influence of soil stress history on the behaviour of the soil. As a result, these interpretation methods may only be used to analyse the pressuremeter test in normally or lightly overconsolidated clays where the soil strength does not vary significantly with loading history. For heavily overconsolidated clays, the methods based on the total stress analysis are not strictly correct as the variation of soil strength with effective stresses is ignored.

Although Carter *et al.* (1979) and Randolph *et al.* (1979) have used finite element methods to study the expansion of cavities in critical state soils, the yield surface of the modified Cam clay model used by these authors is considered to be unsuitable for modelling heavily overconsolidated clays as it tends to overestimate the soil strength significantly. Recently, Collins and Yu (1996) presented analytical solutions for undrained expansion of cavities in both normally and overconsolidated clays using a variety of critical state soil models. These analytical solutions for cavities expanding from a finite initial radius can be used to analyse the self-boring pressuremeter test in soil.

In the present effective stress analysis of the self-boring pressuremeter test in overconsolidated clays, the pressuremeter test is simulated as an undrained cylindrical cavity expansion process. The soil is modelled using the critical state theory (Schofield and Wroth, 1968; Atkinson and Bransby, 1978; Muir Wood, 1990).

As presented in Chapter 4, Collins and Yu (1996) derived large strain analytical solutions for cavity expansion in soils with various critical state plasticity models. In this section, the combined Cam clay and Hvorslev yield surface shown in Figure 4.5 is used in the analysis of the self-boring pressuremeter tests in clays. This is because this combined yield surface adequately models the stress-strain behaviour of both normally consolidated and heavily overconsolidated clays (see Collins and Yu, 1996 for more discussion and detailed formulations). The variables q and p' denote the effective shear and mean stresses respectively. The definitions for the critical state soil properties $N, M, \varkappa, \lambda$ are well known and can also be easily found in Schofield and Wroth (1968), Atkinson and Bransby (1978) or Muir Wood (1990).

The values of the critical state parameters chosen are those relevant to London clay: $\Gamma = 2.759$, $\lambda = 0.161$, $\varkappa = 0.062$, the critical state friction angle $\phi'_{cs} = 22.75°$ and the Hvorslev friction angle $\phi'_{hc} = 19.7°$ (Atkinson and Bransby, 1978; Muir Wood, 1990). The Poisson's ratio μ is assumed to be 0.3. If the critical state friction angle of the soil is assumed to be the same for both the triaxial and plane strain loading conditions (see Muir Wood, 1990 for experimental evidence and further discussion), the values of M and h for cylindrical cavity expansion can be determined using $M = 2\sin\phi'_{cs}$ and $h = 2\sin\phi'_{hc}$ respectively.

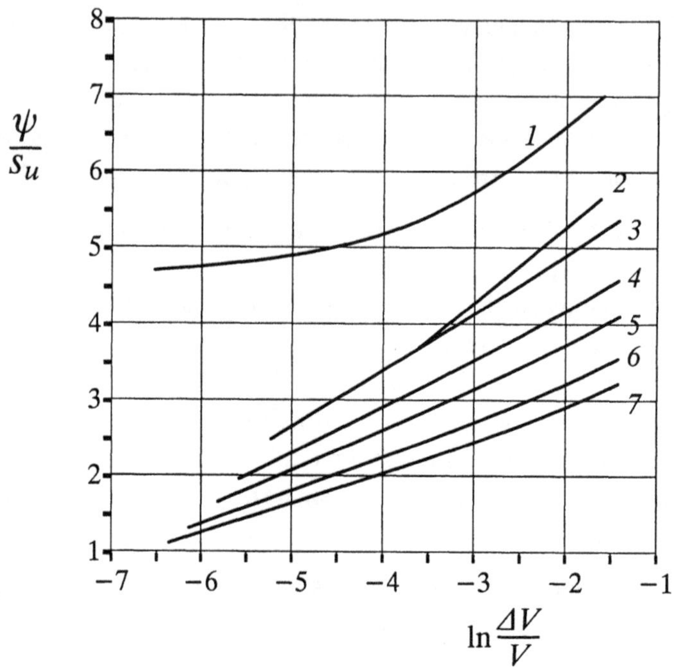

Figure 8.6: Pressuremeter loading curves for overconsolidation ratios

$n_p = 1.001$(curve1), 2.5(curve2)

and 5(curve3), 7.5(curve4), 10(curve5), 15(curve6), 20(curve7)

The first set of results relates to self-boring pressuremeter tests in London clay with an overconsolidation ratio of $n_p = 5$ and three different values of initial specific volume $v = 1.5,\ 2.0,\ 2.5$. These results, as presented in Yu and Collins (1998), show that although the initial specific volume has some effect on the location of the pressuremeter curves, it has very little effect on the slope of pressuremeter curves. In other words, the initial specific volume has no influence on the undrained shear strength derived from a pressuremeter test. For this reason, the following discussion will be based on the results obtained when the specific volume of the soil v is equal to 2.0.

Results showing the pressuremeter curves for seven different overconsolidation ratios of $n_p = 1.001,\ 2.5,\ 5,\ 7.5,\ 10,\ 15,\ 20$ are presented in Figure 8.6. Note that only the plastic part of the pressuremeter curve is presented in the figure. The reason for using $n_p = 1.001$ to represent a normally consolidated clay is that when

$n_p = 1$ the shear strain required to reach the yield surface is zero which causes the radius of the elastic-plastic boundary to be indeterminate (see Chapter 4 for details). The pressuremeter pressure (i.e. cavity pressure) has been normalized by the theoretical triaxial undrained shear strength of the soil which is related to the soil properties by $s_u = 0.5M\exp((\Gamma - v)/\lambda)$. The plastic portion of the pressuremeter curve for $n_p = 2.5$ is shorter than those for other values of the overconsolidation ratio n_p. This is because the case $n_p = 2.5$ is very close to the situation where under undrained loading conditions the soil behaves purely elastically before reaching the critical state (i.e. $n_p = 2.718$ corresponds to the critical state for the original Cam clay model). As a result, the elastic cavity strain attained before the plastic portion of the pressuremeter curve is larger than that for the other values of overconsolidation ratio n_p used in the calculation.

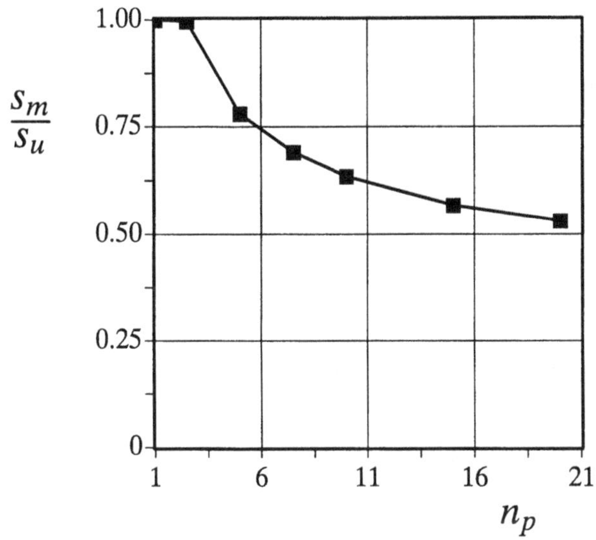

Figure 8.7: Ratio of pressuremeter strength s_m to triaxial undrained shear strength s_u versus n_p

The interpretation procedure of Gibson and Anderson (1961) is used to derive undrained shear strength s_m from the pressuremeter curves in the range of cavity strains between 5–15%. The application of Gibson and Anderson's method to the theoretical pressuremeter curves (shown in Figure 8.6) obtained from the present effective stress analysis suggests that for normally consolidated and lightly overconsolidated clays, the derived pressuremeter undrained shear strength is equal or close to the theoretical undrained shear strength s_u used in the calculation. For heavily overconsolidated clays, however, the shear strength derived from the pressuremeter curves is significantly less than the actual value with the difference increasing with the value of overconsolidation ratio. The variation of the ratio of pressuremeter to actual undrained shear strengths with the value of overconsolidation ratio n_p is presented in Figure 8.7.

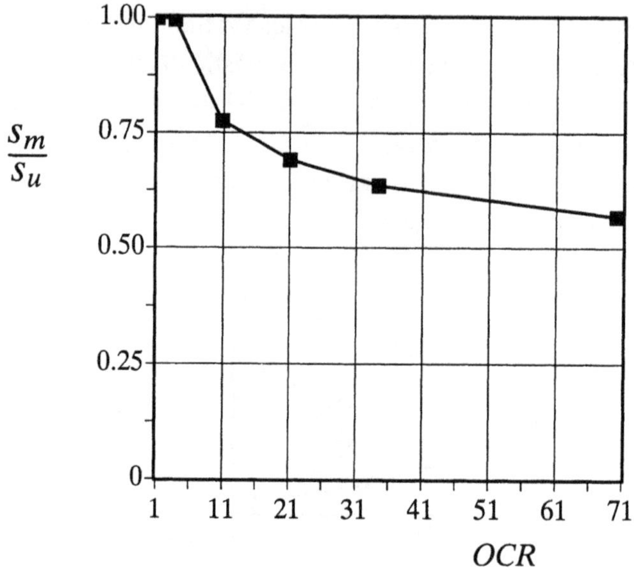

Figure 8.8: Ratio of pressuremeter strength to triaxial undrained shear strength versus OCR

As presented in Collins and Yu (1996), the overconsolidation ratio n_p is defined in terms of the mean effective stress. In practice, the overconsolidation ratio often refers to the one dimensional definition, OCR, that is in terms of the vertical effective stress. As discussed by Muir Wood (1990), the overconsolidation ratio n_p can be converted to the one-dimensional overconsolidation ratio OCR. Although the relationship between n_p and OCR is slightly dependent on the value of the coefficient of earth pressure at rest for a normally consolidated sample K_{onc}, the average correlation may be adequately represented by :

$$\frac{OCR}{n_p} = \frac{4 + n_p + \sqrt{8n_p + {n_p}^2}}{8} \tag{8.4}$$

With equation (8.4), the result presented in Figure 8.7 is re-plotted in Figure 8.8 in terms of the ratio of pressuremeter to actual undrained shear strengths against OCR. It is interesting to note from Figure 8.7 and Figure 8.8 that for heavily overconsolidated clays the use of the total stress analysis of Gibson and Anderson (1961) significantly underestimates the actual shear strength.

The application of the present effective analysis to London clay suggests that although Gibson and Anderson's total stress analysis may be adequately used to analyse undrained pressuremeter tests in normally consolidated and lightly overconsolidated clays, it is not suitable for interpreting the results for tests in heavily overconsolidated clays. Because of the significant effect of pressuremeter geometry, a reliable interpretation method must take account of the pressuremeter geometry as well as use an effective stress analysis with a realistic critical state soil model.

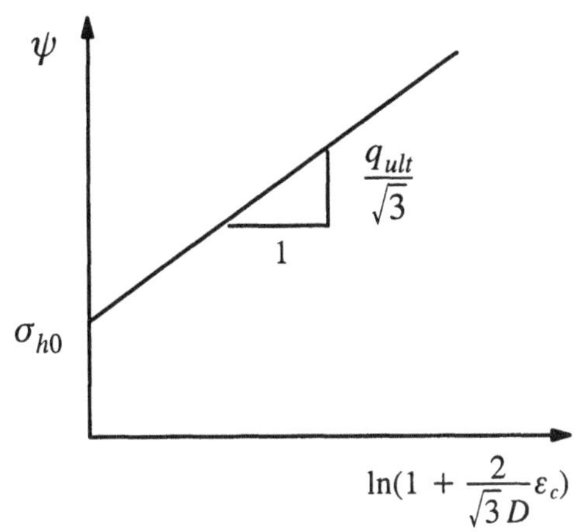

Figure 8.9: Graphical method using the strain hardening analsyis of Provest and Hoeg (1975)

Shear strength derived from total stress analysis with a hyperbolic soil model

As detailed in Chapter 5, neglecting elastic deformation results in a small strain closed form solution for the undrained expansion of a cylindrical cavity in soils modelled by a hyperbolic stress-strain model (Prevost and Hoeg, 1975).

According to the theoretical solution, if the soil stress-strain relation for a strain hardening soil can be represented by

$$q = \frac{\gamma^p}{D + \gamma^p} q_{ult} \tag{8.5}$$

where, as defined in Chapter 5, q is the shear stress; γ^p is the plastic shear strain; D is a material constant and the second material constant q_{ult} is the ultimate shear stress. It then follows that the pressuremeter loading curve can be described as a function of the soil parameters D and q_{ult}:

$$\psi = \sigma_{h0} + \frac{q_{ult}}{\sqrt{3}} \ln(1 + \frac{2}{\sqrt{3}D}\varepsilon_c) \tag{8.6}$$

where $\varepsilon_c = (a - a_0)/a_0$ is the cavity strain. In practice, the material constant D may be easily chosen for a given soil. If this is the case, pressuremeter loading curves may be used to estimate the ultimate shear strength of the soil (i.e. residual shear strength). This can be achieved by plotting the pressuremeter loading results in terms of cavity pressure versus cavity strain as shown in Figure 8.9.

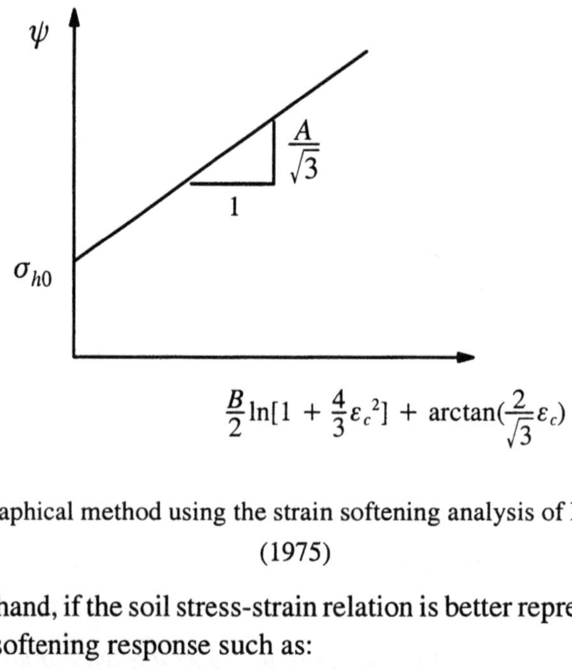

Figure 8.10: Graphical method using the strain softening analysis of Provest and Hoeg (1975)

On the other hand, if the soil stress-strain relation is better represented by a strain hardening and softening response such as:

$$q = A\frac{B(\gamma^p)^2 + \gamma^p}{1 + (\gamma^p)^2} \tag{8.7}$$

where A and B are two soil properties, the theoretical pressuremeter loading curve may be obtained as follows:

$$\psi = \sigma_{h0} + \frac{A}{\sqrt{3}}\left\{\frac{B}{2}\ln[1 + (\frac{2}{\sqrt{3}}\varepsilon_c)^2] + \arctan(\frac{2}{\sqrt{3}}\varepsilon_c)\right\} \tag{8.8}$$

The above plastic pressuremeter curve is graphically shown in Figure 8.10. As a result, if the constant B is known then the pressuremeter test can be used to determine the constant A using Figure 8.10.

Stress-strain curves derived from pressuremeter loading tests

The inverse cavity expansion in an undrained clay was first studied by Palmer (1972), Baguelin *et al.* (1972) and Ladanyi (1972). A detailed description of the solution procedure for both undrained and drained cavity expansion problems is given in Section 5.5 of Chapter 5.

For undrained clay, the following definition is assumed for soil shear strength:

$$s_u(\varepsilon) = \frac{\sigma_r - \sigma_\theta}{2} \tag{8.9}$$

where σ_r and σ_θ are the radial and hoop stresses respectively. Note the undrained shear strength $s_u(\varepsilon)$ in equation (8.9) is not a constant, but rather is an unknown function of the strain ε.

Combining equation (8.9) and the incompressibility condition leads to the following cavity expansion curve:

$$\psi = \sigma_{h0} + \int_0^{\varepsilon_c} \frac{s_u(\varepsilon)}{\varepsilon} d\varepsilon \tag{8.10}$$

where ψ is the cavity pressure. The above equation can be used to give the following function for the undrained shear strength:

$$s_u(\varepsilon_c) = \varepsilon_c \frac{d\psi}{d\varepsilon_c} \tag{8.11}$$

in which the derivative $d\psi/d\varepsilon_c$ is readily obtained as the cavity expansion curve $\psi - \varepsilon_c$ is known from pressuremeter tests.

8.2.4 Consolidation coefficient

Another soil property that can be estimated with a self-boring pressuremeter capable of measuring pore pressures is the horizontal consolidation coefficient c_h. The method followed in such a measurement is the so-called 'holding test' proposed by Clarke *et al.* (1979). When a pressuremeter is expanded in a clay under undrained conditions, excess pore pressures are generated in the surrounding soil in which soils are deforming plastically. If at this stage the diameter of the cavity is held constant, relaxation of soil is observed by the decrease of the measured excess pore pressure and the total cavity pressure. If the total pressure is held constant, relaxation occurs in the decrease of the measured excess pore pressure and the continuing increase in cavity diameter.

It can be shown (Mair and Wood, 1987) that at any plastic stage of pressuremeter loading tests, the excess pore pressure at the cavity wall is linked to the cavity volumetric strain by:

$$U = s_u \ln(\frac{G}{s_u}) + s_u \ln\frac{\Delta V}{V} \tag{8.12}$$

If the cavity radius is held constant, the excess pore pressure dissipates. The consolidation coefficient may be estimated using a dimensionless time factor $T_{50} = c_h t_{50}/a^2$, where t_{50} is actual time taken for the excess pore pressure to fall to half its maximum value.

Using an analytical method, Randolph and Wroth (1979) derived a relationship between the normalised maximum excess pore pressure U_{max}/s_u and the time factor T_{50}. This correlation is shown in Figure 8.11.

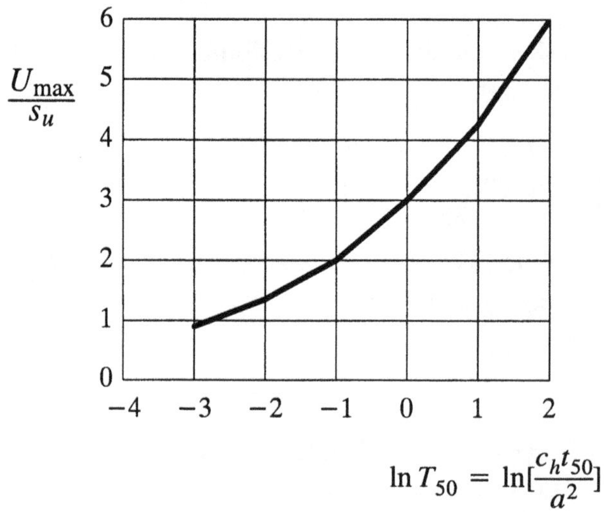

Figure 8.11: Theoretical correlation between maximum excess pore pressure and time factor by Randolph and Wroth (1979)

With the actual time t_{50} and normalised maximum excess pore pressure at the cavity wall measured in pressuremeter holding tests, the correlation shown in Figure 8.11 may be used to determine the horizontal consolidation coefficient c_h.

8.2.5 Effects of finite pressuremeter length and initial stress state

Effect of finite pressuremeter length

The interpretation methods presented so far assume that the pressuremeter is infinitely long so that it can be simulated as a cylindrical cavity expansion process. Most self-boring pressuremeters have a length that is about 6 times the size of their diameters. As a result, the two dimensional effects in the pressuremeter tests may be significant. The most convenient way of determining such an effect is to use axisymmetric finite element analysis that accounts for the actual geometry of a pressuremeter.

Such numerical studies have been carried out by Yu (1990), Yeung and Carter (1990), Houlsby and Carter (1993), and Charles *et al.* (1999). While all these studies suggest that ignoring finite pressuremeter length will significantly overestimate undrained shear strength, the details of these results differ slightly. This is mainly due to the fact that different soil models and computer programs were used by these authors. For example, the study carried out by Yu (1990) suggested that the following expression may be used to represent the finite pressuremeter length effects on the derived undrained shear strength:

$$\alpha = \frac{s_u}{s_u^6} = 1\text{–}0.02\ln\frac{G}{s_u^6} \tag{8.13}$$

where s_u and s_u^6 are undrained shear strengths derived from pressuremeters with the length to diameter ratios of ∞ and 6 respectively. G is the shear modulus of soil. The undrained shear strength of soil can be estimated by multiplying the undrained shear strength s_u^6 derived from a self-boring pressuremeter test with a correction factor α given in (8.13).

The effective stress analysis reported by Charles *et al.* (1999) on pressuremeter geometry effects in undrained clay modelled by various critical state soil models is an advanced numerical study. The study concluded that the overestimation of strength predicted by the total stress analysis with perfectly plastic soil models tended to be higher than that predicted by an effective stress analysis. It was also found that the effect of pressuremeter geometry decreased rapidly with increasing values of OCR.

As pointed out by Clarke (1993), experience with the use of Gibson and Anderson's method suggests that in soft clays (normally consolidated clays) the pressuremeter tends to give higher values of undrained shear strength when compared with those from triaxial tests. On the other hand, in stiff overconsolidated clays (such as London clay) the results from the pressuremeter are similar to triaxial results. Since ignoring the finite pressuremeter length tends to overestimate the shear strength significantly (Yu, 1990; Yu, 1993; Yeung and Carter, 1990; Houlsby and Carter, 1993), it is logical that the pressuremeter would give a higher shear strength which is the case for soft clays. Using total stress analysis, it is expected that a similar pattern for stiff clays will be obtained, but in reality the pressuremeter shear strength is similar to the triaxial undrained shear strength in stiff clays. To provide a theoretical explanation for this difference, the analytical solutions for undrained cavity expansion in a critical state soil developed by Collins and Yu (1996) can be used to analyse the results of pressuremeter tests in both soft and stiff clays.

To explain why the shear strength derived from a pressuremeter test is similar to triaxial results for stiff clays (i.e. overconsolidated clays), but significantly higher than triaxial strength for soft clay (i.e. normally consolidated or lightly overconsolidated clays), it is useful to recall recent research on the effect of pressuremeter geometry on undrained shear strength derived from a pressuremeter test in clay. Recent theoretical research, carried out by Yu (1990, 1993) and Houlsby and Carter (1993) using finite element methods, suggested that neglecting the pressuremeter geometry tended to give much higher values of undrained shear strength. The overestimation increases with increasing value of stiffness index I_r (which is defined as the ratio of shear modulus G to undrained shear strength). For stiffness indices of I_r=100–500, the undrained shear strength was overestimated by about 20 to 40% if the actual pressuremeter geometry was not accounted for. As reported by Bond and Jardine (1991), the one dimensional overconsolidation ratio of London clay was in the range of OCR=20–50. From the result presented in Figure 8.8, it can be concluded that the use of the total stress analysis of Gibson and Anderson (1961) would underestimate the undrained shear strength of London clay by about 30–40%. If both the effect of pressuremeter geometry (overestimating undrained shear strength) and the effect of OCR (underestimating undrained shear strength) are taken into account, the net effect is the pressuremeter shear strength from the total stress analysis of Gibson and Anderson (1961) would give similar strength values to those from triaxial tests for heavily overconsolidated clay. For a normally and lightly overconsolidated clay, however, the effect of OCR is negligible (see Figure 8.8). Therefore the method of Gibson and Anderson (1961) overestimates the triaxial shear strength as it ignores the effect of pressuremeter geometry. As discussed by Clarke (1993), this is the general trend observed in recent years with the use of the self-boring pressuremeter test in both soft and stiff clays.

Effect of initial stress states

All analytical pressuremeter interpretation methods described in the previous sections are based on cavity expansion in soils with an isotropic *in-situ* stress state. In other words, the initial horizontal stress is assumed to be equal to the vertical stress. In reality, however, this is unlikely to be the case. It is necessary to assess the accuracy of the pressuremeter interpretation methods for soils with anisotropic initial stress states.

To quantify the effects of *in-situ* stress states on the soil properties derived from the self-boring pressuremeter tests, the finite element program CAVEXP developed by the author (Yu, 1990, 1994, 1996) is used. The detailed formulation used in CA-

VEXP has been briefly described in Chapter 7. A finite element parametric study carried out by the author suggested that the initial stress state has negligible effects on the undrained shear strength derived from self-boring pressuremeter tests.

8.3 SELF-BORING PRESSUREMETER TESTS IN SAND

Like tests in clay, the self-boring pressuremeter test in sand can be used to measure shear modulus and total horizontal *in-situ* stress. In addition, the soil strength (e.g. the angles of friction and dilation) and *in-situ* state parameters can also be estimated from the results of drained self-boring pressuremeter tests in sand.

In this section, some of the major interpretation procedures developed using cavity expansion solutions are presented. As self-boring pressuremeter testing assumes zero installation disturbance, only small strain cavity expansion solutions are needed in the interpretation. It is noted that all the analyses for self-boring pressuremeter tests in sand are based on the assumption that tests are carried out slowly so that the fully drained condition is likely to be valid. All the stress or pressure variables (e.g. cavity pressure) referred to in this section are effective stresses.

8.3.1 Shear modulus

As for the case in clays, if the pressuremeter results are presented in terms of effective cavity pressure ψ' (i.e. the difference of cavity pressure and initial pore water pressure) against cavity strain ε_c, the cylindrical cavity expansion solution suggests that the shear modulus of soil G is equal to half the slope of an unloading-reloading loop. As most soils do not behave exactly as a linear elastic plastic material, the pressuremeter modulus is a function of both pressure and unloading-reloading loop strain levels. In other words, the measured modulus does not depend on the location where and how deep the loop is made. While carrying out unloading-reloading loops, it is important to ensure that the loop remains in an elastic region. For an elastic-perfectly plastic Mohr-Coulomb soil, cavity expansion theory can be used (Wroth, 1982) to show that for an entirely elastic unloading the maximum cavity pressure reduction must be less than the following value:

$$(\Delta\psi')_{\max} = \frac{2\sin\phi'}{1+\sin\phi'}\psi'_{un} \tag{8.14}$$

where ψ'_{un} is the effective cavity pressure at the start of pressuremeter unloading, and ϕ' is the drained angle of internal friction.

Because of inevitable soil disturbance due to the installation of a pressuremeter in sand, the shear modulus obtained from the initial pressuremeter curve is likely to be unreliable and is therefore not recommended for use in geotechnical design.

8.3.2 *In-situ* total horizontal stress

If a self-boring pressuremeter is installed in the ground with no or little disturbance to the surrounding soil, the initial pressuremeter pressure corresponding to zero cavity strain should theoretically be equal to the *in-situ* total horizontal stress.

In practice, however, extreme care must be taken in order to obtain a reliable result. Limited experience to date seems to indicate that the *in-situ* total horizontal stress σ_{h0} in sand estimated from the 'lift off' approach tends to be too low. This is because self-boring pressuremeter tests in sand tend to cause soil disturbance. Although the detection of 'lift off' is the only reasonable way of estimating σ_{h0} in sand, the ability of self-boring pressuremeters to measure σ_{h0} is not yet fully proven (see, for example, Mair and Wood, 1987; Clarke, 1995). Improvement in the ability of estimating σ_{h0} depends on further improvement in drilling and installation techniques.

8.3.3 Drained shear strength

For drained pressuremeter tests, the Gibson and Anderson analysis has been modified by Hughes *et al.* (1977) to account for the effect of soil dilation during the test. To develop a simple closed form solution, Hughes *et. al.* (1977) assumed that both the friction angle and dilation angle were constant during the pressuremeter test. From the analysis, the angles of internal friction and dilation, ϕ' and ν' were deduced from the results of a pressuremeter test. Although the assumption of a perfectly–plastic model has proven to be reasonable for very dense sands (Jewell *et. al.*, 1980; Fahey, 1980), it is less satisfactory for medium and loose sands. A relatively new drained analysis, which is similar to Palmer's undrained analysis, has been developed by Manassero (1989). With Manassero's analysis, a stress ratio-shear strain relationship can be derived from the pressuremeter test provided a plastic flow rule is assumed.

Shear strength derived from pressuremeter loading tests

Hughes *et al.* (1977) developed a small strain cavity expansion solution that can be used to deduce the angles of soil friction and dilation from the pressuremeter loading test results. In the analysis of Hughes *et al.*, the sand is assumed to behave as an elastic-perfectly plastic Mohr-Coulomb material. The pressuremeter test is idealised as the expansion of an infinitely long cylindrical cavity in soils under fully drained conditions.

By ignoring elastic deformation in the plastically deforming zone, the analytical solution for the cavity expansion curve in the plastic stage can be approximated as follows:

$$\ln \psi' = s \ln \varepsilon_c + A \tag{8.15}$$

where $s = (1 + \sin\nu')\sin\phi'/(1 + \sin\phi')$ and A is a constant.

The theoretical pressuremeter curve, as defined by (8.15), shows that if the pressuremeter results are plotted as the effective cavity pressure versus the volumetric strain in logarithmic scales, the slope of the plastic portion (which is a straight line) is equal to s, that is a function of the angles of soil friction and dilation, see Figure 8.12.

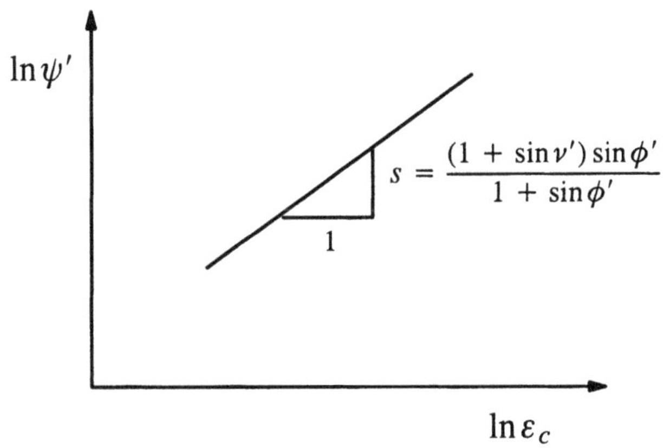

Figure 8.12: Graphical method using the analysis of Hughes *et al.* (1977)

To obtain the values of ϕ' and ν', further information is required which may be provided by Rowe's stress-dilatancy relation between the angles of soil friction and dilation as well as the critical state friction angle ϕ'_{cv}. The latter quantity ϕ'_{cv} is a constant for a given sand which can be easily measured from disturbed samples.

The use of Rowe's stress dilatancy law leads to the following expressions for determining the angles of soil friction and dilation from the measured pressuremeter slope s and soil critical state friction angle ϕ'_{cv}:

$$\sin\phi' = \frac{s}{1 + (s-1)\sin\phi'_{cv}} \tag{8.16}$$

$$\sin\nu' = s + (s-1)\sin\phi'_{cv} \tag{8.17}$$

Shear strength derived from pressuremeter unloading tests

Using the same assumptions, Houlsby *et al.* (1986) extended the analysis of Hughes *et al.* (1977) to interpret the pressuremeter unloading test results. In the analysis of Houslby *et al.*, the sand is assumed to behave as an elastic-perfectly plastic Mohr-

Coulomb material. Elastic deformation in the plastically deforming region was again neglected. More recently, a much more rigorous unloading analysis that takes into account of both large strains and elastic deformation in the plastic region was developed by Yu (1990). The results of a comparison given by Yu and Houlsby (1995) indicate that while the effect of large strains is considerable, the influence of elastic deformation in the plastic zone on the slope of cavity unloading curves is very small.

The small strain unloading analysis of Houlsby *et al.* (1986) is summarized in Figure 8.13, which shows that the plastic unloading slope in a plot of $\ln \psi'$ against $-\ln[(\varepsilon_c)_{max} - \varepsilon_c]$ is primarily controlled by the soil strength parameters and to a small extent by soil stiffness. Note that $(\varepsilon_c)_{max}$ is the maximum cavity strain at the start of unloading tests. The slope s_d of the pressuremeter unloading curve in this plot can be approximated as a function of the angles of soil friction and dilation as follows:

$$s_d = -\frac{N - \frac{1}{N}}{N + \frac{1}{n}} \tag{8.18}$$

where $N = (1 - \sin\phi')/(1 + \sin\phi')$ and $n = (1 - \sin\nu')/(1 + \sin\nu')$.

Once again, the use of Rowe's stress dilatancy law can be used to derive the following expressions for determining the angles of soil friction and dilation from measured pressuremeter slope s and soil critical state friction angle ϕ'_{cv}:

$$\sin\phi' = m - \sqrt{m^2 - 1} \tag{8.19}$$

$$\sin\nu' = \frac{\sin\phi' - \sin\phi'_{cv}}{1 - \sin\phi' \sin\phi'_{cv}} \tag{8.20}$$

where m is given by:

$$m = \sin\phi'_{cv} + \frac{1 + \sin\phi'_{cv}}{s_d} \tag{8.21}$$

Limited experience in using the unloading analysis of Houlsby *et al.* (1986) suggests that the angles of friction and dilation obtained using this approach tend to be much lower than those measured from other tests and interpretation methods (Withers *et al.*, 1989; Houlsby and Yu, 1990; Yu and Houlsby, 1995). This is probably due to the fact that the elastic-perfectly plastic Mohr-Coulomb model is grossly unrealistic for modelling the complex behaviour of sand during unloading conditions. As a result, this method has not been widely used in practice.

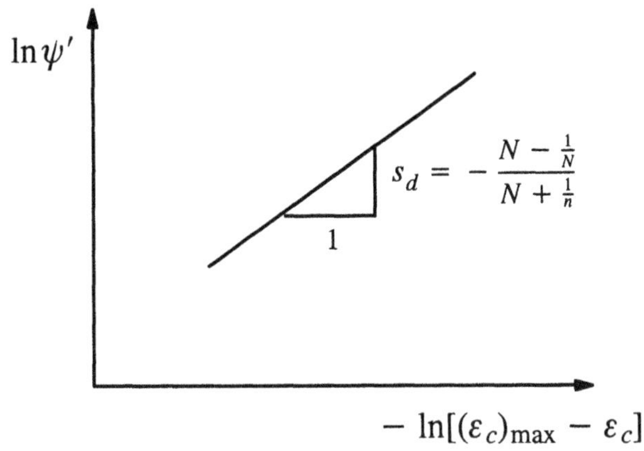

Figure 8.13: Graphical method using the analysis of Houlsby *et al.* (1986)

Stress-strain curves derived from pressuremeter loading tests (Manassero, 89)

The interpretation methods described so far are developed by assuming a complete soil stress-strain relation for sand. As in clays, an alternative approach is to solve the pressuremeter test as an inverse cavity expansion problem. In this second approach, by assuming a plastic flow rule for sand, the pressuremeter loading curve can be used to deduce a complete shear stress-strain curve for soils.

The undrained inverse cavity expansion analysis developed by Palmer (1972) may also be applied to cavity expansion problems in sand under fully drained conditions. Because of dilation, the analysis for sand is not as simple as for undrained clay. In particular, no closed form solutions can be obtained and instead a finite difference method is needed to obtain the stress-strain relationship from a given cavity expansion curve. The pressuremeter analysis using this approach has been derived by Manassero (1989) and Sousa Coutinho (1989).

For dilatant sands, the relationship between strains may be assumed to be related by an unknown function *f*:

$$\varepsilon_r = f(\varepsilon_\theta) \tag{8.22}$$

with a condition that $\varepsilon_r = f = 0$ when $\varepsilon_\theta = 0$. The function *f* must be determined.

Equation (5.165) can be used to give:

$$\frac{d\varepsilon_r}{d\varepsilon_\theta} = f' \tag{8.23}$$

where f' denotes the derivative of *f* with respect to ε_θ.

The strain compatibility condition can be written as:

$$\frac{d\varepsilon_\theta}{dr} = \frac{\varepsilon_r - \varepsilon_r}{r} \tag{8.24}$$

Using equation (5.165), the above equation gives:

$$\frac{dr}{r} = \frac{d\varepsilon_\theta}{f - \varepsilon_\theta} \tag{8.25}$$

By ignoring elastic deformation, Rowe's stress-dilatancy relationship can be written as follows:

$$\frac{\sigma_r}{\sigma_\theta} = -K\frac{d\varepsilon_\theta}{d\varepsilon_r} \tag{8.26}$$

where $K = (1 + \sin\phi_{cv})/(1 - \sin\phi_{cv})$ and ϕ_{cv} is the critical state friction angle.

For the cylindrical cavity, the total stresses have to satisfy the following equilibrium equation:

$$r\frac{d\sigma_r}{dr} + \sigma_r - \sigma_\theta = 0 \tag{8.27}$$

which can be rewritten as:

$$d\sigma_r = -(\sigma_r - \sigma_\theta)\frac{dr}{r} \tag{8.28}$$

By using equations (5.168) and (5.169), the above equation becomes:

$$\frac{1}{\sigma_r} \times \frac{d\sigma_r}{d\varepsilon_\theta} = -\frac{1 + \frac{f'}{K}}{f - \varepsilon_\theta} \tag{8.29}$$

Unfortunately, the above equation cannot be integrated analytically. But when applying it at the cavity wall, the finite difference method can be used to solve for a numerical function f and therefore the relationship between ε_r and ε_θ. This is possible because at the cavity wall both σ_r and the derivative $d\sigma_r/d\varepsilon_\theta$ are given.

Once the function f is determined, all other variables (i.e., stresses and displacement) can be easily determined using the equations presented before. It is noted that in the above analysis, small strain is assumed and more importantly, elastic deformation in the plastically deforming zone is neglected. As illustrated in Yu and Houlsby (1991), however, this latter assumption may have a significant effect on the derived soil properties derived from self-boring pressuremeter tests in sand.

Like Palmer's undrained analysis for clays, limited experience suggests that Manassero's drained pressuremeter analysis seems to be very sensitive to initial conditions (e.g. possible disturbance during installation, uncertainty concerning the true reference condition).

8.3.4 State parameter

Since its introduction (Been and Jefferies, 1985), the state parameter has found many applications in geotechnical engineering. This is because the state parameter combines the effects of both relative density and stress level, and thus represents an important step forward as compared to the relative density concept in characterising sand behaviour.

It has been demonstrated that many commonly used sand properties, such as angles of friction and dilation, normalise quite well to the state parameter. This is the utility of the concept to the practising engineer. The practical application of the state parameter concept is dependent on the ability to measure the state parameter *in-situ*. This section presents methods that can be used to determine the initial sand state from self-boring pressuremeter test results.

Soil state derived from pressuremeter loading tests

Yu (1994) developed an interpretation method by which the results of the self–boring pressuremeter test can be correlated with the initial sand state. It was found that for a particular sand, a linear correlation exists between the pressuremeter loading slope s (see Figure 8.12) and the initial state parameter of the soil. In addition, this correlation was found to be largely independent of initial stress state and soil stiffness, and was therefore considered to be unique.

The soil model used in Yu (1994) was a state parameter based critical state model. A brief description of this model can be found in Chapter 7 (see section 7.2.2 (d)). With this strain hardening/softening sand model, no closed form solution is available for cavity expansion problems. As a result, a numerical method is needed. In the study presented by Yu (1994), the finite element program, CAVEXP, has been used to simulate the self-boring pressuremeter test as the expansion of a cylindrical cavity in sand.

A typical arrangement of finite element mesh used in the analysis is shown in Figure 8.14. For all the analyses presented in Yu (1994), 100 two-noded elements in combination with an infinite spring element are used to simulate the behaviour of an infinite medium. The radius of the interface between the last annular element and the spring element was selected to be 300 times the initial cavity radius (i.e. $r_n = 300r_1$).

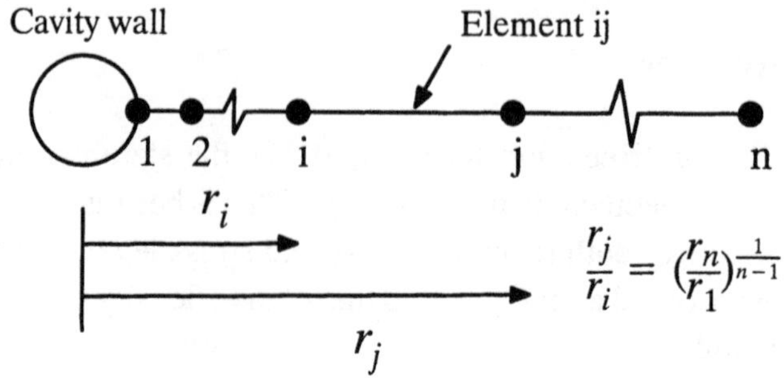

Figure 8.14: A typical finite element mesh used for cavity expansion analysis (Yu, 1994)

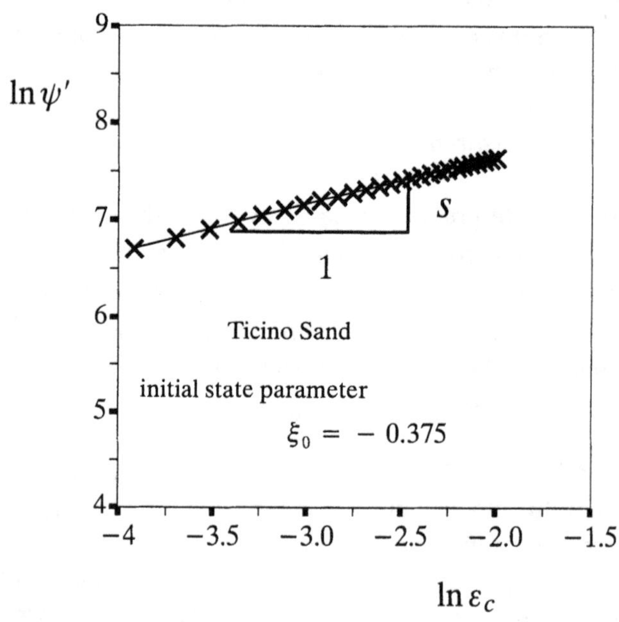

Figure 8.15: A typical numerical pressuremeter loading curve in Yu (1994)

The interpretation of the pressuremeter test using the state parameter model presented in Yu (1994) follows the method of Hughes *et al.* (1977) in the sense that only the loading portion of the pressuremeter curve is used to derive the soil properties.

Figure 8.15 shows typical results of numerical simulations of the self-boring pressuremeter tests in Ticino sand. As presented by Been *et al* (1987), the critical state properties of Ticino sand are $\Gamma = 0.986$, $\lambda = 0.024$, $\phi_{cv} = 31^0$. The pressure–expansion curve from the numerical analysis is interpreted as if it were derived from a real field test. It is interesting to note that the loading curve of the pressuremeter test in the log-log plot is approximately straight. The slope of the pressuremeter curve in such a plot is estimated using the portion of the loading curve from the cavity strain of 2% to 15%. With the known values of s and the initial state parameter of the soil ξ_0, the numerical correlation between these two parameters can therefore be established.

Sand	Monterey No.0 sand	Hokksund sand	Kogyuk sand	Ottawa sand	Reid Bedford sand	Ticino sand
e_{max}	0.82	0.91	0.83	0.79	0.87	0.89
e_{min}	0.54	0.55	0.47	0.49	0.55	0.6
A	0.83	0.80	0.75	0.95	0.63	0.6
Γ	0.878	0.934	0.849	0.754	1.014	0.986
λ	0.013	0.024	0.029	0.012	0.028	0.024
ϕ_{cv}	32	32	31	28.5	32	31

Figure 8.16: Soil properties for six chamber sands (friction angles are in degrees)

Yu's loading analysis concentrated mainly on those types of sand that are commonly used in calibration chamber tests and therefore, a direct comparison between the analysis and the experimental data can be made. Examples of those calibration chamber materials are Monterey No.0 sand, Hokksund sand, Kogyuk 350/2 sand, Ottawa sand, Reid Bedford sand, and Ticino sand. The critical state properties $(\lambda, \Gamma, \phi_{cv})$ for these materials are listed in Figure 8.16, which are compiled by Collins *et al.* (1992) using the data from Been and Jefferies (1985) for Kogyuk 350/2 sand, and Been *et al.* (1987) for other sands.

To develop a sensible numerical correlation between the loading slope of pressuremeter curve s and the initial state parameter of the material ξ_0, it is also necessary to study the influence of other key parameters. Previous numerical studies on the self-boring pressuremeter test in sand by Yu and Houlsby (1994) suggested that the stiffness index of the material plays an important role in the analysis of the pressuremeter test. Thus the effect of the stiffness index (which is defined as the ratio of the shear modulus and the initial mean stress level) can be studied by varying the value of the stiffness index from 500 to 2000. In addition, the variation of the initial lateral stress ratio K_0 (which is defined as the ratio of the initial horizontal and vertical stresses) from 0.5 to 2.0 is also used in the analysis so that any possible effect of an anisotropic initial stress state can be identified.

Like all the existing analyses, the pressuremeter test is simulated by the expansion of a cylindrical cavity in the soil. Each test was continued until the cavity strain reached 15% and the least square method was then applied to the strain range of between 2% and 15% to estimate the slope s of the pressuremeter expansion curve on a logarithmic scale.

The numerical correlations between the pressuremeter loading slope s and initial sand state parameter were shown to be largely independent of the stiffness index and initial stress states of the material. This is because the difference between the numerical correlations for the stiffness index of 500 and 2000 is very small and may be considered to be negligible for all practical applications. The numerical correlations for Monterey No. 0 sand with different initial lateral stress ratios suggest that the anisotropic initial stress state has very little effect on the relationship between the pressuremeter loading slope and the initial state parameter of the soil.

In summary, the two important conclusions emerging from the above numerical results are: (a) the loading slope of the pressuremeter curve for a particular material is mainly controlled by the initial state parameter of the soil; (b) the numerical correlations between s and ξ_0 are approximately linear. For practical purposes, the following linear equation may be used to express this correlation (see Figure 8.17):

$$\xi_0 = 0.59 - 1.85\, s \tag{8.30}$$

where ξ_0 and s denote initial sand state parameter and the derived loading slopes of pressuremeter tests with an infinitely long probe.

Once the state parameter is known, the soil friction angles can then be estimated using an average correlation between the friction angle and state parameter (e.g. Been *et al.*, 1987). This leads to the following expression for determining plane strain friction angle ϕ_0^{ps} from pressuremeter loading slope:

$$\phi_0^{ps} = 0.6 + 107.8\, s \tag{8.31}$$

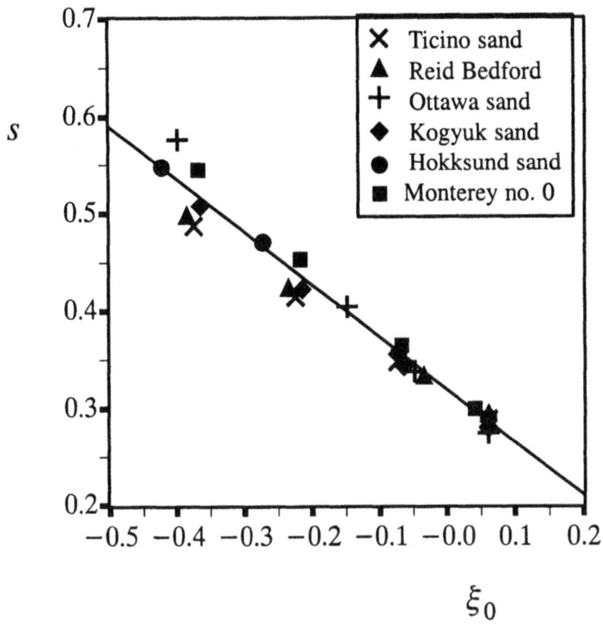

Figure 8.17: A linear theoretical correlation between s and ξ_0

Soil state derived from pressuremeter unloading tests

Although the self-boring pressuremeter may be used to minimise soil disturbance, the process of installation of the self-boring pressuremeter is not always efficient (Clough *et al.*, 1990). As pointed out by Jamiolkowski *et al.* (1985), soil disturbance during the installation of the self-boring pressuremeter may have significant effects on the shape of the initial loading portion of the pressuremeter curve. It is therefore necessary, wherever possible, to place less reliance on interpretation methods that are purely based on the initial loading portion of the test results.

Experimental studies on both self-boring and full-displacement pressuremeters (Hughes and Robertson, 1985; Bellotti *et al.*, 1986; Schnaid and Houlsby, 1992) have shown that the unloading portion of pressuremeter tests is less sensitive to initial soil disturbance. This experimental observation has, to a large extent, been supported by a theoretical study carried out by Whittle and Aubeny (1993) on the effects of installation disturbance on *in-situ* test results using the Strain Path Method. For this reason, there has been some development in the analysis of the unloading of a pressuremeter test in both sand (Houlsby *et al.*, 1986; Withers *et al.*, 1989; Yu, 1990; Yu and Houlsby, 1994) and clay (Jefferies, 1988; Houlsby and Withers, 1988; Ferreira and Robertson, 1992). While the methods developed by these authors may be used to derive realistic soil properties from the unloading portion of the pressuremeter test in clay, the analysis for sand, as presented by Houlsby *et al.* (1986) is known to be unrealistic as it gives a much lower value of the friction angle (Withers *et al.*, 1989; Yu, 1990).

In view of this, Yu (1996) developed an interpretation method for the unloading stage of a pressuremeter test in terms of the state parameter. The method uses the unloading portion of a pressuremeter test to derive the soil state parameter, and therefore represents an attractive alternative to the loading analysis presented in the previous section (Yu, 1994). Yu's unloading analysis is based on a strain-hardening (or softening) plasticity model in which the angles of friction and dilation are assumed to be a function of the state parameter. As the proposed method only uses the unloading portion of the test results, it does not require pressuremeter data with perfect self-boring installation. In other words, this unloading analysis may also be applicable to the results of a full-displacement or cone pressuremeter test in sand.

The soil model used in Yu's unloading analysis is essentially the same as that used in Yu's loading analysis for the pressuremeter test. The only difference is that the variation of shear modulus with void ratio and mean pressure has been included in the unloading analysis.

The first unloading analysis for pressuremeter tests in sand was an approximate small strain solution developed by Houlsby *et al.* (1986), that was later extended by Withers *et al.* (1989) to include the case of spherical cavity. The analysis was based on an elastic-perfectly plastic Mohr-Coulomb model and the elastic deformation in the plastically deforming zone was ignored. The main conclusion of this analysis was that the plastic unloading curve is a straight line in a plot of $\ln \psi'$ against $-\ln[(\varepsilon_c)_{max} - \varepsilon_c]$, where ψ' = effective cavity pressure, and $(\varepsilon_c)_{max}$ and ε_c are the maximum and current cavity strains respectively. The slope of this curve was found to be mainly controlled by the strength parameters of the soil. Unfortunately application of this analysis to results of field pressuremeter tests has shown that it gives much lower values for the friction angle when compared with those obtained from the loading portion of the test.

Using the same soil model, Yu (1990) and Yu and Houlsby (1994) presented a more rigorous unloading analysis that accounted for both large strains and elastic deformations in the plastically deforming zone. It was found that the plastic unloading curves from the large strain analysis were approximately straight on a logarithmic plot. A preliminary comparison indicated that the plastic unloading slope from Houlsby's small strain solution was slightly higher than that from the large strain analysis for loose and medium soils, but less than the large strain solution for a dense soil.

Experimental data from pressuremeter tests suggests that although a linear unloading curve on a logarithmic plot is normally observed, the slope of the pressuremeter unloading curve does not generally compare well with the theoretical prediction (Withers *et al.*, 1989). Broadly speaking, the numerical unloading slopes obtained from the small strain solution are significantly higher than those measured. The large strain unloading analysis predicts smaller unloading slopes for both loose and medium sands, and tends to compare better with the measured values. Nevertheless, the angles of friction and dilation derived from the measured unloading slopes of a pressuremeter test are still much lower than the actual values. This disagreement is not entirely surprising considering that an elastic-perfectly plastic Mohr-Coulomb soil model may not be sufficiently accurate for modelling the complex soil behaviour during the unloading of a pressuremeter test.

The study reported in Yu (1996) aimed to develop a more realistic unloading analysis that accounts for the dependence of strength parameters on soil deformation history. Using the state parameter-based constitutive model outlined in the previous section, the pressuremeter test has been simulated by a cylindrical cavity expansion process in a strain-hardening (or softening) dilatant soil. As no closed form cavity expansion solution exists with this soil model, the finite element program, CAVEXP, has been used to correlate the measured pressuremeter curve with the initial state parameter of the soil.

As described earlier, the existing unloading analyses using the Mohr-Coulomb soil model suggest that the plastic unloading curve is a straight line in a plot of $\ln \psi'$ against $-\ln[(\varepsilon_c)_{\max} - \varepsilon_c]$ and that the slope of the curve is mainly controlled by the strength parameters of the soil. The interpretation of the pressuremeter test using the state parameter model, as presented in Yu (1996), follows a similar approach by using the unloading portion of the pressuremeter test to derive the soil properties.

Figure 8.18 shows a typical result of finite element simulations of self-boring pressuremeter tests in Ticino sand. As presented by Been *et al.* (1987), the critical state properties of Ticino sand are $\Gamma = 0.986$, $\lambda = 0.024$, $\phi_{cv} = 31^0$. The pressure-contraction curve from the numerical analysis is interpreted as if it were derived from a real field test. It is interesting to note that a significant part of the unloading curve of the pressuremeter test in the log-log plot is approximately straight. The slope of the pressuremeter unloading curve in such a plot is estimated using the portion of the unloading curve that is obviously linear. With the known values of s_d and the initial state parameter of the soil ξ_0, the numerical correlation between these two parameters can be established.

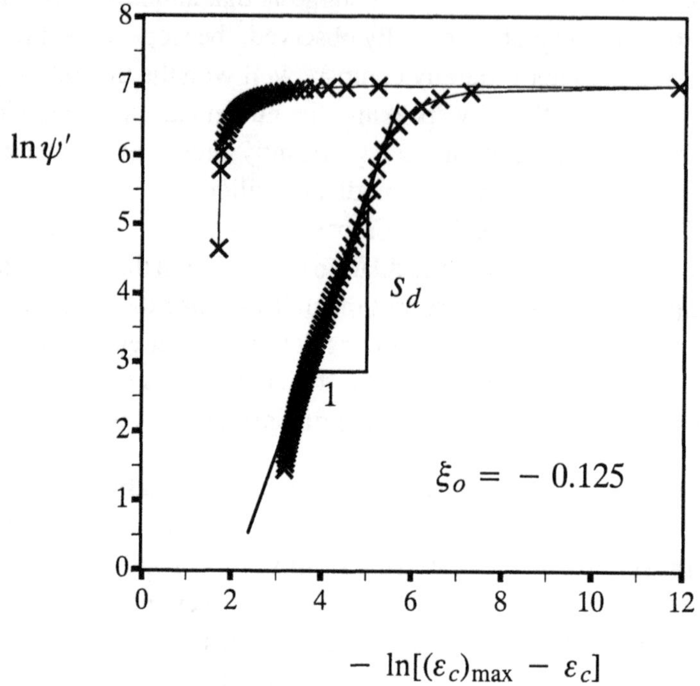

Figure 8.18: A typical numerical pressuremeter unloading curve in Yu (1996)

Figure 8.19 presents the changes of state in the soil adjacent to the pressuremeter membrane during loading and unloading of a pressuremeter test in Ticino sand. The critical state line for the same soil is also plotted in the figure so that the change of soil state parameter during the pressuremeter testing can be easily observed.

It can be concluded that during the loading test, both medium and dense sands dilate significantly whilst loose sand tends to compress. A very small reduction of void ratio is noticed when the soil unloads elastically and once plastic unloading occurs, soil of all densities starts to dilate. The test path for loose sand indicated in the figure suggests that it is possible that the soil may change from a 'loose state' to a 'dense state' during the pressuremeter test. Collins *et al.* (1992) also noticed this behaviour when using state parameter models to analyse cavity expansion problems.

Like Yu's loading analysis, the unloading analysis of Yu (1996) concentrated mainly on those types of sand that are commonly used in calibration chamber tests so that realistic material properties could be used in the numerical simulation. Examples of those calibration chamber materials are Monterey No.0, Hokksund, Kogyuk 350/2, Ottawa, Reid Bedford, and Ticino sands. The critical state properties

$(\lambda, \Gamma, \phi_{cv})$ for these sands are given in Figure 8.16. This data was obtained from Been *et al.* (1987), Been and Jefferies (1985) and Collins *et al.* (1992).

As mentioned previously, the pressuremeter loading and unloading test is simulated by the expansion of a cylindrical cavity followed by the contraction of the same cylindrical cavity in the soil.

For simplicity, it is assumed that the membrane remains in contact with the sand throughout the test. It must be emphasized that in practice this may not be the case, especially at very low stresses. Following most existing pressuremeter analyses, the possible effect of loose or saturated sand collapsing under some loading conditions in pressuremeter testing is ignored in the present analysis.

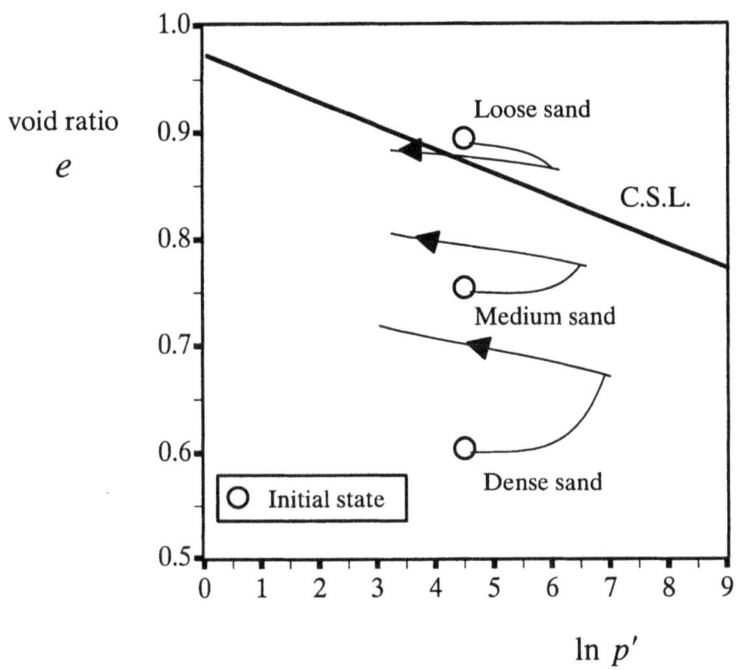

Figure 8.19: A typical numerical pressuremeter unloading curve in Yu (1996)

In general, each test continues until the cavity strain reaches $(\varepsilon_c)_{\max} = 18.23\%$ (which corresponds to cavity expansion ratio of 1.2) and then unloading of the cavity is carried out. The pressuremeter loading and unloading curves are plotted as $\ln \psi'$ against $-\ln[(\varepsilon_c)_{\max} - \varepsilon_c]$, and the slope s_d of the pressuremeter unloading curve on the logarithmic scale is then estimated.

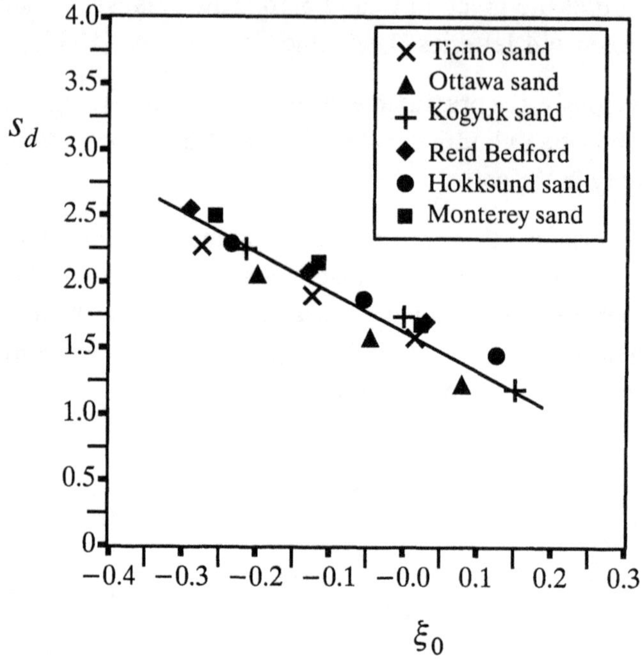

Figure 8.20: Theoretical correlation between pressuremeter unloading slope and initial state parameter

Figure 8.20 and Figure 8.21 present the numerical correlations between the derived pressuremeter unloading slope s_d, the initial state parameter ξ_0 and initial plane strain friction angle ϕ_0^{ps} for the six calibration chamber sands mentioned earlier. These results were obtained by choosing a range of initial void ratios and stress states. As the soil model used is based on a triaxial friction angle, a correlation is needed to convert it to a plane strain friction angle. The main reason for presenting the results in terms of plane strain friction angle is that it is more convenient to compare the present analysis with previous interpretation methods. Although a few alternative expressions may be used to convert the triaxial friction angle to the plane strain friction angle, the simple relationship $\phi_0^{ps} = 1.1\phi_0^{tr}$, suggested by Wroth (1984), has been adopted in Yu (1996).

It is interesting to note that the numerical correlations obtained are largely independent of sand type. This means the difference between the correlations for sands with different values of λ and ϕ_{cv} is quite small and may be considered negligible for most practical applications.

The average relationship between the pressuremeter unloading slope and the initial soil state parameter shown in Figure 8.20 is considered to be linear. The following equation may be used to express this correlation:

$$\xi_0 = -0.33 s_d + 0.53 \tag{8.32}$$

Similarly the correlation between the unloading slope and initial plane strain friction angle shown in Figure 8.21 can be described by the following linear approximation:

$$\phi^{ps}{}_0 = 18.4 s_d + 6.6 \tag{8.33}$$

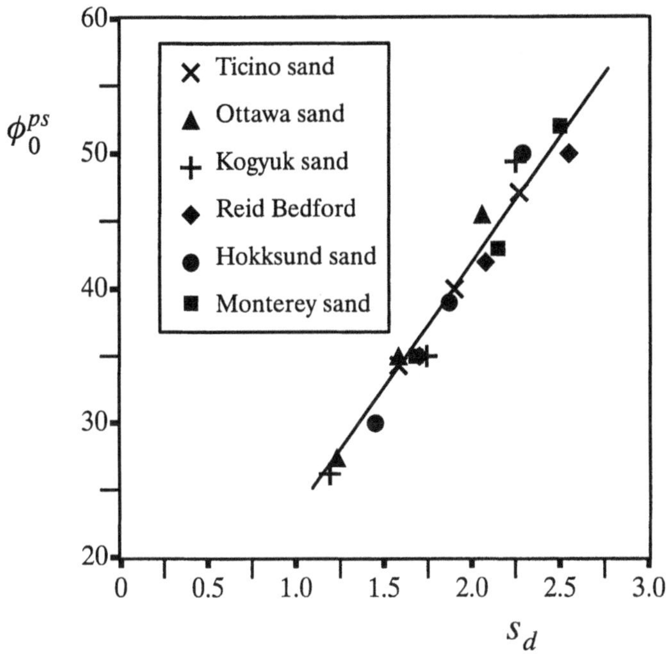

Figure 8.21: Theoretical correlation between pressuremeter unloading slope with initial sand friction angle

8.3.5 Effect of finite pressuremeter length

All the interpretation methods presented so far in sand are based on the assumption that the pressuremeter membrane is infinitely long, meaning that the pressuremeter loading and unloading tests can be simulated as the expansion of an infinitely long cylindrical cavity in soils.

In reality, however, the typical length to diameter ratio of a self-boring pressuremeter is about 6 and it is expected that the two-dimensional nature of the pressuremeter problem has some effects on the derived soil properties. To quantify this possible effect, it is necessary to resort to numerical methods and experimental chamber testing. Research on finite element analysis of finite pressuremeter length effects on sand has been conducted and reported by Yu (1990), Yu and Houlsby (1994) and Yu (1993). A recent experimental investigation using chamber testing by Ajalloeian and Yu (1998) on pressuremeter length effects supports the findings of the numerical study conducted by Yu (1990).

While numerical studies have not been carried out on the effects of finite pressuremeter length on unloading analysis of pressuremeter tests in sand, experimental results obtained by Ajalloeian (1996) suggested that the finite pressuremeter length has a less effect on unloading as compared to the loading portion of tests.

As expected, both the numerical and chamber studies suggest that the finite length of the pressuremeter probe results in a stiffer pressuremeter loading response. In particular, the real pressuremeter loading slopes s^6 with a length to diameter ratio of 6 were typically 10–20% higher than those obtained from the cylindrical cavity expansion process s. In agreement with the chamber test results reported by Laier *et al.* (1975) and Ajalloeian and Yu (1998), the numerical study suggests that the effect of the finite pressuremeter length is largely independent of the soil density. It is however noticed that the overestimation of the loading slope due to the finite length of the pressuremeter probe increases slightly with the stiffness index of the soil. According to Yu and Houlsby (1994), the effect of the finite pressuremeter length on the loading slope s for the case when the length to diameter ratio is 6 can be taken into account using the following equation:

$$a = \frac{s}{s^6} = 1.19 - 0.058 \ln \frac{G}{p'_0} \leq 1 \tag{8.34}$$

where s and s^6 represent the derived loading slopes of pressuremeter tests with the length to diameter ratios of ∞ and 6, respectively; G and p'_0 are soil shear modulus and initial mean effective stress of the soil.

So with the interpretation methods using the pressuremeter loading slope, it is possible to account for the finite pressuremeter length effects simply by applying a correction factor to the derived pressuremeter slope, before being used to derive soil strength and state parameter. The following procedure may be used in practice:

(1) Present field test data in a plot of pressure against central membrane strain.
(2) Estimate the value of shear modulus (G) from the slope of an unloading–reloading loop.
(3) Estimate the value of *in-situ* horizontal stress, and therefore the mean effective stress p'_0 using, for example, a 'lift–off pressure' approach.

(4) Calculate the value of stiffness index (G/p'_0) from the known values of G and p'_0.
(5) Present pressuremeter loading curves in a log–log plot from which the value of s^6 can be estimated.
(6) Using equation (8.34), calculate the finite length (or 2D) correction factor α.
(7) Estimate the value of s, that corresponds to an infinitely long pressuremeter, by $s = \alpha \times s^6$.
(8) After s is determined, the soil properties such as state parameter and angles of internal friction and dilation can be estimated using cavity expansion correlations such as those derived by Yu (1994, 1996) and Hughes *et al.* (1977).

8.4 CONE PRESSUREMETER TESTS IN CLAY AND SAND

The cone pressuremeter (also called the full-displacement pressuremeter) is an *in-situ* testing device that combines the standard cone penetrometer with a pressuremeter module incorporated behind the cone tip. The idea of mounting a pressuremeter module behind the tip of the cone penetrometer was first introduced in the early 1980s. The cone penetration test (CPT) and pressuremeter test are now used very widely in geotechnical engineering as the former can be used to quickly obtain approximate soil profiles while the latter is capable of accurately measuring soil stiffness and strength. The development of the cone pressuremeter aims to combine the merits of both the standard cone and the pressuremeter into a single instrument. The cone pressuremeter can be installed by standard CPT jacking equipment and enables pressuremeter tests to be carried out as part of routine CPT operations.

Compared with self-boring pressuremeter tests, cone pressuremeter tests are much more difficult to analyse because the tests are carried out in a soil which has already been disturbed by the penetration of the cone. As a result, the interpretation of cone pressuremeter tests must account for the effect of installation process. This is why the development of equipment for the cone pressuremeter was, for a long time, more advanced than its interpretation methods.

8.4.1 Cone pressuremeter testing in clay

Undrained shear strength, shear modulus and in-situ total horizontal stress from pressuremeter loading and unloading curves

Using cavity expansion theory, Houlsby and Withers (1988) developed the first theoretical interpretation method for deriving soil properties from cone pressuremeter tests in clay. Although only valid for undrained clay, the study by Houlsby and Withers (1988) may be viewed as a key development in the analysis of cone pressuremeter tests in soil.

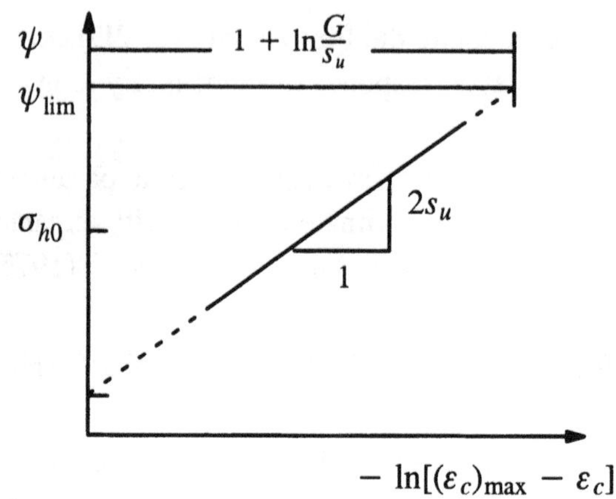

Figure 8.22: Graphical method using the unloading analysis of Houlsby and Withers (1988)

In the analysis of Houlsby and Withers (1988), the initial installation of the cone pressuremeter is modelled theoretically as the expansion of a cylindrical cavity within the soil. The expansion phase of the pressuremeter test is modelled as a continued expansion of the same cylindrical cavity, and the contraction phase as a cylindrical cavity contraction. Although this one-dimensional simulation of the two dimensional cone pressuremeter tests is somewhat in error, more rigorous analyses of cone penetration problems (eg. Baligh 1986; Teh, 1987; Yu *et al.*, 2000) show that the stress distribution far behind the cone tip is similar to the distribution created by the expansion of a cylindrical cavity from zero initial radius. Considering the pressuremeter membrane is located some distance behind the cone tip in the cone pressuremeter, it seems logical to use simple cavity expansion theory for the analysis of cone pressuremeter tests.

As the installation and subsequent expansion of the cone pressuremeter is simulated as the expansion of a cylindrical cavity from zero initial radius, the solutions developed in Section 3.2.2(c) suggest that the cavity pressure remains constant during any stage of installation and expansion tests. The constant cavity pressure is the same as the limit pressure for a cavity expansion from a finite initial radius:

$$\psi_{\lim} = \sigma_{h0} + s_u[1 + \ln\frac{G}{s_u}] \tag{8.35}$$

The complete analytical solution for cavity contraction from this limiting plastic state is derived in Section 3.2.3(b). For the case of cylindrical cavity, the unloading pressure-displacement curve is defined by:

$$\psi = \psi_{\lim} - 2s_u[1 + \ln((\varepsilon_c)_{\max} - \varepsilon_c) - \ln(\frac{s_u}{G})] \tag{8.36}$$

where $(\varepsilon_c)_{\max}$ is the maximum cavity strain at the start of the unloading phase.

The large strain unloading solution of Houlsby and Withers (1988), as defined by equation (8.36), is summarized in Figure 8.22, which shows that the plastic unloading slope in a plot of ψagainst $-\ln[(\varepsilon_c)_{\max} - \varepsilon_c]$ is controlled by the soil strength parameter. The slope of the pressuremeter unloading curve in this plot is equal to twice the undrained shear strength of the soil s_u. From the figure, it is also possible to estimate the shear modulus and initial total horizontal stress.

Shear modulus from unloading-reloading loops

Like self-boring pressuremeters, it is reasonable to deduce some useful information about soil shear modulus from the slope of an unload-reload loop. However, the stress and strain levels from which the shear modulus is obtained must be recorded, as the pressure level for the cone pressuremeter tests is generally much higher than the self-boring pressuremeter. This is due to the effect of initial cone installation process on soil stresses.

8.4.2 Cone pressuremeter testing in sand

Despite initial success in the analysis of cone pressuremeter tests in clay, progress on the analysis of cone pressuremeter tests in sand has been very slow, largely due to the more complex behaviour of granular materials. For a long time, empirical correlations were adopted to derive soil properties from the results of cone pressuremeter tests in sand, and this situation cast doubt on the potential and applicability of the cone pressuremeter for frictional soils.

The development of cone pressuremeter testing in sands improved significantly when, almost eight years after the publication of Houlsby and Withers (1988) analysis for clays, Yu *et al.* (1996) published an equivalent, theoretically based, interpretation method for cone pressuremeter tests in sand. This method has been applied successfully by various investigators and is recommended by Lunne *et al.* (1997) in their book on cone penetration testing in geotechnical practice. The procedure outlined in Yu *et al.* (1996) represents a major development for deriving soil properties from cone pressuremeter tests in sand. It may have paved the way for a much wider application of the cone pressuremeter in geotechnical engineering practice.

Experimental results of calibration chamber testing of the Fugro cone pressuremeter presented by Schnaid (1990) clearly showed that good correlations exist between the ratio of cone tip resistance and pressuremeter limit pressure to many soil properties such as relative density and friction angle. Although both the cone resistance and the pressuremeter limit pressure are known to be dependent upon the size of the calibration chamber used, it has been shown that the ratio of these two quantities is largely unaffected by the chamber size (Schnaid and Houlsby, 1991). It is therefore reasonable to suggest that the correlations between soil properties and the ratio of cone resistance and pressuremeter limit pressure obtained from the chamber testing may be directly applied to field conditions. However, it should be stressed that so far the validity of this hypothesis has been verified only with the results of a limited number of field tests (Schnaid, 1994). In addition, as stressed by Wroth (1984), empirical correlations can only be used with confidence if the interpretation procedure is set against a background of theory, however idealised this may be. It is therefore necessary to develop some theoretical correlations between the soil properties and the cone pressuremeter results.

Yu *et al.* (1996) presented some theoretical correlations between the ratio of cone resistance and pressuremeter limit pressure to strength parameters of the soil. In developing these theoretical correlations, it was assumed that both the cone resistance and the pressuremeter limit pressure were strongly related to the limit pressures of spherical and cylindrical cavities. In particular, the solutions of cavity expansion in an elastic-perfectly plastic Mohr-Coulomb soil were used to correlate the ratio of cone resistance and pressuremeter limit pressure to peak friction angle of the soil. In addition, the limit pressure solutions for cavity expansion in a strain hardening/softening soil using a state parameter-based soil model were used to correlate the ratio of cone resistance and pressuremeter limit pressure to *in-situ* sand state parameter. The validity of these theoretical correlations was verified by comparing them with the results of the calibration chamber testing. Since there is no large body of experimental data available for cone pressuremeter tests in sand, validation of the proposed correlations was also explored by making use of limit pressure values obtained from the self-boring pressuremeter and the cone resistance from independent cone penetration tests. Application of the proposed correlations to the results of a large number of chamber and field tests suggested that they can be confidently used to determine the soil friction angle and *in-situ* state parameter.

Cavity expansion modelling of cone pressuremeter test

The nature of large strains associated with the cone penetration has made it extremely difficult to develop a rigorous analysis for the cone penetration test in sand. Although some success has been made in using the strain path method to analyze the cone penetration test in clay (Baligh, 1985), the analysis for sands is still in its infancy despite the fact that a lot of effort has been devoted to this problem over the last two decades. As discussed by Robertson and Campanella (1983), the current analyses for the cone penetration in sands may be divided into two categories, namely those based on bearing capacity theory (e.g. Durgunoglu and Mitchell, 1975) and those based on cavity expansion theory (e.g. Vesic, 1972; Baligh, 1976; Yu and Houlsby, 1991; Wang, 1992; Salgado, 1993). Despite much of its success, it is likely that the bearing capacity theory cannot provide a reliable prediction for the cone resistance, as it does not take into account soil compressibility. On the other hand, experimental evidence provided by a comprehensive set of calibration chamber tests of the cone penetrometer (Baldi *et al.*, 1981) has shown that cavity expansion theory is capable of modelling the measured response extremely well.

The analogy between cavity expansion and the cone penetration process was first noticed by Bishop *et al.* (1945) and Hill (1950) after observing that the pressure required to produce a deep hole in an elastic-plastic medium was proportional to that necessary to expand a cavity of the same volume under the same conditions. Since then, a tremendous amount of research has been carried out in an attempt to correlate the cone tip resistance to cavity expansion limit pressures (see, for example, Ladanyi and Johnston, 1974; Vesic, 1972, 1977; Baligh, 1976; Mitchell and Tseng, 1990; Yu and Houlsby, 1991; Collins *et al.*, 1992; Salgado, 1993). Due to the approximate nature of these correlations, it is not possible to determine which one would be the most suitable. A theoretical study carried out by Wang (1992) has suggested that the correlation between the cone resistance and the spherical cavity limit pressure proposed by Ladanyi and Johnston (1974) can be successfully used to predict the cone resistance measured in calibration chambers. It is for these reasons that the simple correlation of Ladanyi and Johnston (1974) and Ladanyi (1975) will be used in conjunction with the cavity expansion solutions developed by Yu and Houlsby (1991) and Collins *et al.* (1992) to determine the cone resistance in sands. For the geometry of a standard cone penetrometer, Ladanyi and Johnston (1974) suggested the following relationship between the cone tip resistance q_c and the spherical cavity limit pressure σ_s :

$$q_c = \sigma_s \, (1 + \sqrt{3} \, \tan\phi_{ps}') \tag{8.37}$$

where ϕ_{ps}' denotes the plane strain peak friction angle of the soil, as the cone penetration is assumed to occur under plane strain conditions. It should also be noted that for the above equation to be valid, a perfectly rough soil-cone interface must be assumed.

On the other hand, theoretical and experimental studies of the cone pressuremeter test in sand clearly indicate that the pressuremeter limit pressures obtained from the cone pressuremeter test can be accurately predicted by the cylindrical cavity limit pressures (Schnaid, 1990; Yu, 1990; Schnaid and Houlsby, 1991; Schnaid and Houlsby, 1992). This leads to:

$$\psi_l = \sigma_c \tag{8.38}$$

where ψ_l and σ_c are the pressuremeter limit pressure measured from the cone pressuremeter and the cylindrical cavity limit pressure respectively.

Combining equations (8.37) and (8.38) gives:

$$\frac{q_c}{\psi_l} = (1 + \sqrt{3}\ \tan\phi_{ps}')\ \frac{\sigma_s}{\sigma_c} \tag{8.39}$$

The above equation suggests that the ratio of cone resistance to pressuremeter limit pressure is directly related to the ratio of spherical and cylindrical cavity limit pressures. Because σ_s/σ_c is controlled by the strength and stiffness properties of the soil, q_c/ψ_l should also be a good indicator of these soil properties.

Yu *et al.* (1996) assumed that the cone pressuremeter test is carried out under drained loading conditions and all the stresses used are therefore effective stresses. In other words, q_c was used to represent the difference of the total cone resistance and initial pore pressure, and ψ_l denoted the difference of the total pressuremeter limit pressure and initial pore pressure.

Evaluation of friction angle

In this section, the analytical solutions for limit pressure of cavity expansion in a linear-elastic and perfectly-plastic dilatant soil presented by Yu and Houlsby (1991) are used to correlate q_c/ψ_l with soil strength parameters. In developing the analytical solution for the expansion of both cylindrical and spherical cavities in dilatant soil, Yu and Houlsby (1991) used the Mohr-Coulomb yield criterion with a non-associated flow rule. The dilatancy of the soil can therefore be properly taken into account. For the case of cylindrical cavity expansion, the axial or vertical stress is assumed to be the intermediate stress and plane strain conditions in the axial direction are assumed. An explicit expression for the pressure-expansion relation has been derived without any restriction imposed on the magnitude of the deformations. Consequently, the limit cavity pressure when the radius of the cavity approaches infinity can be determined analytically.

For given angles of internal friction and dilation and values of shear modulus, cavity expansion theory can be used to calculate the ratio of spherical and cylindrical cavity limit pressures σ_s/σ_c. The ratio of cone resistance to pressuremeter limit pressure q_c/ψ_l can then be readily determined from equation (8.39). Using the cavity expansion solutions of Yu and Houlsby (1991), the variation of q_c/ψ_l with friction angle and shear modulus has been studied in detail and the results of this study are presented in Figure 8.23. In the analysis, Rowe's stress-dilatancy relationship was used to relate the angles of internal friction and dilation. As expected, the ratio of q_c/ψ_l was found to increase with value of of friction angle. Figure 8.23 also suggests that q_c/ψ_l is not only a function of soil strength parameters but also strongly dependent upon the soil stiffness. This means that any sensible method for deriving soil friction angle from the value of q_c/ψ_l must take into account the shear modulus of the soil. It is interesting to note that the results presented in Figure 8.23 can also be represented in a modified plot so that all numerical data lies on a unified line as shown in Figure 8.24. For the case when the stiffness index is in the range of 200 to 1500, the numerical data can be accurately represented by a straight line, namely:

$$\phi_{ps} = \frac{14.7}{\ln I_s}\frac{q_c}{\psi_l} + 22.7 \tag{8.40}$$

where the stiffness index $I_s = G/p_0$, and G and p_0 represent the shear modulus and initial effective mean pressure respectively.

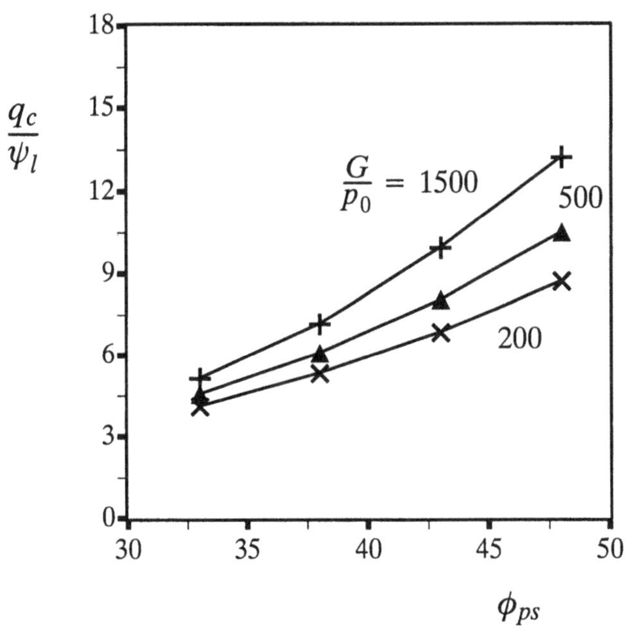

Figure 8.23: Variation of ratio of cone resistance to pressuremeter limit pressure with friction angle and stiffness index

The theoretical correlation (8.40) can be used to derive plane strain friction angles from the measured ratio of q_c/ψ_l, provided a reasonable estimate of the stiffness index I_s can be made. To demonstrate the relevance of this correlation, equation (8.40) was applied in Yu *et al.* (1996) to a large number of cone and pressuremeter tests to predict the angle of internal soil friction. The validity of this correlation was confirmed by comparing these estimates with other measurements.

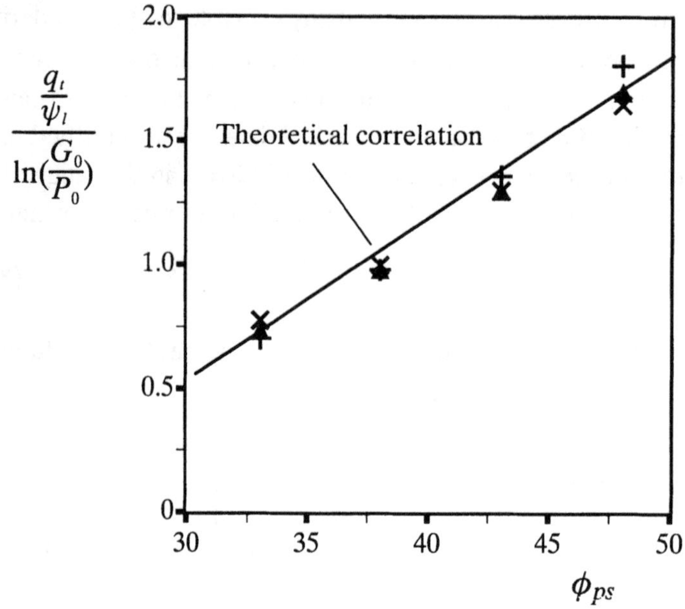

Figure 8.24: Theoretical correlation between the ratio of cone resistance to pressuremeter limit pressure with friction angle

Evaluation of state parameter

The cavity expansion solutions developed by Collins (1990), Collins *et al.* (1992) and Wang (1992) using a state parameter soil model are used in this section to develop a theoretical correlation between the ratio of q_c/ψ_l and *in-situ* state parameter of the soil.

The state parameter model used by Collins (1990) is an elastic-plastic strain hardening (or softening) model which postulates that a sand specimen can be considered as an isotropic continuum. Collins' model is very similar to that used by Yu (1994, 1996) for the analysis of self-boring pressuremeter tests in sand. The Mohr-Coulomb yield function was used to model the sand behaviour. It was further assumed that the angles of friction and dilation are a function of the state parameter.

To account for the variation of the shear modulus on void ratios and stress levels, the following empirical formula of Richard *et al.* (1970) has been used:

$$\frac{G}{p_r} = S\,\frac{(e_c - e)^2}{1 + e}\,(\frac{p}{p_r})^{0.5} \tag{8.41}$$

where the reference pressure $p_r = 100$ kPa and e_c is a dimensionless constant with values ranging from 2.17 for round-grained sands to 2.97 for sands with angular grains. The corresponding range for the constant S is 690-323.

It is noted that the above expression, combined with a constant value of Poisson's ratio, leads to a non-conservative model in the sense that some unloading-reloading cycles result in a continuous dissipation of energy (Zytynski *et al.*, 1978). However, as pointed out by Gens and Potts (1988), this fact may not be too important if monotonic loading is considered, and it may only become significant if the loading involves many stress reversals. It is true that a pressure dependent shear modulus together with a constant value of Poisson's ratio has been successfully used in the past to model the non-linear elastic behaviour of granular materials for static geotechnical problems.

Using the state parameter soil model mentioned above, Collins *et al.* (1992) and Wang (1992) presented solutions for the limit pressure of the expansion of both spherical and cylindrical cavities in six different calibration chamber sands. These numerical solutions suggested that the ratio of spherical and cylindrical cavity limit pressures σ_s/σ_c could be accurately calculated by the following equation:

$$\frac{\sigma_s}{\sigma_c} = m_1\, p_0^{m_2 + m_3(1+e_0)}\, \exp[m_4(1 + e_0)] \tag{8.42}$$

where e_0 , p_0 denote the initial void ratio and mean effective stress before cone pressuremeter testing; and m_1, m_2, m_3, m_4 are material constants, the values of which are given in Figure 8.25 for the six calibration chamber sands. The ratio of cone resistance to pressuremeter limit pressure q_c/ψ_l can then be determined from equation (8.39). In doing this, the plane strain friction angle in (8.39) can be converted from the triaxial friction angle employed in the state parameter soil model using the simple expression of Bishop (1966):

(a) for $\phi_{tc} < 33^0$, $\phi_{ps} = \phi_{tc}$;
(b) for $33^0 \leq \phi_{tc} < 36^0$, $\ln\phi_{ps} = 1.666\ln\phi_{tc} - 2.336$;
(c) for $\phi_{tc} \geq 36^0$, $\ln\phi_{ps} = 1.293\ln\phi_{tc} - 1.002$.

The variation of q_c/ψ_l with initial state parameter was evaluated with different initial effective mean stresses and it was found that the theoretical correlations between q_c/ψ_l and initial state parameter were largely independent of initial mean effective stress. The correlations of q_c/ψ_l and initial state parameter for the six calibration chamber sands are presented in Figure 8.26. As expected, it was found that

q_c/ψ_l decreased with initial state parameter. It was also interesting to note that the difference between the correlations for different sand types seemed to be reasonably small and this was particularly true for medium and loose sands. For the convenience of applications, the dependence of the theoretical correlations upon sand types can be ignored. The average correlation for these six different sands can be adequately represented by the following linear equation:

$$\xi_0 = 0.46 - 0.3 \ln \frac{q_c}{\psi_l} \tag{8.43}$$

The correlation (8.43) can be used to derive the initial state parameter from the measured ratio of q_c/ψ_l. To demonstrate the relevance of this correlation, equation (8.43) was applied with success by Yu *et al.* (1996) to a number of cone and pressuremeter tests to predict the state parameter of the soil.

Sand	Monterey No.0 sand	Hokksund sand	Kogyuk sand	Ottawa sand	Reid Bedford sand	Ticino sand
m_1	1087	560	237	1163	342	376
m_2	−0.47	−0.424	−0.359	−0.469	−0.385	−0.387
m_3	0.225	0.195	0.167	0.24	0.172	0.175
m_4	−3.214	−2.84	−2.485	−3.483	−2.521	−2.604

Figure 8.25: Constants m_1, m_2, m_3, m_4 for six chamber sands

Evaluation of shear modulus

As for clay, the shear modulus of sands can be estimated from the slope of an unloading-reloading loop. As detailed in Schnaid (1990) and Schnaid and Houlsby (1994), the stress level at which the unloading-reloading loop is carried out is very important and must be recorded, as the cone pressuremeter modulus is sensitive to the pressure level.

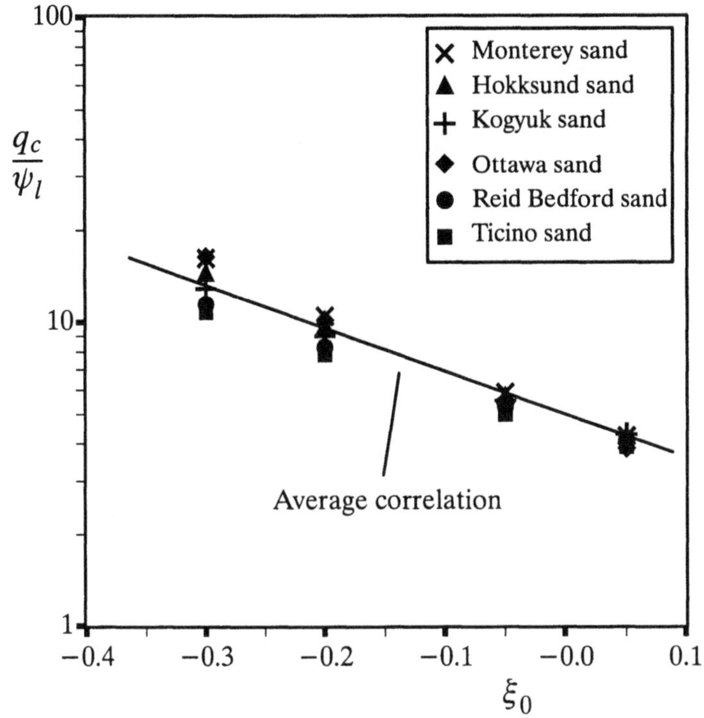

Figure 8.26: Theoretical correlation between the ratio of cone resistance to pressurmeter limit pressure with state parameter

8.5 CONE PENETRATION TESTS IN SOIL

The analogy between cavity expansion and cone penetration was first pointed out by Bishop *et al.* (1945) after observing that the pressure required to produce a deep hole in an elastic-plastic medium was proportional to that necessary to expand a cavity of the same volume under the same conditions. Two steps must be followed in order to use a cavity expansion approach to predict cone resistance: (a) to develop theoretical (analytical or numerical) limit pressure solutions for cavity expansion in soils, and (b) to relate cavity expansion limit pressures to cone resistance. As concluded by Yu and Mitchell (1998) in a review of the methods for analysis of cone resistance, the cavity expansion approach provides a more accurate prediction of cone resistance than that obtained using bearing capacity theory. This is because the influences of soil stiffness, compressibility (or dilatancy) and penetration-induced horizontal stress increase can all be adequately taken into account. This section is based closely on the review published by Yu and Mitchell (1998). For clarity, the theories for clay and sand will be presented separately.

8.5.1 Cone penetration in cohesive soils

This section describes some of the well known correlations for relating the cone resistance and cavity limit pressures in undrained clay soils.

Solution by Ladanyi and Johnson (1974)

Ladanyi and Johnson (1974) assumed that the normal stress acting on the cone face was equal to that required to expand a spherical cavity from zero radius. If the cone-soil interface is perfectly rough, the equilibrium equation in the vertical direction can be used to give the cone resistance as follows:

$$q_c = \psi_s + \sqrt{3} s_u \tag{8.44}$$

where ψ_s is the spherical cavity limit pressure. If the clay is modelled by the Tresca yield criterion, then the spherical cavity limit pressure can be shown to be (Hill, 1950; Collins and Yu, 1996):

$$\psi_s = \frac{4}{3} s_u (1 + \ln \frac{G}{s_u}) + p_0 \tag{8.45}$$

in which G is the shear modulus and p_0 is the *in-situ* mean total stress in the soil.

Combining equations (8.44) and (8.45) leads to the following cone resistance expression:

$$q_c = N_c s_u + p_0 \tag{8.46}$$

where the cone factor is :

$$N_c = 3.16 + 1.33 \ln \frac{G}{s_u} \tag{8.47}$$

Solution by Vesic (1972, 1977)

Vesic (1972, 1977) developed a theory that relates pile resistance to the spherical cavity limit pressure. The failure mechanism assumed by Vesic was based on observations of both model and full-size pile testing. Piles are analogous to a cone penetrometer. In particular, Vesic assumed that radial shear zone was identical to that emanating from the transverse section of an infinitely long wedge (plane strain case). After taking account of the shear resistance along the failure surface, Vesic obtained the following expression for the cone factor :

$$N_c = 3.90 + 1.33 \ln \frac{G}{s_u} \tag{8.48}$$

which is slightly higher than that given by equation (8.47).

Solution by Baligh (1975)

Baligh (1975) interpreted the total cone resistance as a combination of the work required to give the tip of the cone a virtual vertical displacement and the work done to expand a cylindrical cavity around the shaft of the cone in the radial direction. Using this approach, the cone factor was found to be:

$$N_c = 12.0 + \ln\frac{G}{s_u} \tag{8.49}$$

However a simple summation of both contributions cannot be entirely correct for a complex nonlinear problem. The assumption that a renewed cylindrical cavity expansion takes place behind the tip of the cone from zero radius tends to overestimate the total work required, and this leads to an overestimation of the cone resistance.

Solution by Yu (1993)

According to the experimental work of Lehane and Jardine (1992) on displacement piles, the normal pressure along the shaft after pile penetration is assumed to be equal to the average of the three stresses (i.e. radial, hoop and vertical stresses) required to expand a cylindrical cavity. Using this experimental finding, Yu (1993) showed that the rigorous axisymmetric steady state solutions of a rigid cone presented by Durban and Fleck (1992) and Sagaseta and Houlsby (1992) can be used to arrive at the following expression for the cone factor:

$$N_c = \frac{2}{\sqrt{3}}[\pi + \alpha + \arcsin(\lambda) + \lambda \cot\frac{\alpha}{2} - \sqrt{1-\lambda^2} + \frac{D}{2} + \ln\frac{\sqrt{3}G}{2s_u}] \tag{8.50}$$

where the parameter D, which is a function of the cone roughness parameter λ (i.e., $\lambda = 0$ for a perfectly smooth cone and $\lambda = 1$ for a perfectly rough cone) and the cone apex angle α, can be obtained from the following equations:

$$D = \frac{\sin\frac{\beta}{2} + \lambda \sin\beta}{\cos\frac{\beta}{2} - \cos\beta}; \qquad \beta = 180^0 - \alpha/2$$

For a standard 60 degree cone, the solution (8.50) proposed by Yu (1993) can be further simplified to:

$$N_c = 4.18 + 1.155\ln\frac{\sqrt{3}}{2}\frac{G}{s_u} \tag{8.51}$$

for a perfectly smooth cone, and to:

$$N_c = 9.4 + 1.155\ln\frac{\sqrt{3}}{2}\frac{G}{s_u} \tag{8.52}$$

for a perfectly rough cone.

Comparison with Field Testing

In normally consolidated clays, the cone resistance is often correlated with vane shear strength to gain information about the value of the cone factor. Although the vane tends to overestimate shear strength in OC clays (Meigh, 1987), comparative experimental research (e.g. Nash *et al.*, 1992) suggests that the vane can be satisfactorily used to measure undrained shear strength in NC clays as it gives similar strengths to those obtained from self-boring pressuremeter and triaxial compression tests. The results of an extensive review by Lunne and Kleven (1981) for NC clays indicated that the average value of the cone factor was about 15, with the majority of results falling between 11 and 19. It is more difficult to establish similar correlations for stiff overconsolidated clays because of the effects of fabric and fissures on the response of the clay. Nevertheless, experimental and theoretical research suggests that the cone factors for OC clays are generally higher than those of normally consolidated clays (Meigh, 1987, Yu *et al.*, 2000).

Some of the theoretical cone factors derived from the methods of bearing capacity theory, cavity expansion theory, strain path method and large strain finite element method are given and compared in Yu and Mitchell (1998). The bearing capacity theory indicates that the cone factor does not depend on the soil stiffness index as in bearing capacity theory elastic deformations are not taken into account. The bearing capacity cone factors, which range from 8.3 to 10.4, are much lower than the average cone factor of 15 as observed in field testing. It is interesting to note that the rest of the cone factors listed in Yu and Mitchell (1998) all show a tendency to increase with the stiffness index. Although slightly higher, the cavity expansion solutions of Ladanyi and Johnston (1974) and Vesic (1977) for a rough cone are still lower than the average cone factor obtained from field correlations. On the other hand, as indicated previously, the cavity expansion solution suggested by Baligh (1975) tends to overestimate the cone factor. This is confirmed by the fact that the cone factor of Baligh (1975) is even higher than the finite element solution of van den Berg (1994) which is known to be higher than the exact solution due to the effect of incompressibility constraints. For a smooth cone, the strain path solutions of Whittle (1992) and Teh and Houlsby (1991) are similar to the finite element solution of van den Berg (1994).

For a perfectly rough cone, the cavity expansion solution of Yu (1993) gives cone factors of between 13.8 and 16.4 for stiffness indices ranging from 50 to 500. These cone factors, which agree well with the average cone factor of 15 observed in the field, are slightly lower than those predicted by the finite element method and therefore should be closer to the exact theoretical values assuming perfect plasticity. The cone factors of Yu (1993) for a smooth cone were between 8.5 and 11.2 for stiffness indices ranging from 50 to 500. Once again, it was found that these cone factors were about 10-15% lower than the large strain finite element solutions and, therefore, may be considered to be very close to the exact theoretical cone factors assuming perfect plasticity.

All the cone factors presented so far were obtained from a total stress analysis using the Tresca or von Mises yield criterion. Strictly speaking, this is only valid for normally consolidated clays. For accurate modelling of an overconsolidated clay, an effective stress analysis with more realistic soil models such as Cam clay has to be carried out so that the variation of soil strength with stress history can be considered.

Comparison with Chamber Testing

Kurup *et al.* (1994) presented the results of a calibration chamber study on piezocone tests in cohesive soils. These chamber testing results provide a valuable set of experimental data for verification of theoretical predictions of cone resistance in clay. Using two miniature cone penetrometers (11.28 mm and 12.72mm in diameter), eight chamber tests were carried out on three isotropically consolidated specimens. The calibration chamber used has an internal diameter of 525mm which is over 40 times as large as the diameters of the miniature cone penetrometers. For such a large value of chamber to cone diameter ratio, the chamber size would have a very small effect on the measured cone resistance in clay. Specimens 1 and 2 were normally consolidated samples, while specimen 3 was a lightly overconsolidated sample with an overconsolidation ratio of 5. The clay specimens were prepared by mixing kaolinite and fine sand with deionized water at a water content of twice the liquid limit. A mixture of 50% kaolinite and 50% fine sand by weight was used to prepare specimens 1 and 3. Specimen 2 was prepared from a mixture of 33% kaolinite and 67% fine sand.

The value of cone factor for each cone penetration test was determined from the measured cone resistance, initial stresses applied on the soil specimen and the undrained shear strength measured from undrained triaxial compression tests. The ratio of the shear modulus at 50% peak shear stress to the undrained shear strength was used to arrive at the value of stiffness index. The stiffness indices obtained for specimens 1, 2 and 3 were 267, 100 and 150 respectively. Comparison of the calibration chamber cone factors and some of the theoretical cone factors is summarized in Yu and Mitchell (1998). As observed by Yu and Mitchell (1998), the cone factors from bearing capacity theory, Vesic's cavity expansion solution and the strain path method were all 10 to 34% lower than the measured values. The large strain finite element analysis of van den Berg (1994) and Baligh's cavity expansion solution overestimated the measured cone factors to some extent. Yu's cavity expansion solution (Yu, 1993) for a rough cone predicted the measured cone factors within 10%.

8.5.2 Cone penetration in cohesionless soils

Some of the available correlations for calculating cone resistance from cavity limit pressures in cohesionless soils are summarised in this section. The cone penetration is assumed to occur under fully drained conditions so that all the stresses are effective stresses.

Solution by Ladanyi and Johnson (1974)

For cohesionless soils, Ladanyi and Johnson (1974) assumed that the normal stress acting on the cone face is equal to that required to expand a spherical cavity. Following this assumption, the cone resistance in a cohesionless soil can be obtained as:

$$q_c' = [1 + \sqrt{3}\tan(\lambda\phi')]\ \psi_s' \tag{8.53}$$

Cavity expansion solutions (e.g. Vesic, 1972; Baligh, 1976; Carter *et al.*, 1986; Yu and Houlsby, 1991; Yu, 1992, and Collins *et al.*, 1992) can then be used to determine the effective spherical cavity limit pressure ψ_s' in the following form :

$$\psi_s' = A\ p_0' = \frac{(1 + 2K_0)}{3} A\sigma_{v0}' \tag{8.54}$$

where p_0' is the initial mean effective stress. The parameter A denotes the ratio of the effective spherical cavity limit pressure to the initial mean effective stress and is normally a function of soil strength and stiffness. $K_0 = \sigma_{h0}'/\sigma_{v0}'$ is the coefficient of earth pressure at rest. For normally consolidated deposits, the coefficient of earth pressure at rest may be estimated by $K_0 = 1 - \sin\phi'$.

Combining equations (8.53) and (8.54) gives:

$$q_c' = N_q\sigma_{v0}' \tag{8.55}$$

where the cone factor for cohesionless soil is :

$$N_q = \frac{(1 + 2K_0)A}{3}[1 + \sqrt{3}\tan(\lambda\phi')] \tag{8.56}$$

Equation (8.56) suggests that the cone factor for sand is mainly controlled by the value of A, which is dependent on the particular cavity expansion theory used in the interpretation. Analytical expressions for A are not available for most cavity expansion theories in sand, but as will be discussed in a later section such expressions can be obtained numerically. The correlations (8.53)-(8.56) combined with the cavity expansion solutions of Collins *et al.* (1992) and Yu and Houlsby (1991) have been successfully used by Wang (1992), Collins *et al.* (1994) and Yu *et al.* (1996) to interpret the results of cone penetrometers and cone pressuremeters.

Solution by Vesic (1972, 1977)

The failure mechanism proposed by Vesic (1977) for the soil around the cone can be used to relate the cone resistance to the spherical cavity expansion limit pressure obtained by Vesic (1972). After some derivations, it can be shown that the Vesic model gives the following expression for the cone factor in sand :

$$N_q = \left(\frac{1 + 2K_0}{3 - \sin\phi'}\right) \exp[(\pi/2 - \phi')\tan\phi'] \tan^2(45^0 + \phi'/2)\ (I_{rr})^n \quad (8.57)$$

in which the reduced rigidity index is defined as $I_{rr} = I_r/(1 + I_r\varepsilon_v)$, where ε_v is the average volumetric strain in the plastically deformed region, and the rigidity index I_r and parameter n are given respectively by :

$$I_r = \frac{G}{p_0' \tan\phi'} \quad (8.58)$$

$$n = \frac{4\sin\phi'}{3(1 + \sin\phi')} \quad (8.59)$$

After applying the Vesic correlation to the results of a number of chamber tests, Mitchell and Keaveny (1986) concluded that the cone penetration test can be closely modelled by the spherical cavity limit pressure for sands with a low value of the reduced rigidity index (i.e. more compressible soils). However, for sands with a high reduced rigidity index (i.e. less compressible soils), use of the cylindrical cavity limit pressure gives better predictions.

To use equation (8.57), the average volumetric strain ε_v in the plastically deformed region must be estimated, using either laboratory testing or empirical correlations. This adds uncertainty to the accuracy of the interpretation process and must therefore be considered a limitation in using Vesic's cavity expansion solutions (Vesic, 1972) to correlate the cone resistance with soil properties. In addition, the dilatancy of soil during shear cannot be modelled by Vesic's cavity expansion solution. Consequently, this solution cannot be used to model cone penetration in medium to dense sands where dilation is known to be significant.

Solution by Salgado (1993)

Salgado (1993) used a stress rotation analysis to relate the cone resistance to a cylindrical cavity limit pressure. The cavity expansion solution developed by Salgado (1993) is an extension of the analytical solution of Yu (1990) and Yu and Houlsby

(1991) for finite cavity expansion in non-associated, dilatant soils. Unlike the solution of Yu and Houlsby (1991) which is for perfectly plastic soils, the cavity expansion solutions presented by Salgado (1993) account for the effects of the variable friction and dilation angles, as well as the dependence of the shear modulus on pressure and void ratio.

Based on a number of simplifying assumptions, cone resistance is related to the effective cylindrical cavity limit pressure ψ_c' by Salgado (1993) as follows :

$$q_c' = 2\exp(\pi\tan\phi')\frac{[(1+C)^{1+\beta_T} - (1+\beta_T)C - 1]}{C^2\,\beta_T(1+\beta_T)}\,\psi_c' \tag{8.60}$$

where the quantity β_T must be determined numerically (Salgado, 1993) and for a standard 60 degree cone the parameter C is :

$$C = \sqrt{3}\exp(\frac{\pi}{2}\tan\psi) \tag{8.61}$$

and ψ is the angle of soil dilation.

Salgado (1993) and Salgado *et al.* (1997) applied the correlations (8.60)-(8.61) to predict the cone resistances for a large number of chamber tests and concluded that the correlations worked reasonably well. Typically, the measured cone resistance can be predicted to within 30%.

Solution by Yasufuku and Hyde (1995)

Based on the observation of model pile tests on crushable sand, Yasufuku and Hyde (1995) proposed a simple failure mechanism for relating cone resistance to a spherical cavity limit pressure. The failure mechanism assumed by Yasufuku and Hyde (1995) can be used to obtain the following expression for the cone resistance :

$$q_c' = \frac{\psi_s'}{1-\sin\phi'} = \frac{A}{1-\sin\phi'}\,p_0' = \frac{(1+2K_0)A}{3(1-\sin\phi')}\sigma_{v0}' \tag{8.62}$$

From equation (8.55), the cone factor suggested by equation (8.62) is:

$$N_q = \frac{(1+2K_0)A}{3(1-\sin\phi')} \tag{8.63}$$

Although the correlation (8.62) has been successfully used to predict pile end-bearing capacity in crushable sands, the applicability of this correlation to other less compressible soils remains to be seen.

Comparison with Chamber Testing

As the values of cone factor for sand obtained using various methods differ significantly, they should be compared with experimental data wherever possible. Field testing is not generally suitable for validating theoretical solutions because of soil variability and the number of unknown quantities associated with a field test. By comparison, the results of chamber testing provide a useful database for verifying various theoretical solutions for the cone resistance of sand. However, as noted earlier, the effect of chamber size and boundary conditions on the test data must be corrected before the test data can be compared with theoretical solutions. This is because most theoretical solutions relate to the cone penetration test in an infinite soil mass.

Effect of Chamber Size. The chamber size effects, which increase with increasing relative density, are not identical for all sands. For example, Ghionna (1984) indicated that for Ticino sand the measured cone resistance is relatively independent of chamber to cone diameter ratio for ratios above 34. In contrast, the chamber size effect for Hokksund sand was found to be more significant. This difference may be related to the fact that Ticino sand is more compressible than Hokksund sand. Yu (1990) suggested that the chamber size effect may be accounted for by comparing the cavity limit pressure in a finite soil medium (i.e. in a large calibration chamber) and that in an infinite soil medium (i.e. in the field). This approach was later followed by Houlsby and Yu (1990) and Schnaid and Houlsby (1991). By combining cavity expansion theory and stress rotation analysis, Salgado (1993) was also able to analyse chamber size effects. All these cavity expansion studies indicate that the cavity expansion approach works reasonably well for predicting the chamber size effects on dense sands. However, for loose and medium sands, a larger chamber size effect than observed in the laboratory was predicted (Parkin and Lunne, 1982; Eid, 1987; Parkin, 1988; Ghionna and Jamiolkowski, 1991). In addition, as shown by Salgado (1993) the cavity expansion approach suggests a rather significant chamber size effect even when the chamber to cone diameter ratio is up to 100. This contrasts with the current belief that the chamber size effect can be neglected provided the chamber to cone diameter ratio is above 60 or 70 (Ghionna and Jamolkowski, 1991 and Mayne and Kulhawy, 1991). More experimental research is required on chamber testing of cone penetrometers, in particular of those with high chamber to cone diameter ratios, to resolve the issue of chamber size effect.

In view of this, the calibration chamber data for cone tests on normally consolidated Ticino sand are used for comparison, because (i) sufficient experimental data exists for cone tests on Ticino sand with three different chamber to cone diameter ratios of 33.6, 47.2, 60; in addition, some tests were carried out with a chamber to cone diameter ratio of 120 (see, for example, Parkin, 1988; Been *et al.*, 1987; Salgado, 1993); (ii) the chamber size effect for Ticino sand was found to be relatively small even for samples of high relative density (Ghionna, 1984; Parkin, 1988); (iii) the test data on Ticino sand has been regrouped by Sladen (1989) according to mean stress level so that the scatter is significantly reduced.

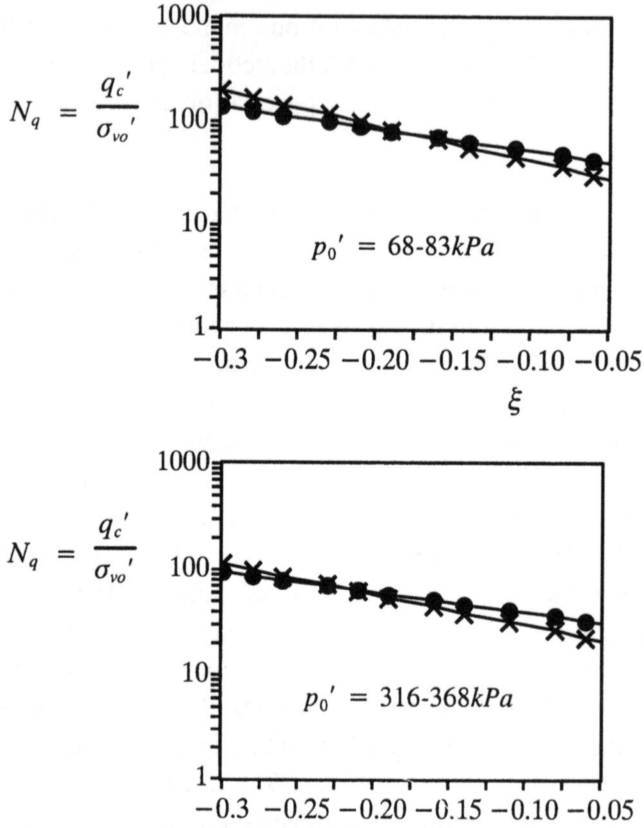

Figure 8.27: Comparison between experimental data (cross) and cavity expansion prediction (solid circle) for Ticino sand (adopted from Yu and Mitchell, 1998)

Correlation between Cone Resistance and State Parameter. Unlike what was assumed in Been *et al.* (1987), a detailed analysis of the cone test data on normally consolidated Ticino sand by Sladen (1989) shows that there is *no unique correlation* between the normalised cone resistance (i.e. the cone factor) and the state parameter ξ (which is defined as the difference between current void ratio and the void ratio at the critical state at the same mean effective stress). Although experimental evidence suggests that this correlation varies systematically with the mean stress level, theoretical solutions must be developed to predict such a dependence.

The dependence of the cone factor–state parameter correlation on the mean stress can be accurately predicted by the cavity expansion solution of Collins *et al.* (1992). Using a state parameter soil model, Collins *et al.* (1992) found that the spherical cavity expansion limit pressure ψ_s' can be expressed as follows:

$$\frac{\psi_s'}{p_0'} = A = m_1(p_0')^{(m_2+m_3v_0)} \exp(-m_4v_0) \tag{8.64}$$

where v_0 is the initial specific volume of the soil (=1 + void ratio); the constants m_1, m_2, m_3, m_4 depend on the critical state properties. Their values for Ticino sand are: $m_1 = 2.012 \times 10^7, m_2 = -0.875, m_3 = 0.326, m_4 = 6.481$ (Collins *et al.*, 1992).

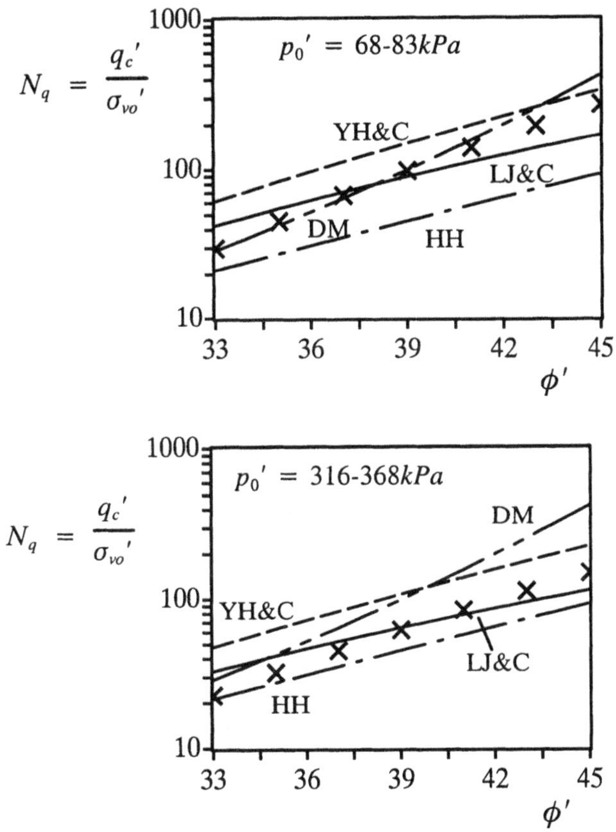

Figure 8.28: Comparison between experimental data (cross) and prediction of cone factors by several methods for Ticino sand (adopted from Yu and Mitchell, 1998)

Equation (8.64) can be used in conjunction with Ladanyi and Johntson's correlation to obtain a mean stress dependent expression for the cone factor in sand. Presented in Figure 8.27 are comparisons between the cavity expansion prediction and experimental data for two different mean stress levels. The experimental curves were obtained by applying a chamber size factor (given by Been *et al.*, 1987) to the best-fit lines presented by Sladen (1989). The theoretical solution was derived for a partly rough cone by assuming the friction angle at the soil-cone interface is half the soil friction angle (i.e. $\lambda = 0.5$). It can be seen from these comparisons that it is possible to predict the measured cone factors very accurately.

Correlation between Cone Resistance and Soil Friction Angle. To evaluate the performance of other solutions, Figure 8.28 shows comparisons of four theoretical correlations with the experimental data in terms of cone factor against friction angle. As expected, the bearing capacity solution of Durgunoglu and Mitchell (DM) (1975) compares well with the data for low stress levels (i.e. shallow penetration). For the cases of high mean stresses, the Durgunoglu and Mitchell solution significantly overestimates the cone factor. For all stress levels considered, the average chamber correlation suggested by Houlsby and Hitchman (HH) (1988) gives a much lower cone factor than the experimental data. In other words, if the average correlation of Houlsby and Hitchman is used to back-calculate the friction angle for Ticino sand, a much higher friction angle (i.e. unconservative value) is obtained. Compared with the experimental data, the cavity expansion correlation proposed by Yasufuku and Hyde (YH&C) (1995) seems to give consistently higher values of cone factor for all values of friction angle and mean stress levels. On the other hand, the correlation suggested by Ladanyi and Johnston (LJ&C) (1974), when used in conjunction with the cavity expansion solution of Collins *et al.* (1992), agrees with the measured cone resistance. The agreement is best for the case of high initial mean stresses. For very dense sand with a very low mean stress (i.e. shallow penetration in dense sand), cavity expansion theory is less satisfactory. This is not surprising considering shallow cone penetration cannot be accurately modelled by a cavity expansion process.

8.5.3 General Remarks

Many relationships have been proposed to relate cone resistance to cavity expansion limit pressures. The cavity expansion approach is more realistic than bearing capacity theory for the following two reasons:

(a) Both elastic and plastic deformations of the soil during the cone penetration test can be properly taken into account using cavity expansion theory.

(b) The cavity expansion approach is able to consider, in an approximate manner, both the influence of the cone penetration process on initial stress states and the effect of stress rotations that occur around the cone tip.

8.6 SUMMARY

1. Cavity expansion theory has been used by the geotechnical community as the most important theoretical basis for interpretation of pressuremeter and cone penetration tests in soils. The two-dimensional nature of pressuremeter tests can be readily accounted for using correction factors derived from two dimensional finite element analysis.
2. For self-boring pressuremeter tests in clay, interpretation methods based on cavity expansion theory are developed to derive shear modulus, undrained shear strength and shear stress-strain relationship from pressuremeter curves.
3. For cone pressuremeter tests in clay, cavity expansion based interpretation methods can be used to derive shear modulus, *in-situ* horizontal stress, and undrained shear strength.
4. For self-boring pressuremeter tests in sand, interpretation methods are developed to derive shear modulus, angles of soil friction and dilation, and shear stress-strain relationship from pressuremeter curves. There are also methods that can be used to measure *in-situ* state parameter of sand deposits.
5. Less progress has been made in the rigorous analysis of cone pressuremeter tests in sand, although semi-analytical methods based on cavity expansion theory have been developed for measuring soil strength and state parameter from measured cone tip resistance and pressuremeter limit pressure.
6. The analogy between cavity expansion and cone penetration is based on the experimental observation that the pressure required to produce a deep hole in an elastic-plastic medium is proportional to that necessary to expand a cavity of the same volume under the same conditions. Two steps must be followed in order to use cavity expansion theory to predict cone resistance: (a) to develop theoretical (analytical or numerical) limit pressure solutions for cavity expansion in soils, and (b) to relate cavity expansion limit pressures to cone resistance. The cavity expansion approach provides a more accurate prediction of cone resistance than bearing capacity theory because the influences of soil stiffness, compressibility (or dilatancy) and penetration-induced horizontal stress increase can all be adequately taken into account.

REFERENCES

Ajalloeian, R. and Yu, H.S. (1998). Chamber studies of the effects of pressuremeter geometry on test results in sand. *Geotechnique,* 48(5), 621–636.

Atkinson, J.H. and Bransby, P.L. (1978). *The Mechanics of Soils*. McGraw–Hill.

Baguelin, F., Jezequel, J.F., Lemee, E. and Mehause, A. (1972). Expansion of cylindrical probes in cohesive soils. *Journal of the Soil Mechanics and Foundations Division*, ASCE, 98(SM11), 1129–1142.

Baldi, G., Bellotti, R., Ghionna, V., Jamiolkowski, M. and Pasqualini, E. (1981). Cone resistance in dry N.C. and O.C. sands. *Cone Penetration Testing and Experience,* ASCE, National Convention, St. Louis, Missouri (Ed. Norris and Holtz), 145–177.

Baligh, M.M. (1975). Theory of deep static cone penetration resistance. *Report No. R75–56, Department of Civil and Environmental Engineering,* MIT, USA.

Baligh, M.M. (1976). Cavity expansion in sands with curved envelopes. *Journal of the Geotechnical Engineering Division,* ASCE, 102, 1131–1146.

Baligh, M.M. (1985). Strain path method. *Journal of Geotechnical Engineering,* ASCE, 111(9), 1108–1136 .

Been, K and Jefferies, M.G. (1985). A state parameter for sands. *Geotechnique,* 35(2), 99–112.

Been, K., Crooks, J.H.A., Becker, D.E. and Jefferies, M.G. (1987). The cone penetration test in sands: II, general inference of state. *Geotechnique,* 37(3), 285–299.

Bellotti, R., Ghionna, V., Jamiolkowski, M., Lancellotta, R. and Manfredini, G. (1986). Deformation characteristics of cohesionless soils from in-situ tests. *In-situ 86 – Use of In-situ Tests in Geotechnical Engineering,* ASCE, Blackburg, 47–73.

Bishop, A.W. (1966). The strength of soils as engineering materials. *Geotechnique,* 16, 91–128.

Bishop, R.F., Hill, R and Mott, N.F. (1945). The theory of indentation and hardness tests. *Proceedings of Physics Society,* 57, 147–159.

Bond, A.J. and Jardine, R.J. (1991). Effect of installing displacement piles in a high OCR clay. *Geotechnique*, 41, 341–363.

Carter, J.P., Booker, J.R. and Yeung, S.K. (1986). Cavity expansion in cohesive frictional soils. *Geotechnique*, 36, 349–358.

Carter, J.P., Randolph, M.F. and Wroth, C.P. (1979). Stress and pore pressure changes in clay during and after the expansion of a cylindrical cavity. *International Journal for Numerical and Analytical Methods in Geomechanics*, 3, 305–323.

Charles, M., Yu, H.S. and Sheng, D. (1999). Finite element analysis of pressuremeter tests using critical state soil models. *Proceedings of the 7th International Symposium on Numerical Models in Geomechanics* (NUMOG7), Graz, 645–650.

Clarke, B.G. (1993). The interpretation of self-boring pressuremeter tests to produce design parameters. *Predictive Soil Mechanics,* (Editors: G.T. Houlsby and A.N. Schofield), Thomas Telford, London, 156–172.

Clarke, B.G. (1995). *Pressuremeter in Geotechnical Design*. Chapman and Hall, London.

Clough, G.W., Briaud, J.L. and Hughes, J.M.O. (1990). The development of pressuremeter testing. *Proceedings of the 3rd International Symposium on Pressuremeters*, Oxford, 25–45.

Collins, I.F. (1990). On the mechanics of state parameter models for sands. *Proc. 7th International Conference on Computer Methods and Advances in Geomechanics*, Cairns, Australia, Vol 1, 593–598.

Collins, I.F. and Yu, H.S. (1996). Undrained cavity expansions in critical state soils. *International Journal for Numerical and Analytical Methods in Geomechanics*, 20(7), 489–516.

Collins, I.F., Pender, M.J. and Wang, Y. (1992). Cavity expansion in sands under drained loading conditions. *International Journal for Numerical and Analytical Methods in Geomechanics*, 16(1), 3–23.

Collins, I.F., Pender, M.J. and Wang, Y. (1994). Critical state models and the interpretation of penetrometer tests. *Computer Methods and Advances in Geomechanics*, Siriwardane and Zaman (eds) Balkema, Rotterdam, 2, 1725–1730.

Durban, D. and Fleck, N.A. (1992). Singular plastic fields in steady penetration of a rigid cone. *Journal of Applied Mechanics, ASME*, 59, 706–710.

Durgunoglu, H.T. and Mitchell, J.K. (1975). Static penetration resistance of soils. *Proceedings of the ASCE Specialty Conference on In-situ Measurements of Soil Properties,* Vol 1, 151–189.

Eid, W.K. (1987). *Scaling Effect on Cone Penetration Testing in Sand.* PhD thesis, Virginia Tech, USA.

Fahey, M. (1980). *A Study of the Pressuremeter Test in Dense Sand.* PhD Thesis, University of Cambridge, England.

Ferreira, R.S. and Robertson, P.K. (1992). Interpretation of undrained self–boring pressuremeter test results incorporating unloading. *Canadian Geotechnical Journal,* 29, 918–928.

Gens, A. and Potts, D.M. (1988). Critical state models in computational geomechanics. *Engineering Computations,* 5, 178–197.

Ghionna, V, (1984). Influence of chamber size and boundary conditions on the measured cone resistance. *Seminar on Cone Penetration Testing in the Laboratory,* University of Southampton.

Ghionna, V, and Jamiolkowski, M. (1991). A critical apraisal of calibration chamber testing of sands. *Proceedings of the 1st International Symposium on Calibration Chamber Testing,* Potsdam, New York, 13–39.

Gibson,R.E. and Anderson, W.F. (1961). *In-situ* measurement of soil properties with the pressuremeter. *Civil Engineering Public Works Review.*, 56, 615–618.

Hill, R. (1950). *The Mathematical Theory of Plasticity.* Oxford University Press.

Houlsby, G.T. and Carter, J.P. (1993). The effect of pressuremeter geometry on the results of tests in clays. *Geotechnique*, 43, 567–576.

Houlsby, G.T. and Hitchman, R. (1988). Calibration chamber tests of a cone penetrometer in sand. *Geotechnique,* 38(1), 39–44.

Houlsby, G.T. and Withers, N.J. (1988). Analysis of the cone pressuremeter test in clay. *Geotechnique*, 38, 575–587.

Houlsby, G.T. and Yu, H.S. (1990). Finite element analysis of the cone pressuremeter test. *Proceedings of the 3rd International Symposium on Pressuremeters,* Oxford, 201–208.

Houlsby, G.T., Clarke, B.G. and Wroth, C.P. (1986). Analysis of the unloading of a pressuremeter in sand. *Proceedings of the 2nd International Symposium on Pressuremeter- and its Marine Applications*, ASTM SPT950, 245–262.

Hughes, J.M.O. and Robertson, P.K. (1985). Full–displacement pressuremeter testing in sands. *Canadian Geotechnical Journal,* 22, 298–307.

Hughes, J.M.O, Wroth, C.P. and Windle, D. (1977). Pressuremeter tests in sands. *Geotechnique,* 27(4), 455–477.

Jamiolkowski, M., Ladd, C.C., Germaine, J.T. and Lancellotta, R. (1985). New developments in field and laboratory testing of soils. *Proceedings of the 11th International Conference on Soil Mechanics and Foundation Engineering,* Vol. 1, 57–154.

Jefferies, M.G. (1988). Determination of horizontal geostatic stress in clay with self-bored pressuremeter. *Canadian Geotechnical Journal*, 25, 559–573.

Jewell, R.J., Fahey, M. and Wroth, C.P. (1980). Laboratory studies of the pessuremeter test in sand. *Geotechnique*, 30(4), 507–531.

Kurup, P.U., Voyiadjis, G.Z. and Tumay, M.T. (1994). Calibration chamber studies of piezocone test in cohesive soils. *Journal of Geotechnical Engineering,* ASCE, 120(1), 81–107.

Ladanyi, B. (1963). Evaluation of pressuremeter tests in granular soils. *Proceedings of the 2nd Pan American Conference on Soil Mechanics*, San Paulo, 1, 3–20.

Ladanyi, B. (1975). Bearing capacity of strip footings in frozen soils. *Canadian Geotechnical Journal,* 12, 393–407.

Ladanyi, B. and Johnston, G.H. (1974). Behaviour of circular footings and plate anchors embedded in permafrost. *Canadian Geotechnical Journal,* 11, 531–553.

Laier, J.E., Schmertmann, J.H. and Schaub, J.H. (1975). Effects of finite pressuremeter length in dry sand. *Proceedings of ASCE Speciality Conference on In-situ Measurement of Soil Properties*, Raleigh, Vol 1, 241–259.

Lehane, B. and Jardine, R. (1992). The behaviour of a displacement pile in Bothkennar clay. *Predictive Soil Mechanics* (Editors: G.T. Houlsby and A.N. Schofield), Thomas Telford, London, 421–435.

Lunne, T. and Kleven, A. (1981). Role of CPT in North Sea foundation engineering. *In: Cone Penetration Testing and Experience.* (Editors: G.M. Norris and R.D. Holtz), ASCE, 76–107.

Lunne, T., Robertson, P.K. and Powell, J.J.M. (1997). *Cone Penetration Testing in Geotechnical Practice.* Blackie Academic and Professional, London.

Mair, R.J. and Wood, D.M. (1987). *Pressuremeter Testing, Methods and Interpretation,* CIRIA report, Butterworths, London.

Manassero, M. (1989). Stress-strain relationships from drained self-boring pressuremeter tests in sand. *Geotechnique,* 39(2), 293–308.

Mayne, P.W, and Kulhawy, F.H. (1991). Calibration chamber database and boundary effects correction for CPT data. *Proceedings of the 1st International Symposium on Calibration Chamber Testing,* Potsdam, New York, 257–264.

Meigh, A.C. (1987). *Cone Penetration Testing.* CIRIA Report, Butterworths, London.

Mitchell, J.K. and Tseng, D.J. (1990). Assessment of liquefaction potential by cone penetration resistance. *Proceedings of the H.B. Seed Memorial Symposium,* Vol 2, 335–350.

Muir Wood, D. (1990). *Soil Behaviour and Critical State Soil Mechanics.* Cambridge University Press.

Nash, D.F.T., Powell, J.J.M. and Lloyd, I.M. (1992). Initial investigation of the soft clay test site at Bothkennar. *Geotechnique,* 42(2), 163–181.

Palmer, A.C. (1972). Undrained plane strain expansion of a cylindrical cavity in clay: a simple interpretation of the pressuremeter test. *Geotechnique,* 22(3), 451–457.

Parkin, A.K. (1988). The calibration of cone penetrometers. *Proceedings of the 1st International Symposium on Penetration Testing,* Vol 1, 221–243.

Parkin, A. and Lunne, T. (1982). Boundary effects in the laboratory calibration of a cone penetrometer in sand. *Proceedings of the 2nd European Symposium on Penetration Testing,* Vol 2, 761–768.

Prevost, J.H. and Hoeg, K. (1975). Analysis of pressuremeter in strain softening soil. *Journal of the Geotechnical Engineering Division,* ASCE, 101(GT8), 717–732.

Randolph, M.F., Carter, J.P. and Wroth, C.P. (1979). Driven piles in clay – the effects of installation and subsequent consolidation. *Geotechnique,* 29, 361–393.

Robertson, P.K. and Campanella, R.G. (1983). Interpretation of cone penetration tests: part I: sand. *Canadian Geotechnical Journal,* 20(4), 718–733.

Richart, Jr. F.E., Hall, J.R. and Woods, R.D. (1970). *Vibrations of Soils and Foundations.* Prentice-Hall, Engewood Cliffs, New Jersey.

Sagaseta, C. and Houlsby, G.T. (1992). Stresses near the shoulder of a cone penetrometer in clay. *Proceedings of the 3rd Int. Conference on Computational Plasticity: Fundamentals and Applications,* Vol 2, 895–906.

Salgado, R. (1993). *Analysis of Penetration Resistance in Sands.* PhD Thesis, University of California at Berkeley, USA.

Salgado, R., Mitchell, J.K. and Jamiolkowski, M. (1997). Cavity expansion and penetration resistance in sand. *Journal of Geotechnical and Geoenvironmental Engineering, ASCE*, 123 (4), 344–354.

Schnaid, F. (1990). *A Study of the Cone Pressuremeter Test in Sand.* DPhil Thesis, Oxford University, England.

Schnaid, F. (1994). Relating cone and pressuremeter tests to assess properties and stresses in sand. *Proceedings of XIII International Conference on Soil Mechanics and Foundation Engineering,* New Delhi, India.

Schnaid, F. and Houlsby, G.T. (1991). An assessment of chamber size effects in the calibration of *in-situ* tests in sand. *Geotechnique,* 41(3), 437–445.

Schnaid, F. and Houlsby, G.T. (1992). Measurement of the properties of sand by the cone–pressuremeter test. *Geotechnique,* 42(4), 587–601.

Schofield, A.N. and Wroth, C.P. (1968). *Critical State Soil Mechanics*. McGraw–Hill.

Sladen, J.A. (1989). Problems with interpretation of sand state from cone penetration test. *Geotechnique,* 39 (2), 323–332.

Sousa Coutinho, A.G.F. (1990). Radial expansion of cylindrical cavities in sandy soils: application to pressuremeter tests. *Canadian Geotechnical Journal,* 27, 737–748.

Teh, C.I. (1987). *An Analytical Study of the Cone Penetration Test.* PhD thesis, Oxford University, England.

Teh, C.I. and Houlsby, G.T. (1991). An analytical study of the cone penetration test in clay. *Geotechnique*, 41(1), 17–34.

van den Berg (1994). *Analysis of Soil Penetration*. PhD thesis, Delft University.

Vesic, A.S. (1972). Expansion of cavities in infinite soil mass. *Journal of the Soil Mechanics and Foundations Division,* ASCE, 98, 265–290.

Vesic, A.S. (1977). Design of pile foundations. *National Cooperative Highway Research Program, Synthesis of Highway Practice 42,* Transportation Research Board, National Research Council, Washington, D.C.

Wang, Y. (1992). *Cavity Expansion in Sands with Applications to Cone Penetrometer Tests*. PhD Thesis, University of Auckland, New Zealand.

Whittle, A.J. (1992). Constitutive modelling for deep penetration problems in clay. *Proceedings of the 3rd International Conference on Computational Plasticity: Fundamentals and Applications*, Vol 2, 883–894.

Whittle, A.J. and Aubeny, C.P. (1993). The effects of installation disturbance on interpretation of in–situ tests in clays. *Predictive Soil Mechanics* (Editors: G.T. Houlsby and A.N. Schofield), Thomas Telford, London, 742–767.

Withers, N.J., Howie, J., Hughes, J.M.O. and Robertson, P.K. (1989). Performance and analysis of cone pressuremeter tests in sands. *Geotechnique,* 39(3), 433–454.

Wroth, C.P. (1982). British experience with the self–boring pressuremeter. *Proceedings of the Symposium on Pressuremeter and its Marine Applications,* Paris, Editions Technip, 143–164.

Wroth, C.P. (1984). The interpretation of in–situ soil tests. *Geotechnique,* 34, 449–489.

Yasufuku, N. and Hyde, A.F.L. (1995). Pile end bearing capacity in crushable sands. *Geotechnique*, 45(4), 663–676.

Yeung, S.K. and Carter, J.P. (1990). Interpretation of the pressuremeter test in clay allowing for membrane end effects and material non-homogeneity. *Proceedings of the 3rd International Symposium on Pressuremeters,* Oxford, 199–208.

Yu, H.S. (1990). *Cavity Expansion Theory and its Application to the Analysis of Pressuremeters.* DPhil Thesis, University of Oxford, England.

Yu, H.S. (1992). Expansion of a thick cylinder of soils. *Computers and Geotechnics*, 14, 21–41.

Yu, H.S. (1993). A new procedure for obtaining design parameters from pressuremeter tests. *Transactions of Civil Engineering*, Institution of Engineers, Australia, CE35, No. 4, 353–359.

Yu, H.S. (1993). Discussion on: singular plastic fields in steady penetration of a rigid cone. *Journal of Applied Mechanics,* ASME, 60, 1061–1062.

Yu, H.S. (1994). State parameter from self–boring pressuremeter tests in sand. *Journal of Geotechnical Engineering,* ASCE, 120(12), 2118–2135.

Yu, H.S. (1996). Interpretation of pressuremeter unloading tests in sands. *Geotechnique,* 46(1), 17–31.

Yu, H.S. and Collins, I.F. (1998). Analysis of self–boring pressuremeter tests in overconsolidated clays. *Geotechnique*, 48(5), 689–693.

Yu, H.S. and Houlsby, G.T. (1990). A new finite element formulation for one–dimensional analysis of elastic–plastic materials. *Computers and Geotechnics,* 9(4), 241–256.

Yu, H.S. and Houlsby, G.T. (1991). Finite cavity expansion in dilatant soil: loading analysis. *Geotechnique,* 41(2), 173–183.

Yu, H.S. and Houlsby, G.T. (1992). Effect of finite pressuremeter length on strength measurement in sand. *Civil Engineering Research Report No.* 074.04.1992, ISBN No. 0 7259 0753 3, The University of Newcastle, NSW 2308, Australia.

Yu, H.S. and Houlsby, G.T. (1995). A large strain analytical solution for cavity contraction in dilatant soils. *International Journal for Numerical and Analytical Methods in Geomechanics,* 19(11), 793–811.

Yu, H.S. and Mitchell, J.K. (1998). Analysis of cone resistance: review of methods. *Journal of Geotechnical and Geoenvironmental Engineering,* ASCE, 124(2), 140–149.

Yu, H.S., Schanid, F. and Collins, I.F. (1996). Analysis of cone pressuremeter tests in sand. *Journal of Geotechnical Engineering,* ASCE, 122 (8), 623–632.

Yu, H.S., Herrmann, L.R. and Boulanger, R.W. (2000). Analysis of steady cone penetration in clay. *Journal of Geotechnical and Geoenvironmental Engineering,* ASCE, 126(7).

Zytynski, M., Randolph, M.F., Nova, R. and Wroth, C.P. (1978). On modelling the unloading-reloading behaviour of soils. *International Journal for Numerical and Analytical Methods in Geomechanics,* 2, 87–93

CHAPTER 9

PILE FOUNDATIONS AND EARTH ANCHORS

9.1 INTRODUCTION

It is well established that cavity expansion solutions can be used to estimate the end-bearing capacity of driven piles in soils. Like cone penetration in soils, the rigorous analysis of the behaviour of driven piles is difficult because of its highly material and geometric nonlinear nature. Although considerable progress has been made over the last 20 years in the analysis of driven piles in clays, the analysis and design of piles in sand is still largely empirical.

In this chapter, the main approaches that have been used to apply cavity expansion theory in the analysis of driven piles in soil and rock are discussed. Although attention will be focused on the ultimate capacity of piles, it is worth noting that cavity expansion solutions may also be used to predict the displacement behaviour of tapered piles (Kodikara and Moore, 1993).

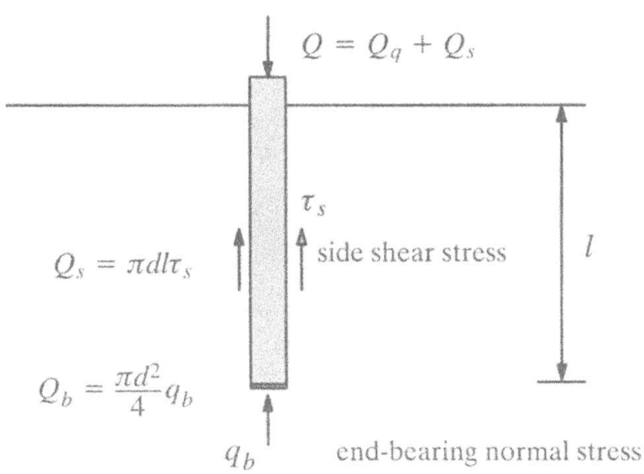

Figure 9.1: An axially loaded driven pile in soil

An axially loaded pile carries the load partly by shear generated along the shaft and partly by the normal stress generated at the base of the pile, Figure 9.1. The relative magnitude of the shaft and base capacities depends on the geometry of the pile and soil profiles. In cohesive soils, the shaft capacity of a friction pile may often represent 80-90% of the total pile capacity. On the other hand, for granular soils the total capacity is generally evenly divided between shaft and base.

Apart from pile modelling, it will be demonstrated in the latter part of this chapter that cavity expansion solutions can also be used to predict the uplift capacity of plate anchors buried in clay and sand.

9.2 AXIAL CAPACITY OF DRIVEN PILES IN CLAY

As a very high proportion of the overall capacity of a driven pile in clay is due to the shaft friction, considerably more effort has been devoted to developing reliable methods for estimating values of skin friction for piles in clay than has been for sand.

9.2.1 Shaft capacity of piles: effect of the installation on soil stress

Over the last three decades, the installation of piles into clay has been studied by many as the expansion of a cylindrical cavity in soils (e.g. Randolph *et al.*, 1979; Davis *et al.*, 1984; Nystrom, 1984; Collins and Yu, 1996). This has stemmed from the observation that when a pile is driven deeply into soil, much of the soil is displaced predominantly outwards in the radial direction, Figure 9.2. In fact, measurements of the radial movements of soil near the pile mid-depth taken from the model tests of Randolph *et al.* (1979) and field measurements of Cooke and Price (1978) show that the radial soil displacements can be accurately predicted by cylindrical cavity expansion solutions.

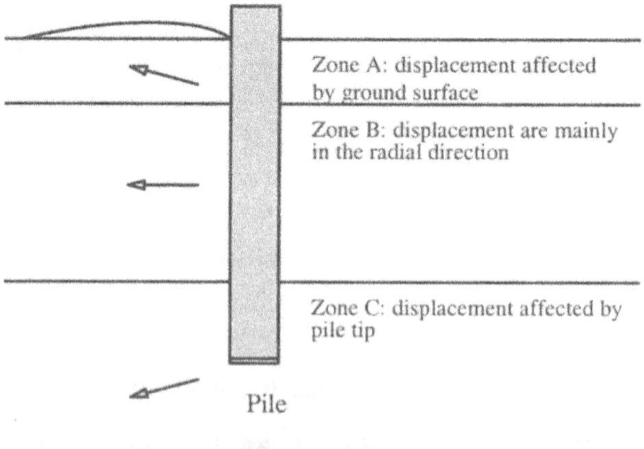

Figure 9.2: Soil displacement due to pile installation

Although cavity expansion theory does not fully model the behaviour of soils surrounding a driven pile in Zones A and C, it provides a useful analytical tool for identifying key parameters that are affecting the pile behaviour. This is supported by a comparative study by Nystrom (1984) indicating that the simple one-dimensional cavity expansion model can be used to predict pile test data as well as the other more complicated two-dimensional finite element methods.

If a pile is driven into the ground quickly, it is reasonable to model the installation as an undrained loading process. Cavity expansion theory can be used to model pile behaviour in two different ways:

(a) The installation of a pile into clay may be modelled as the expansion of a cylindrical cavity from zero radius to the radius of the pile. This will give the stress change in soils around the pile due to pile installation which can be used to estimate the shaft friction.

(b) End bearing capacity of a driven pile can be correlated to the spherical cavity limit pressure in a semi-empirical way.

Total stress analysis with perfectly plastic soil models

If clay is modelled by the perfectly plastic Tresca yield criterion, closed form solutions exist for both the soil stress changes around the shaft and the cavity expansion limit pressures that must be used to estimate the end bearing capacity of piles.

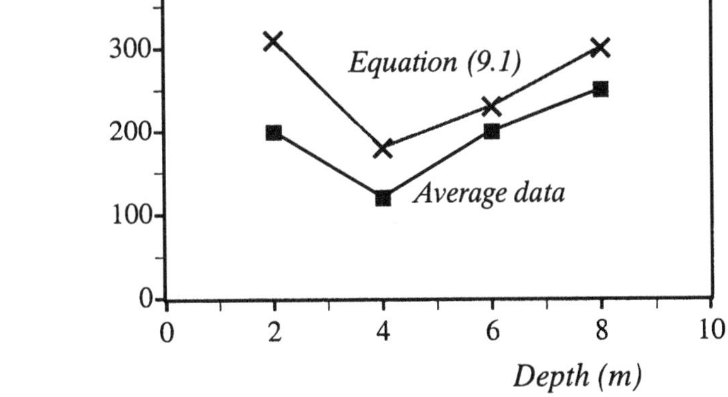

Figure 9.3: Prediction of pile shaft normal stresses for normally consolidated Huntspill clays (Data from Coop and Wroth, 1989)

As discussed in Chapter 3, cavity expansion from zero radius is expected to take place at a constant cavity pressure. This constant cavity pressure is:

$$\psi_c = \sigma_{h0} + s_u[1 + \ln\frac{G}{s_u}] \tag{9.1}$$

for a cylindrical cavity and

$$\psi_s = \sigma_{h0} + \frac{4}{3}s_u[1 + \ln\frac{G}{s_u}] \tag{9.2}$$

for a spherical cavity. σ_{h0} is the *in-situ* total horizontal stress prior to the pile installation; G and s_u are the shear modulus and undrained shear strength of clay respectively.

Figure 9.3 shows the prediction by equation (9.1) for the pile shaft normal stresses measured by Coop and Wroth (1989) for model pile tests in a normally consolidated clay. It is noted that the measured total normal stress at the soil-pile interface is smaller than that predicted using cavity expansion theory. This results from some unloading that occurs in the soil along the pile shaft during pile installation (see, for example, Teh and Houlsby, 1991; Yu *et al.*, 2000) and this unloading reduces the shaft normal stress.

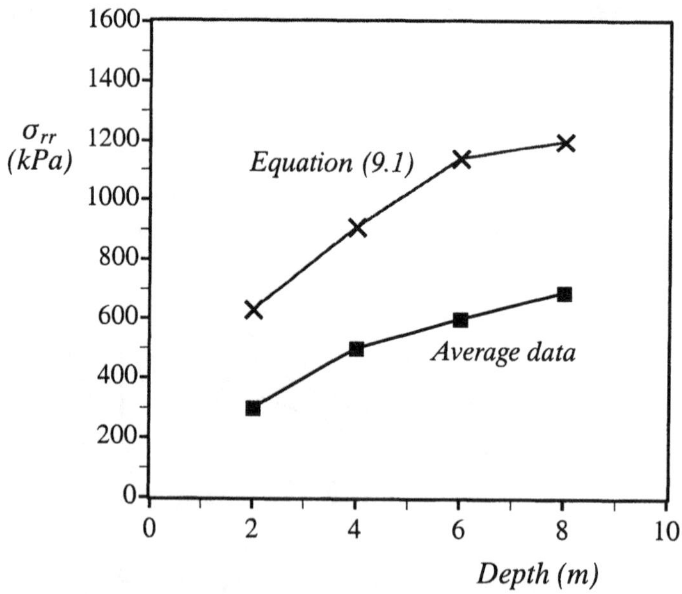

Figure 9.4: Prediction of pile shaft normal stresses for heavily overconsolidated Madingley clays (Data from Coop and Wroth, 1989)

As stressed in Randolph *et al.* (1979) and Collins and Yu (1996), while perfectly plastic soil models may be reasonable for modelling normally or lightly overconsolidated clays, they are not accurate for heavily overconsolidated soils. This is confirmed by the fact the theoretical cylindrical limit pressure is much higher than the measured normal stress in clays with a high OCR (Coop and Wroth, 1989; Bond and Jardine, 1991).

The problem in using the total stress cavity expansion solution (9.1) to estimate the pile normal stress in heavily overconsolidated clays was illustrated by Coop and Wroth (1989) when they applied it to predict the experimental data on Madingley clays. As shown in Figure 9.3 and Figure 9.4, while equation (9.1) provides a reasonable estimate of the pile normal stress in normally consolidated clays, it overestimates the measured normal stresses by over 100% for heavily overconsolidated clays. This suggests that the cavity expansion solutions obtained from a total stress analysis with perfectly plastic soil models are not suitable for estimating the pile normal stresses in high OCR clays.

Once the pile normal stress is estimated, the shaft shear stress may be calculated as follows:

$$\tau_s = \sigma_{rr} \tan \delta \tag{9.3}$$

where δ is the residual friction angle at the soil-pile interface. Experimental tests on soil-pile interface shear tests reported by Coop and Wroth (1989) and Bond and Jardine (1991) indicate that the soil-pile interface friction angle ranges from 0.35 to 0.4 times the critical state friction angle.

Effective stress analysis with critical state soil models

The total stress analysis with a perfectly plastic soil model has the advantage that a closed form solution is possible for cavity expansion problems. However as pointed out by Randolph *et al.* (1979) and Collins and Yu (1996) there are two important shortcomings with the total stress analysis. First of all, no account is taken of pore pressure generated due to pure shear. The second limitation of a perfectly plastic model is its inability to link correctly the strength of the soil and its change with the current effective stress state and stress history of soil. So while it may be reasonable for normally consolidated or lightly overconsolidated soils, the total stress analysis, as presented in the previous section, is not accurate for heavily overconsolidated clays.

An early effective stress cavity expansion analysis using a critical state soil model (i.e. modified Cam clay) was carried out by Randolph *et al.* (1979) using the finite element method. Later closed form solutions were derived by Collins and Yu (1996) for the same problem in soils modelled by various critical state models.

The cavity expansion prediction of radial effective stress is based simply on the assumption that the soil adjacent to the pile shaft is at a critical state under plane strain conditions with a radial major principal stress. The effective radial stress at the cavity wall (i.e. normal stress along the pile shaft) may be expressed as follows:

$$\sigma'_{rr} = [1 + \sqrt{3}/M]s_u \tag{9.4}$$

where M is the slope of the critical state line in the q'–p plot, which is a simple function of the critical state soil friction angle.

The pore pressure results at the cavity wall obtained from the finite element parametric study of Randolph *et al.* (1979) are represented by the following simple equation:

$$\Delta U = U - U_0 = (p'_0 - p'_f) + s_u \ln \frac{G}{s_u} \tag{9.5}$$

where U_0 is the *in-situ* porewater pressure and p'_0 and p'_f denote the mean effective normal stresses at the states of *in-situ* and failure respectively. The second term in equation (9.5) represents the change in the mean normal effective stress as the soil is sheared to failure.

Over the last three decades, many investigators (e.g., Kirby and Esrig, 1979) have measured pore water pressures generated at the pile-soil interface during pile installation in clay soils. Further high quality experimental research on pile testing has been carried out in recent years and notable examples include the work carried out at Oxford University by Coop and Wroth (1989) and Bond and Jardine (1991) at Imperial College.

The work of Coop and Wroth (1989) on instrumented piles in both normally consolidated and heavily overconsolidated clays suggests that although the previous cavity expansion theories (e.g., Randolph *et al,* 1979) may be satisfactorily used to model the installation of a displacement pile in normally consolidated clays, they do not seem to be able to predict the pile behaviour in a heavily overconsolidated clay. In particular, Coop and Wroth (1989) noted that the excess pore pressures predicted by the cavity expansion theory of Randolph *et al.* (1979) for pile installation in high OCR clays are much higher than those measured in the instrumented pile testing. This finding has been confirmed by an independent research programme of Bond and Jardine (1991) on pile testing in a heavily overconsolidated London clay. Using the cavity expansion theory of Randolph *et al.* (1979), Wroth *et al.* (1979) suggested that the normalised excess pore pressure (i.e. the excess pore pressure divided by initial undrained shear strength of the soil) for heavily overconsolidated London clay is in the range of 3.1 to 3.6. However, the results from the instrumented pile tests in London clay indicate that negative pore pressures were recorded during the pile installation.

The relative insensitivity to the overconsolidation ratio of excess pore pressures, as predicted by the cavity expansion theory of Randolph *et al.* (1979), is largely due to a combined effect of their choices of variation of shear modulus G with overconsolidation ratio and the modified Cam clay yield surface for modelling heavily overconsolidated clays.

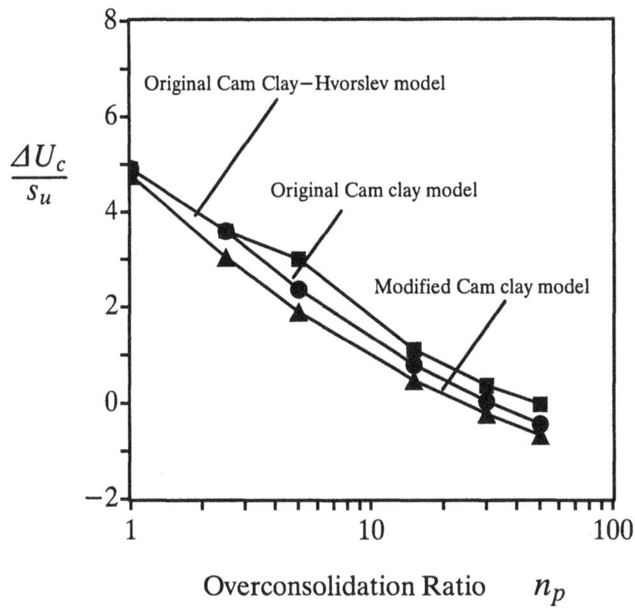

Figure 9.5: Variation of excess porewater pressure at cavity wall with overconsolidation ratio

As a significant development of cavity expansion theory, Collins and Yu (1996) presented an analytical study of the undrained expansion of cylindrical and spherical cavities in soil modelled by various critical state models. The solution procedure is applicable to large strain cavity expansion in any isotropically hardening materials. In all cases only simple quadratures are involved, but in the case of original Cam clay a complete analytical solution has been derived. It has also been shown that the well known perfectly plastic solution can be recovered as a special case of the present solution. The influence of the choice of various critical state models on the cavity expansion behaviour has been investigated in detail by comparing the results of cavity expansion curves and stress distributions derived using different models. A brief application of the present cavity expansion solution to modelling pile installation in cohesive soil confirms its potential and usefulness in geotechnical practice.

To explain the negative excess pore pressures measured at the pile wall during the pile driving in high OCR clays, Collins and Yu's cavity expansion solutions are presented in such a way that the variations of the excess pore pressures with overconsolidation ratio can be studied in detail. The cavity expansion results for London clay predicted by three different critical state soil models when initial specific volume is equal to 2.0 are plotted in Figure 9.5. These are presented in terms of the normalised excess pore pressures at the cavity wall against the overconsolidation ratio. The values of excess pore pressures are obtained at the instant of cavity expansion ratio $a/a_0 = 4$. It is interesting to see that, for all soil models, the normalised excess pore pressures tend to decrease, in a similar manner, with the increase of the overconsolidation ratio. The theoretical trends shown in Figure 9.5 for the case when the overconsolidation ratio is less than 10 are confirmed by the field measurements reported by Kirby and Esrig (1979). According to Figure 9.5, the excess pore pressures become negative if the overconsolidation ratio is greater than 25. As the overconsolidation ratio of London clay is in the range of 20 to 50 (Bond and Jardine, 1991), the results in Figure 9.5 suggests that negative excess pore pressures could be generated at the pile wall during the pile installation in London clay. This is exactly what was observed by Bond and Jardine (1991) in their experimental research on pile testing in London clay for which the specific volume of the soil was slightly less than 2.0. In the study of Collins and Yu (1996), it was also found that the normalised excess pore pressure for a given overconsolidation ratio increases only slightly with the increase of the initial specific volume of the soil.

9.2.2 End bearing capacity of driven piles

It is well known that the long-term, drained end bearing capacity of a pile is much larger than the short term, undrained capacity. However it is also necessary that the pile has a sufficient short term capacity in order to avoid a short term failure immediately after installation. It is also true that the settlements required to mobilise the long term pile capacity may be too large to be tolerated by serviceability requirements. As a result, it is standard practice to calculate the end bearing capacity assuming undrained conditions for clays (Fleming *et al.*, 1985).

The end bearing capacity is traditionally related to the spherical cavity limit pressure. Figure 9.6 shows one of the first analogies between spherical cavity expansion and end bearing failure which was proposed by Gibson (1950). Using this analogy in undrained clay soils, it is possible to demonstrate that the pile end bearing capacity can be expressed as a function of the spherical cavity limit pressure ψ_s and the shear stress mobilised at the interface between the rigid soil and plastic soils $a_1 s_u$ ($a_1 = 0.0 - 1.0$):

$$q_b = \psi_s + \alpha_1 s_u \tag{9.6}$$

For perfectly plastic soils with a Tresca yield criterion, equation (9.2) can be used to determine the spherical cavity limit pressure. In this case, the end bearing capacity of a pile is estimated as follows:

$$q_b = \sigma_{h0} + [\frac{4}{3} + \alpha_1 + \frac{4}{3}\ln\frac{G}{s_u}]\ s_u \tag{9.7}$$

The above equation takes into account the effects of *in-situ* stresses and stiffness index of soils on the end bearing capacity of a pile. It therefore represents a better solution than the existing design methods that mainly use an arbitrary bearing capacity factor (Skempton, 1951; Fleming *et al*, 1985).

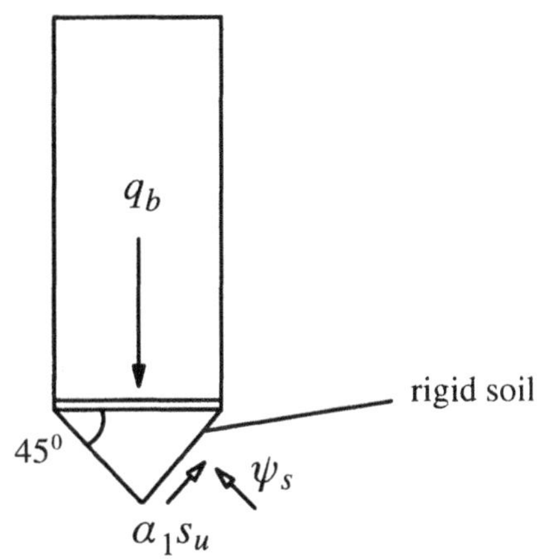

Figure 9.6: Relation between spherical limit pressure and pile end bearing capacity in undrained clay

9.2.3 Increase in capacity of piles with time: effect of consolidation

When piles are driven into clay soils, high excess porewater pressures are generated close to the pile. This high porewater pressure profile reduces the effective stresses along the pile shaft, meaning that the short term shaft capacity is relatively low. After the pile is installed, the excess porewater pressure dissipates, largely by radial flow of water away from the pile. The soil therefore consolidates. The consequence of this consolidation process is that the specific volume of the soil (i.e. water content) reduces and the undrained shear strength increases. Experimental data presented by Seed and Reese (1955) indicates that there was a 7% decrease in water content close to a pile driven into San Fransisco Bay mud. This reduction in water content means a threefold increase in the remoulded soil strength.

As presented before, the excess porewater pressure generated by pile installation may be estimated by simulating it as a cylindrical cavity expansion from zero radius. The solutions are different depending on whether a total stress or an effective stress analysis is used.

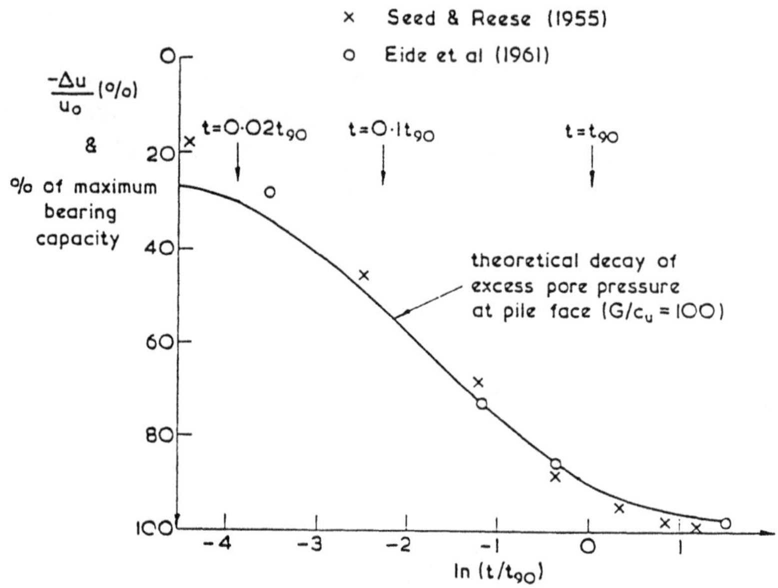

Figure 9.7: Comparison of variation of pile bearing capacity with time and theoretical decay of excess pore pressure (after Randolph and Wroth, 1979)

Using a total stress analysis with the Tresca yield criterion, an analytical solution was derived by Randolph and Wroth (1979). A brief description of this solution was given in Chapter 6. The initial porewater pressure at the pile shaft at the end of pile installation can be shown to be as follows:

$$\Delta U = s_u \ln(\frac{G}{s_u}) \tag{9.8}$$

A key result of Randolph and Wroth's total stress consolidation solution is plotted in Figure 9.7, which details a comparison between the theoretical decay of excess porewater pressure and the measured bearing capacity of driven piles as a percentage of their long term bearing capacity. The time scale has been normalised by dividing by the time taken for 90% of the consolidation to have occurred. The curve is found to be relatively insensitive to the value of stiffness index G/s_u.

As stressed earlier, the total stress analysis is suitable for predicting the behaviour of driven piles into a normally consolidated or lightly overconsolidated clay, but it is not be applicable to heavily overconsolidated clays. If the clay is heavily overconsolidated, it is necessary to use an effective stress analysis method in conjunction with critical state models. The first such study of stress changes caused by pile installation and the consolidation process was carried out by Randolph *et al.* (1979) using finite elements and modified Cam clay. As mentioned before, their study suggests that the porewater pressures generated by pile installation is independent of the value of OCR. This conclusion was not supported by recent field research on model pile testing by Coop and Wroth (1989) and Bond and Jardine (1991). For clay with a high OCR, the measured excess porewater pressures close to the pile shaft may become negative and are generally much lower than those predicted by the analysis of Randolph *et al.* (1979). A complete numerical analysis of the consolidation process presented in Randolph *et al.* (1979) indicates that the undrained shear strength of soil may increase by 50–100% close to the pile due to the decrease in water content caused by consolidation.

When compared with the earlier study of Randolph *et al.* (1979), a more realistic prediction of the excess porewater pressure due to pile installation has been made by Collins and Yu (1996) using more accurate critical state models. In particular, in agreement with field testing, the solutions of Collins and Yu (1996) indicate that for a clay with a very high OCR, the excess porewater pressure close to the pile shaft are negative (see Figure 9.5). Further work is needed to develop consolidation solutions using realistic critical state models such as those used by Collins and Yu (1996).

9.3 AXIAL CAPACITY OF DRIVEN PILES IN SAND

9.3.1 End bearing capacity of piles in sand

Two steps must be followed in order to use cavity expansion theory to predict the end bearing capacity of a driven pile in soil. Firstly, semi-analytical correlations are required to relate end bearing capacity to the cavity limit pressure. Secondly, analytical or numerical solutions must be developed for limit pressures of cavity expansion in soil modelled by a realistic plasticity model. Some details of how these two steps may be addressed are discussed below.

From observations made in deep punching of metals, Bishop *et al.* (1945) and Hill (1950) concluded that the pressure required to produce a deep hole in an elastic-plastic medium is proportional to that necessary to expand a cavity of the same volume and under the same conditions, provided there is no friction. Gibson (1950) was the first to correlate cavity expansion limit pressure to end bearing pressure of a deep foundation. Shown in Figure 9.8 is the correlation proposed by Gibson for sand. It is a simple matter by considering the vertical equilibrium to express the end bearing capacity of the pile as (Ladanyi and Johnston, 1974; Randolph *et al.*, 1994):

$$q_b = (1 + \tan\phi \tan\alpha)\psi_s \tag{9.9}$$

If the soil below the pile base is assumed to have been sheared to its ultimate state, the angle α may be estimated as a function of critical state soil friction angle by $\alpha = 45^0 + \phi_{cv}/2$.

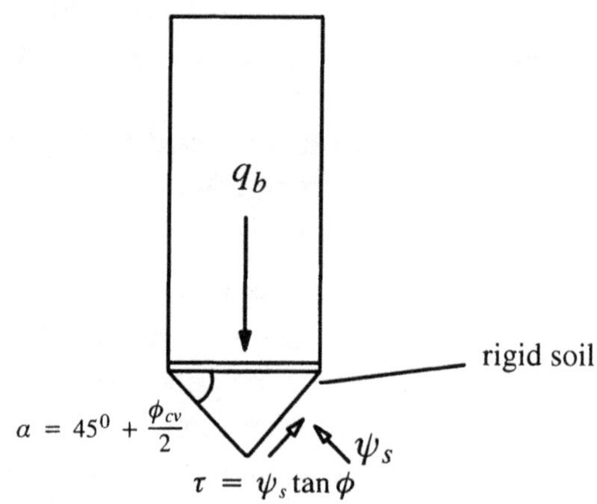

Figure 9.8: Relation between spherical limit pressure and pile end bearing capacity in sand

Randolph *et al.* (1994) proposed a new theoretical framework for predicting the end bearing capacity of a driven pile in sand by combining equation (9.9) and the cavity expansion solutions of Carter *et al.* (1986), Yu and Houlsby (1991) and Collins *et al.* (1992). The spherical cavity limit pressure can be evaluated through the closed form solutions developed by Carter *et al.* (1986) and Yu and Houlsby (1991) for soils modelled by an elastic-perfectly plastic Mohr-Coulomb criterion with constant angles of friction and dilation.

In reality, however, the strength of sands (i.e. angles of friction and dilation) depends on deformation history. To account for such a dependence, a state parameter-based critical state model may be used in which the angles of friction and dilation are assumed to be a function of state parameter (Been and Jefferies, 1985). Using the state parameter model described in Chapter 4 (Section 4.3.1), Collins *et al.* (1992) presented numerical solutions for limit pressures of cavity expansion in sand. It was found that the effect of variable friction and dilation angles was simply accounted for using an average value of strength parameters (friction and dilation angles) at the initial and critical states in an elastic perfectly plastic cavity expansion solution.

Therefore the cavity expansion limit pressure with a state parameter model can be obtained from the closed form solutions of Carter *et al.* (1986) and Yu and Houlsby (1991) using the following expressions to obtain the input strength parameters:

$$\phi = \frac{1}{2}(\phi_i + \phi_{cv}) \tag{9.10}$$

$$\psi = \frac{1}{2}\psi_i \tag{9.11}$$

The critical state friction and dilation angles $(\phi_{cv}, 0)$ are independent of soil disturbance and can be measured easily using routine laboratory testing. The angles of friction and dilation at the initial state (ϕ_i, ψ_i) can be estimated from the initial state parameter using the experimental based expression (Collins *et al*, 1992; Been *et al.*, 1987; Yu, 1994, 1996):

$$\phi_i = \phi_{cv} + A[\exp(-\xi_0) - 1]; \qquad \text{and} \qquad \psi_i = 1.25(\phi_i - \phi_{cv}) \tag{9.12}$$

where ξ_0 is the initial state parameter, which is linked to the initial mean effective stress p'_0 and initial specific volume v_0 as follows:

$$\xi_0 = v_0 + \lambda \ln p'_0 - \Gamma \tag{9.13}$$

λ and Γ are critical state properties of the soil. A is a curve fitting parameter ranging from 0.6 to 0.95 depending on the type of sand.

Alternatively, Randolph *et al.* (1994) followed the work of Bolton (1986, 1987) to estimate initial angles of friction and dilation, linking them to the relative density D_r and the mean effective stress p'_0. The average angles of friction and dilation determined can be shown to be:

$$\phi = \phi_{cv} + 1.5 I_R \tag{9.14}$$

$$\psi = 1.875 I_R \tag{9.15}$$

where I_R is defined by

$$I_R = 5D_r - 1 \qquad \text{for } p'_0 \leq 150kPa \tag{9.16}$$

$$I_R = D_r[5.4 - \ln(p'_0/p_a)] - 1 \qquad \text{for } p'_0 > 150kPa \tag{9.17}$$

and p_a is the atmospheric pressure (100kPa).

Apart from the angles of friction and dilation, soil stiffness (i.e. shear modulus) also plays a key part in the limit pressure solution of cavity expansion in sand. To account for the effect of relative density and mean effective stress on shear modulus, many experimental correlations have been proposed (e.g. Richart *et al.*, 1970; Lo Presti, 1987). For application to pile design, Randolph *et al.* (1994) suggested the following expression for estimating shear modulus of clean sand for input in determining cavity expansion limit pressures:

$$\frac{G}{p_a} = 400\exp(0.7D_r)\left(\frac{p'}{p_a}\right)^{0.5} \tag{9.18}$$

As shown in Randolph *et al.* (1994), the prediction of limiting end bearing pressure obtained from the cavity expansion method is compared with the design charts of Fleming *et al.* (1992) in Figure 9.9.

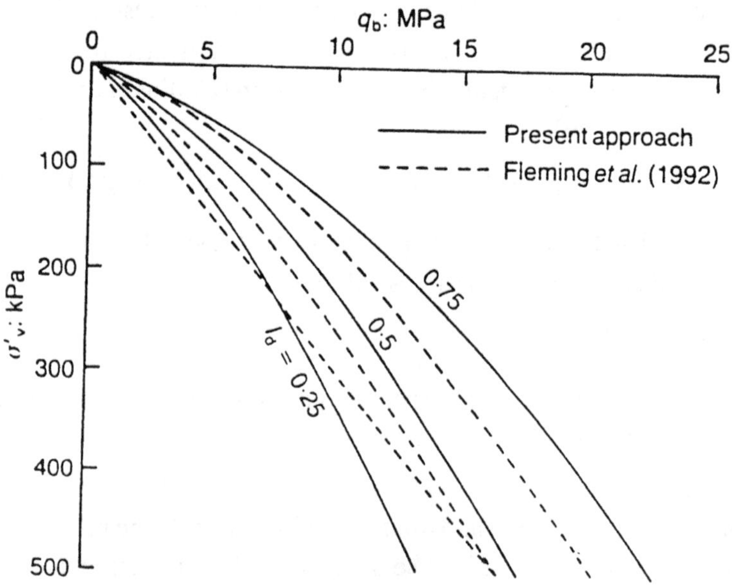

Figure 9.9: Comparsion of end bearing capacity profiles (after Randolph *et al.*, 1994)

The results presented in Figure 9.9 use three different values of relative density. The critical state friction angle is assumed to be 30 degrees. Overall the agreement between the two sets of curves is excellent. It is also noted that the cavity expansion approach gives greater curvature of the end bearing profiles, largely due to the effect of decreasing rigidity index with increasing depth.

When compared with the design charts of Fleming *et al.* (1985) which are based on classical bearing capacity theory, the cavity expansion approach offers the following advantages:

(a) The cavity expansion approach is more flexible and easier to apply. Unlike the bearing capacity approach, it includes the effect of soil compressibility in the solution;

(b) The cavity expansion approach takes into account the variation of soil strength with stress level;

(c) The cavity expansion approach accounts for the dependence of shear modulus on relative density and stress level.

9.3.2 End bearing capacity of piles in crushable sands

It is worth noting that a similar study using cavity expansion theory to predict end bearing capacity of piles has been carried out by Yasufuku and Hyde (1995) for crushable sands. In their study, a slightly different correlation, as defined below, was used to relate the end bearing capacity with spherical cavity limit pressure:

$$q_b = \frac{1}{1-\sin\phi}\psi_s \tag{9.19}$$

In addition, the approximate cavity expansion solutions proposed by Vesic (1972) and Baligh (1976) were used to determine the spherical limit pressures. Both the effect of curvature of failure surface and variable shear modulus are also accounted for in a simple manner.

The use of bearing capacity theory in predicting end bearing pressures of piles would indicate that the end bearing is largely a function of friction angle. However, crushable soils often have relatively high friction angles. The measured end bearing pressures are much lower than predicted by bearing capacity theory. The comparison between cavity expansion prediction and experimental data presented by Yasufuku and Hyde (1995) suggests that the use of spherical cavity expansion methods offers the potential for a much better prediction of end bearing capacities by taking account of soil crushability and compressibility.

9.4 LATERAL CAPACITY OF PILES

In addition to the axial capacity, cavity expansion theory may also be used to estimate the limiting pressure of laterally loaded piles in clay, sand or rock.

9.4.1 Limiting lateral pressures in clay

As discussed by Fleming *et al.* (1985), past experimental research suggests that it is reasonable to assume that the pressure exerted by soil in front of a laterally loaded pile approaches the limiting pressure measured from a pressuremeter test. Behind the pile, the lowest normal stress acting on the pile shaft is a suction of the order of $p_a = -100kPa$. If a gap forms behind the pile, then the normal stress increases to zero (with a dry hole) or to the ambient head of water if free water is available to fill the gap. It is also estimated that the friction along the sides of the pile is in the order of $1.0ds_u$, where d is pile diameter.

According to the above argument, it follows that the ultimate force per unit length resisting the pile lies between the two limiting values defined below:

$$(\psi_c - U_0 + s_u)d < p_u < (\psi_c + p_a + s_u)d \tag{9.20}$$

where the cylindrical cavity limit pressure ψ_c is given by (9.1). Equation (9.20) can be rewritten as follows:

$$\left(\frac{\sigma'_{h0}}{s_u} + 2 + \ln\frac{G}{s_u}\right) < \frac{p_u}{ds_u} < \left(\frac{\sigma_{h0} + p_a}{s_u} + 2 + \ln\frac{G}{s_u}\right) \tag{9.21}$$

It is noted that for typical values of the stiffness index, the above solution compares very well with the more rigorous plasticity solution of Randolph and Houlsby (1984).

9.4.2 Limiting lateral pressures in sand

In practice, piles in sand may fail at some stage by the formation of a plastic hinge at some point down the pile. Lateral movement of the pile to cause such failure is found to be generally in excess of 10% of the pile diameter. Current methods for estimating lateral capacity of piles in sand are largely empirical. At greater depth, pressuremeter testing results (i.e. cavity expansion theory) have been suggested as a possible analytical approach, assuming that the pressuremeter pressure required to cause displacement of over 10 to 15% of the pile diameter is similar to pile lateral pressure needed to cause failure of piles (Fleming *et al.*, 1985). So far, however, no analytical expressions have been proposed to calculate such a pressure for pressuremeter tests.

Using the closed form solution of large strain cylindrical cavity expansion in sand developed in Chapter 3 (Section 3.3.3), a simple analytical method can be developed for estimating the lateral capacity of piles. By ignoring the elastic deformation in the plastically deforming zone, the cylindrical cavity pressure-expansion relationship can be expressed in closed form:

$$\frac{p_u}{p_0} = \frac{2\alpha}{1+\alpha}\left[\frac{1-(a/a_0)^{-1-1/\beta}}{1-(1-\delta)^{1+\frac{1}{\beta}}}\right]^{1/\gamma} \tag{9.22}$$

where

$$\alpha = \frac{1 + \sin\phi}{1 - \sin\phi} \tag{9.23}$$

$$\beta = \frac{1 + \sin\psi}{1 - \sin\psi} \tag{9.24}$$

$$\gamma = \frac{\alpha(1 + \beta)}{\alpha - 1)\beta} \tag{9.25}$$

$$\delta = \frac{(\alpha - 1)p_0}{2(1 + \alpha)G} \tag{9.26}$$

and a and a_0 are current and initial cavity radii respectively. If the cavity pressure required to cause displacement of over 10 to 15% of the pile (cavity) diameter is assumed to be similar to the pile lateral pressure needed to cause failure of piles, equation (9.22) can be used easily to compute the limiting lateral pressure of a pile.

9.4.3 Limiting lateral pressures in rock

It is true that little attention has been given to predict the lateral capacity of shafts socketed into rock. Carter and Kulhawy (1992) pointed out two possible reasons for the lack of research in this area: (a) Lateral design of piles in rock is governed largely by displacement considerations, and therefore the capacity has been assigned less importance; (b) Theoretically, the problem is very difficult to solve. Nevertheless, Carter and Kulhawy (1992) suggested that even if the displacement consideration controls the design, it is still important to estimate the load capacity so that the likely margin of safety can be determined at working load levels.

An analytical procedure was developed by Carter and Kulhawy (1992) for determining the ultimate lateral loads acting on a pile socketed into rock. This new procedure, as detailed in Carter and Kulhawy (1992), may be summarised as follows (see Figure 9.10):

(1) When a lateral load is applied at the rock surface, the rock mass immediately in front of the shaft undergoes zero vertical stress, while horizontal stress is applied by the leading face of the shaft. Ultimately the horizontal stress may reach the uniaxial compressive strength of the rock mass and, with further increase in the lateral load, the horizontal stress may decrease as the rock mass softens during postpeak deformation. It is therefore reasonable to assume that the reaction stress at the rock mass surface is zero or very small as a result of the postpeak softening in the limiting case of loading of the shaft. Along the sides of the shaft, some shear resistance may be mobilised, which is likely to be same as the maximum unit side resistance under axial compression, $\bar{\tau}_{max}$.

(2) At greater depth, it may be assumed that the stress in front of the shaft increases from the initial in situ horizontal stress level, σ_{h0}, to the limiting pressure, ψ_c, reached during the expansion of a cylindrical cavity in the rock. Behind the pile, a gap is likely to be opened between the pile and rock leading to zero normal stress there. At the sides of the shaft, some shear resistance is mobilised. As a result, at depth, the ultimate force per unit length resisting the lateral loading is likely to be about $d(\psi_c + \bar{\tau}_{max})$.

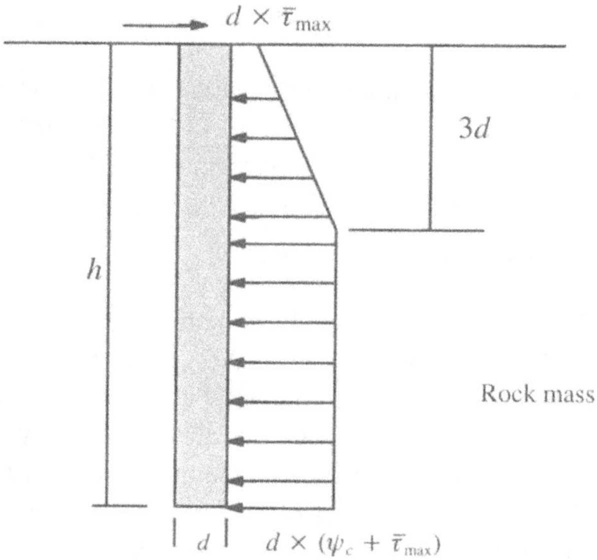

Figure 9.10: Distribution of ultimate lateral force per unit length for piles socketed into rock (Carter and Kulhawy, 1992)

Closed form solutions have been developed by Yu (1991) and Yu and Houlsby (1991) for limit pressures of cavity expansion in materials modelled using the Mohr-Coulomb criterion. It is well known that the plastic behaviour of many rocks can be represented reasonably by the Mohr-Coulomb plasticity model. The limit pressures developed by Yu (1990) and Yu and Houlsby (1991) for cavity expansion in Mohr-Coulomb materials were presented in Chapter 3. In particular, the limit pressure solution ψ_c for a cylindrical cavity can be obtained by finding R_{lim} from the following equation:

$$\sum_{n=0}^{\infty} A_n(R_{lim}, \mu) = \frac{\chi}{\gamma}(1-\delta)^{\frac{\beta+1}{\beta}} \tag{9.27}$$

where A_n is related to R_{lim} by

$$A_n(R_{\text{lim}}, \mu) = \begin{cases} \dfrac{\mu^n}{n!} \ln R_{\text{lim}} & \text{if } n = \gamma \\ \dfrac{\mu^n}{n!(n-\gamma)} ({R_{\text{lim}}}^{n-\gamma} - 1) & \textit{otherwise} \end{cases} \tag{9.28}$$

Once R_{lim} is obtained, the limit cavity pressure ψ_c is readily derived from:

$$R_{\text{lim}} = \frac{(1 + \alpha)[Y + (\alpha-1)\psi_c]}{2\alpha[Y + (\alpha-1)p_0]} \tag{9.29}$$

The parameters α, β, γ and δ used above are defined in equations (9.23) to (9.26) (see Section 3.3.3 in Chapter 3 for further details). The limit pressure depends strongly on the angles of friction and dilation as well as the stiffness properties of the soil.

The problem remains as to how to determine the depth at which the limit pressure from cavity expansion is mobilised. In a study of lateral capacity of piles in clay, Randolph and Houlsby (1984) assumed that this depth would be about 3 times the pile diameter. In the absence of any other data, Carter and Kulhawy (1992) suggested that the assumption of Randolph and Houlsby (1984) may also be adopted for piles in rock, see Figure 9.10.

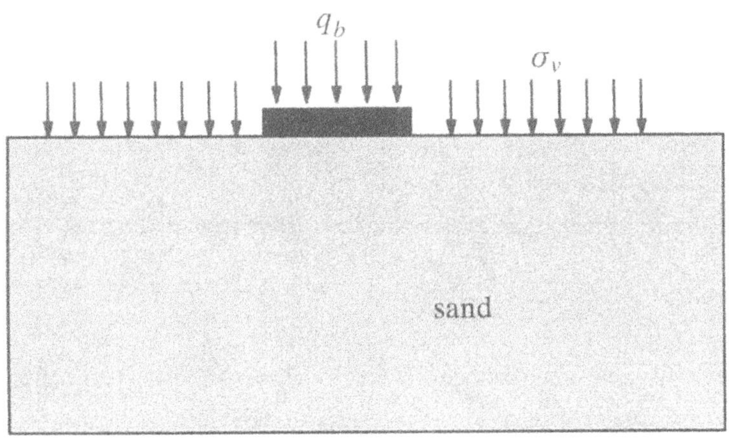

Figure 9.11: Bearing capacity of sand with a surcharge

9.5 BEARING CAPACITY OF SAND WITH A SURCHARGE

A closely related problem to pile design is the determination of the bearing capacity of sand with a surcharge as shown in Figure 9.11. This is because most current design methods for determining the end bearing capacity of piles in sand are based on bearing capacity theory (see Fleming *et al*, 1985; ASCE manual on design of pile foundations, 1993):

$$q_b = N_q \sigma_v \tag{9.30}$$

where σ_v is the overburden stress at the level of pile tip and N_q is a factor derived from bearing capacity theory.

A study was presented by Yeung and Carter (1989) using spherical cavity expansion solutions to estimate the bearing capacity of a circular footing rested on both calcareous and silica sands. Two plasticity models were used. The first was the standard Mohr-Coulomb model with constant strength parameters. For comparison, a more complex, collapse-softening plasticity model was also used.

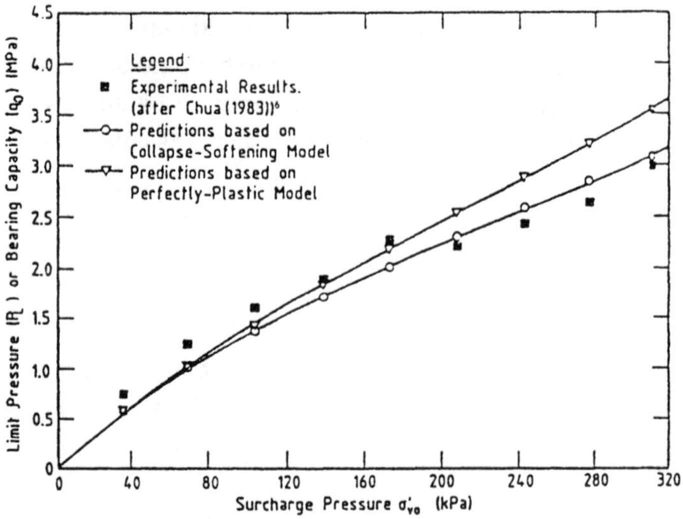

Figure 9.12: Comparison of computed cavity limit pressures and measured bearing capacities from model footing tests in calcareous sand (after Yeung and Carter, 1989)

In Yeung and Carter (1989), the measured bearing capacities of a model circular footing obtained by Chua (1983) were directly compared with the theoretical limit pressures for the expansion of a spherical cavity. Plotted in Figure 9.12 and Figure 9.13 are comparisons between the measured bearing capacities and the theoretical spherical cavity expansion limit pressures for calcareous and silica sands respectively. It can be concluded from these figures that the measured values of bearing capacity can be predicted by spherical cavity expansion solution within 10–20%, with the collapse-softening model producing a slightly better prediction.

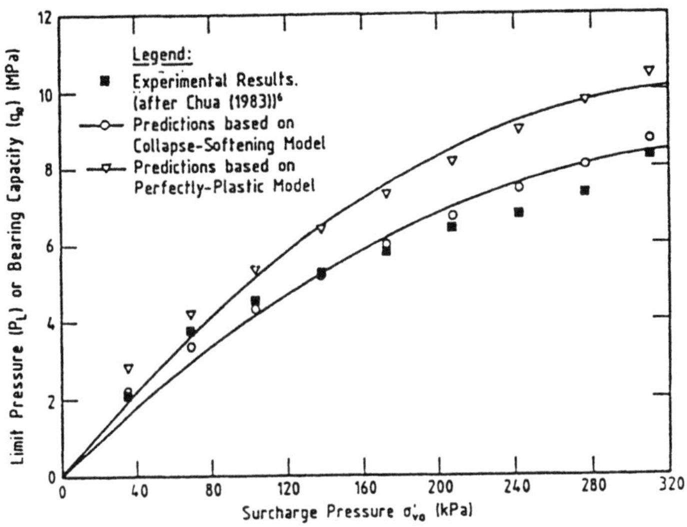

Figure 9.13: Comparison of computed cavity limit pressures and measured bearing capacities from model footing tests in silica sand (after Yeung and Carter, 1989)

9.6 UPLIFT CAPACITY OF PLATE ANCHORS IN SOILS

In this section, procedures that can be used to apply cavity expansion solutions to conduct stability calculations for plate anchors buried in soils are developed. In doing so, the following general assumption is made: the breakout of a plate anchor occurs when the boundary of the plastic zone due to the anchor pull-out action (calculated from cavity expansion theory) is sufficiently close to or on the ground surface. In other words, plate anchors break out when the plastic flow is not confined by the outer elastic zone and becomes free, Figure 9.14. It is assumed that a sufficiently large deformation occurs in the soil above the anchor when the radius of plastic zone $c = mH$. The value of m is less than or equal to 1. It will be shown that due to the incompressible nature of undrained clay, a reasonable value for m is 1 for cohesive soil (i.e. plastic boundary is on the free ground surface). For cohesive frictional soils, however, the best value for m is around 0.5 (i.e. plastic boundary is in the middle of anchor position and free ground surface). This is due to the fact that dilation of sand tends to cause a very large deformation in the soil around the plastic zone and this accelerates the breakout failure of anchors.

9.6.1 Plate anchors in clay

The geometry of a plate anchor, shown in Figure 9.14, is defined by its width B and embedment in the soil H. The initial stress state of soil before pull-out of the anchor is assumed to be p_0 and the pull-out capacity of the anchor is denoted by p. For clay soils under undrained loading conditions, the strength is defined by their undrained shear strength s_u.

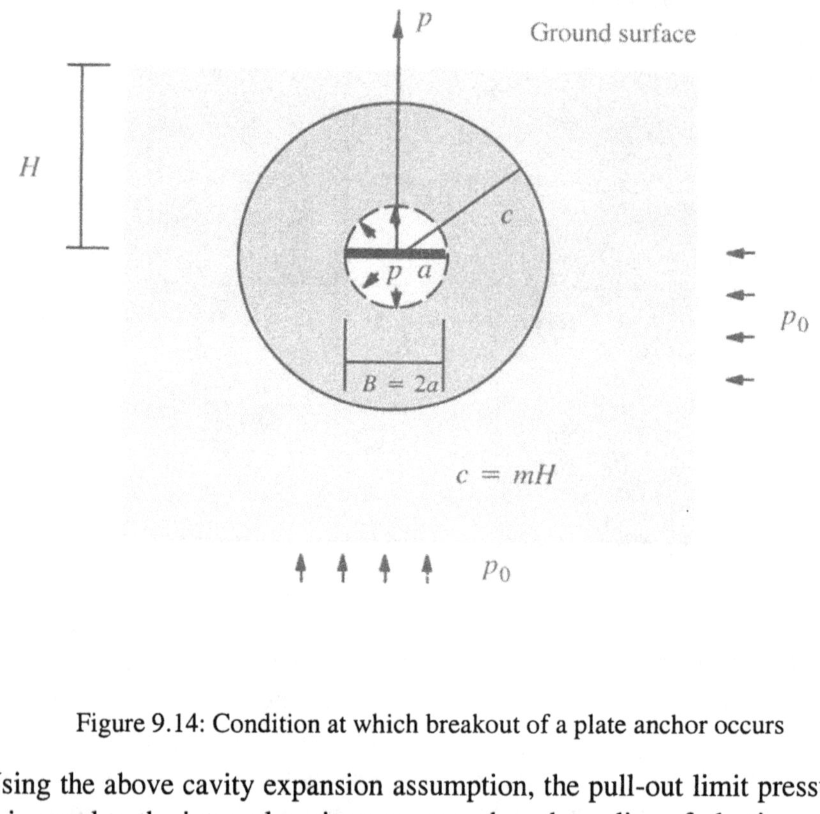

Figure 9.14: Condition at which breakout of a plate anchor occurs

Using the above cavity expansion assumption, the pull-out limit pressure p in clay is equal to the internal cavity pressure when the radius of plastic region c is equal to H (i.e. m=1). The relationship between internal cavity pressure p and plastic radius c has been derived in Chapter 3, and can be expressed as follows:

$$\frac{p-p_0}{s_u} = 2k \ln\frac{c}{a} + \frac{2k}{1+k} \tag{9.31}$$

where s_u is the undrained shear strength of clay. For strip plate anchors (i.e. plane strain anchors) $k=1$ and for circular plate anchors $k=2$. Using $c = mH$ and $a = B/2$, equation (9.31) gives the following 'anchor breakout factor' expression:

$$N_b = \frac{p-p_0}{s_u} = 2\ln(2m\frac{H}{B}) + 1 \tag{9.32}$$

for long strip plates and

$$N_b = \frac{p-p_0}{s_u} = 4\ln(2m\frac{H}{B}) + \frac{4}{3} \tag{9.33}$$

for circular or square plates. Note that for circular anchors, B defined in equation (9.33) should be interpreted as the anchor diameter D.

The initial soil pressure p_0 may be estimated as follows:

$$p_0 = q + \gamma H \tag{9.34}$$

with q as the surcharge acting on the ground surface and γ being the unit weight of clay.

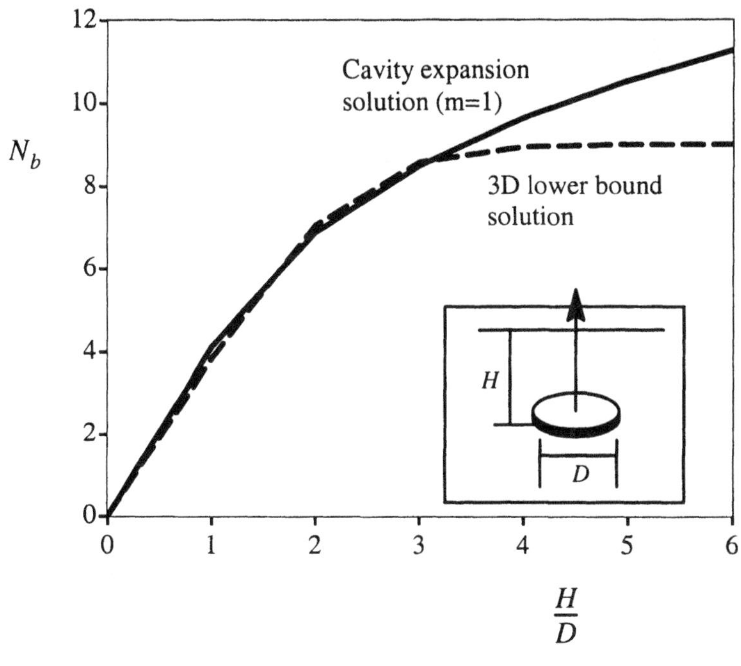

Figure 9.15: Comparison with 3D lower bound solutions for circular anchors

Comparison with other plasticity solutions and experimental data

To illustrate the relevance of this simple cavity expansion solution to the estimation the uplift capacity of plate anchors in clay, a comparison is made with other existing plasticity solutions and experimental data. Figure 9.15 shows the anchor breakout factors predicted by the cavity expansion solution (9.33) and a three-dimensional lower bound analysis reported by Merifield *et al.* (2000) for circular anchors. It is clear from the figure that for shallow anchors ($H/D \leq 4$), the results from the simple cavity expansion method agree very well with the 3D lower bound solutions and therefore, can be used with confidence to predict the anchor pullout capacity.

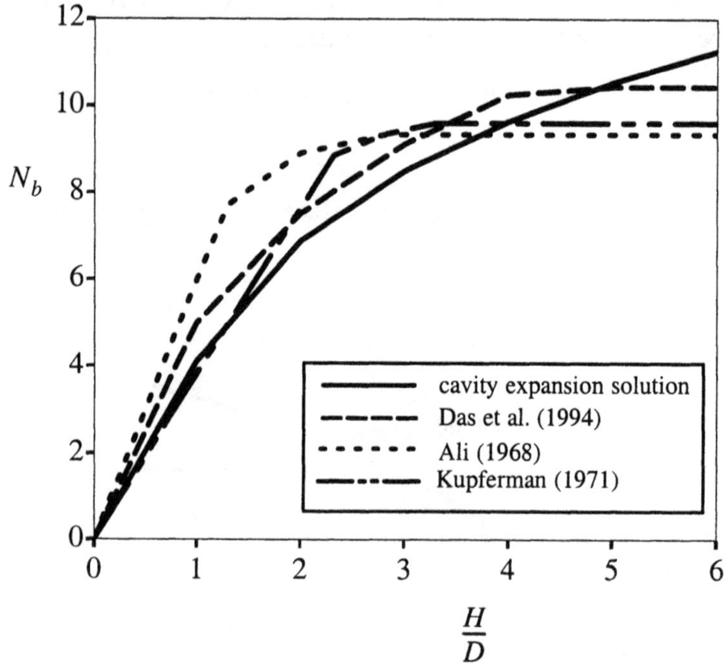

Figure 9.16: Comparison with experimental data for circular anchors

The laboratory model tests on circular anchors performed by Das *et al.* (1994), Ali (1968) and Kupferman (1971) are shown in Figure 9.16. The laboratory results of Das and Kupferman compare reasonably well to the cavity expansion solutions. The greatest variation between cavity expansion solutions and laboratory results is approximately 15%.

The laboratory tests by Ali, which were performed in soft bentonite clay, are difficult to compare with the cavity expansion results because suction forces were allowed to develop below the anchor plate while testing. Corrections to the ultimate capacity were then made by estimating the likely suction forces between the anchor and soil. However, suction forces are likely to be highly variable and may be a function of several variables including the embedment depth, soil permeability, undrained shear strength and loading rate. Therefore, it is reasonable to conclude that much of the difference between the results of Ali and the cavity expansion solutions can be attributed to the uncertainty in estimating the suction forces developed between the soil and the anchor.

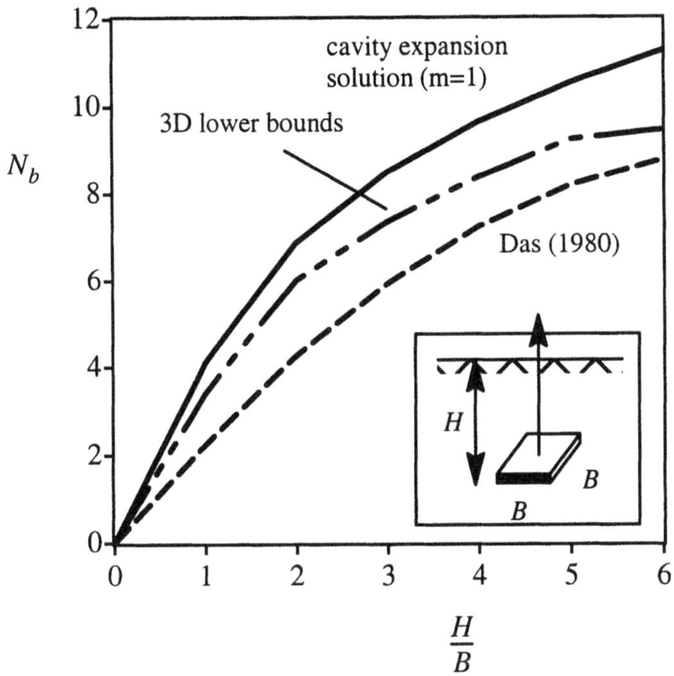

Figure 9.17: Comparison with 3D lower bounds and experimental data for square plate anchors

For square plate anchors, Figure 9.17 compares the cavity expansion solution to the 3D lower bound solution and also limited experimental results obtained by Das (1980). It is clear that the cavity expansion prediction with m=1 is consistently higher than the 3D lower bound results.

To date, very limited experimental work on square anchors has been undertaken. The most comprehensive testing program was undertaken by Das (1980) who performed pullout tests on small model square and rectangular anchors in soft to firm clays. Immediate breakaway was ensured by venting the bottom side of the anchor with hollow copper tubing and filter paper. The break-out factors back figured by Das plot below the limit analysis solutions over the full range of embedment ratios (see Figure 9.17). The difference between the experimental data of Das and the present cavity expansion solutions with m=1 is typically 30%. However, this variation can be removed by using a smaller value of m in the cavity expansion prediction.

It has been established numerically and experimentally that when the length to width ratio is sufficiently large (i.e. greater than 8), rectangular anchors may be analysed as infinitely long strip plate anchors. The advantage of this assumption is that rectangular anchors can be modelled as a two-dimensional problem. The lower and upper bound solutions for infinitely long strip anchors buried in clay have been reported by Merifield *et al.* (1999). As argued earlier in this section, the pullout capacity of strip anchors may also be estimated using cylindrical cavity expansion theory (9.32). Figure 9.18 compares the cavity expansion solutions, as estimated using equation (9.32), with the lower bound solution of Merifield *et al.* (1999). It is clear from the figure that the simple cylindrical cavity expansion solution agrees very well with the rigorous lower bound plasticity results, and can therefore be used with confidence to predict the uplift capacity of strip anchors buried in undrained clay.

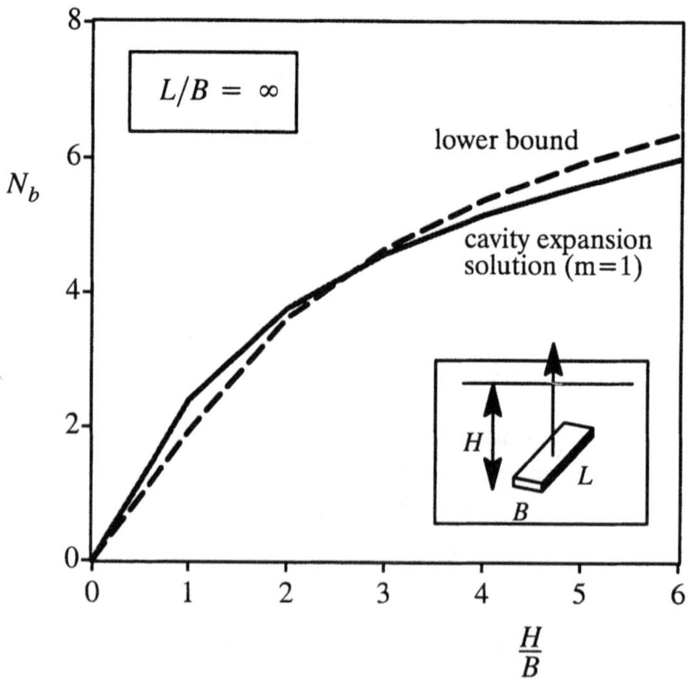

Figure 9.18: Comparison with plane strain lower bounds for long strip plate anchors

9.6.2 Plate anchors in sand

As for cohesive soils, cavity expansion solutions can also be used to conduct stability calculations for plate anchors buried in cohesive-frictional soils. It is also assumed that breakout of a plate anchor occurs as long as the plastic radius of soils, as calculated from cavity expansion theory, is sufficiently close to the ground surface.

The cavity loading solution in cohesive-frictional soils, presented in Chapter 3, gives the relationship between internal pressure p and plastic radius c:

$$\frac{(k+\alpha)[Y+(\alpha-1)p]}{\alpha(1+k)[Y+(\alpha-1)p_0]} = \left(\frac{c}{a}\right)^{k(\alpha-1)/\alpha} \tag{9.35}$$

where α and Y are functions of soil cohesive C and friction angle ϕ defined by

$$Y = \frac{2C\cos\phi}{1-\sin\phi} \quad \text{and} \quad \alpha = \frac{1+\sin\phi}{1-\sin\phi} \tag{9.36}$$

and k is used to indicate strip plate anchors (cylindrical cavity k=1) or circular plate anchors (spherical cavity k=2).

Using $c = mH$ and $a = B/2$, equation (9.35) gives the following 'anchor breakout factor' expression:

$$N_b = \frac{Y+(\alpha-1)p}{Y+(\alpha-1)p_0} = \frac{2\alpha}{1+\alpha}\left(2m\frac{H}{B}\right)^{(\alpha-1)/\alpha} \tag{9.37}$$

for strip plate anchors, and

$$N_b = \frac{Y+(\alpha-1)p}{Y+(\alpha-1)p_0} = \frac{3\alpha}{2+\alpha}\left(2m\frac{H}{B}\right)^{2(\alpha-1)/\alpha} \tag{9.38}$$

for circular or square plate anchors. Note that for circular anchors, B defined in equation (9.38) should be replaced by the anchor diameter D.

The initial soil pressure p_0 may be estimated as follows:

$$p_0 = q + \gamma H \tag{9.39}$$

with q as the surcharge acting on the ground surface, and γ being the unit weight of cohesive-frictional soils.

For anchors buried in a purely frictional soil, the cohesion is zero (Y=0). If it is further assumed that the initial soil pressure is mainly due to self-weight of soils ($q = 0$ and $p_0 = \gamma H$), the stability number can be simplified as follows:

$$N_b = \frac{p}{\gamma H} = \frac{2\alpha}{1+\alpha}\left(2m\frac{H}{B}\right)^{(\alpha-1)/\alpha} \tag{9.40}$$

for strip plate anchors, and

$$N_b = \frac{p}{\gamma H} = \frac{3\alpha}{2+\alpha}\left(2m\frac{H}{B}\right)^{2(\alpha-1)/\alpha} \tag{9.41}$$

for circular or square plate anchors.

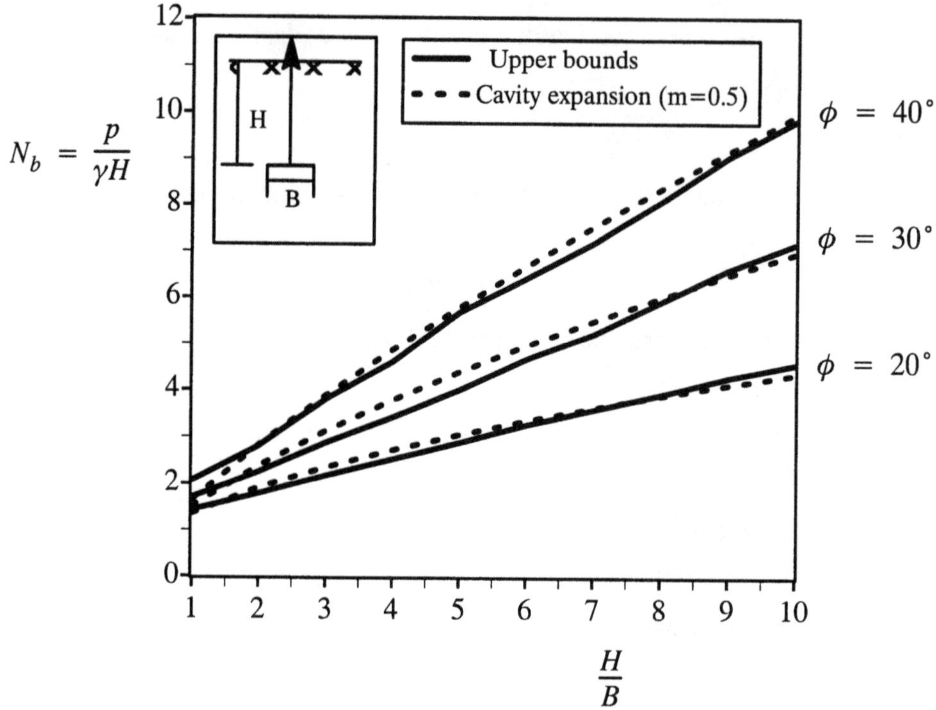

Figure 9.19: Comparison with upper bounds for long strip plate anchors in sand

Comparison with plasticity solutions

To check the validity of the cavity expansion solution, a comparison between the rigorous upper bound solutions and those predicted by equation (9.40) for strip anchors buried in sand is presented. As discussed before, due to the dilatancy of the sand, anchor breakout occurs before the plastic zone reaches the ground surface. In fact, it is reasonable to assume that anchor breakout occurs if the plastic zone extends to the middle surface between the anchor and the ground surface. In other words, a good value to use for m in sand is about 0.5. Figure 9.19 plots the upper bound solutions for three different values of friction angle together with the breakout factors estimated by the cavity expansion equation (9.40) with m=0.5. It is clear from the figure that, despite its simplicity, the cavity expansion results agree very well with the rigorous upper bound plasticity solutions. It can therefore be concluded that the cavity expansion approach provides a useful simple tool for stability analysis of plate anchors.

9.7 SUMMARY

1. The prediction of end-bearing and shaft capacity of driven piles in soils remains a difficult problem in geotechnical engineering. This is because the analysis of pile installation in soil is a large strain problem which involves strong material and geometric nonlinearities. Although some progress has been made in modelling the behaviour of driven piles in clay (e.g., Baligh, 1985 and Yu *et al.*, 2000), little success has been achieved in the rigorous prediction of pile behaviour in sand. Due to the lack of rigorous analysis methods, many semi-analytical or empirically based methods are still widely in use in the analysis and design of pile foundations. Cavity expansion theory has proven to be a very useful theoretical tool in the development of semi-analytical methods for estimating pile capacity.
2. As a very high proportion (80–90%) of the overall capacity of a driven pile in clay is due to the shaft friction, much effort has been devoted to developing reliable methods for estimating values of shaft friction. To estimate the skin friction, it is necessary to know the normal stress acting on the pile shaft after the pile is driven into the ground. Many researchers have assumed that the installation of piles into clay can be modelled as the expansion of a cylindrical cavity from zero radius. This assumption is based on the field and laboratory observation that when a pile is driven deeply into soil, much of the soil is displaced outwards in the radial direction. A comparative study using total stress cavity expansion solutions suggests that this simple cavity expansion approach gives a reasonable prediction for the measured pile normal stresses for normally consolidated or lightly overconsolidated clay. However, for heavily overconsolidated clays, the total stress-based cavity expansion solution tends to over-predict the pile normal stresses. This is because two important shortcomings exist with the total stress analysis. Firstly, no account is taken of pore pressure generated due to pure shear. Secondly, the total stress analysis with a perfectly plastic model is unable to link correctly the strength of the soil and its change with the current effective stress state and stress history of soil. So while it may be reasonable for normally consolidated or lightly overconsolidated soils, the total stress cavity expansion analysis will not be suitable for heavily overconsolidated clays.
3. Many investigators have measured pore water pressures generated at the pile-soil interface during pile installation in clay. For example, high quality experimental research on pile testing has been carried out at Oxford University by Coop and Wroth (1989) and at Imperial College by Bond and Jardine (1991). Both of these studies have observed a very low pore pressures along the pile shaft for overconsolidated clays. In particular, the results from the instrumented pile tests in heavily overconsolidated London clay indicate that negative pore pressures were re-

corded during the pile installation. In this chapter, it is shown that the effective stress cavity expansion solutions of Collins and Yu (1996), as presented in Chapter 4, can be used satisfactorily to explain the measured pore pressures at the pile-soil interface during pile installation in heavily overconsolidated clays.

4. When a pile is driven into clay, high excess porewater pressures are generated close to the pile. This high porewater pressure reduces the effective stresses along the pile shaft, meaning that the short term shaft capacity is relatively low. After the pile is installed, the excess porewater pressures dissipates and the soil therefore consolidates. Accordingly, the pile capacity increases. The cavity expansion consolidation solutions of Randolph and Wroth (1979), as presented in Chapter 6, have been applied with some success to predict such a time-dependent pile behaviour. Work is however needed to develop further consolidation solutions with critical state models that are accurate for both normally consolidated and heavily overconsolidated clays.

5. For driven piles in clay, it is standard practice to calculate the end bearing capacity assuming undrained conditions (Fleming *et al.*, 1985). The end bearing capacity is traditionally related to the spherical cavity limit pressure. Figure 9.6 shows one of the first analogies between spherical cavity expansion and end bearing failure which was proposed by Gibson (1950). Using this analogy in undrained clay soils, it is possible to express the pile end bearing capacity as a function of the spherical cavity limit pressure ψ_s by equation (9.7) assuming the Tresca yield criterion. This equation takes into account the effects of *in-situ* stresses and the stiffness index of soils on the end bearing capacity of a pile, and is therefore a better solution than the existing design methods using an arbitrary bearing capacity factor.

6. Unlike piles in clay, the total capacity of piles in granular soils is generally evenly divided between shaft and base. Two steps must be followed in order to use cavity expansion theory to predict the end bearing capacity of a driven pile in sand. Firstly, semi-analytical correlations are required to relate end bearing capacity to the cavity limit pressure. Secondly, analytical or numerical solutions must be developed for limit pressures of cavity expansion in soil modelled by a realistic plasticity model. Based on the analytical solutions of Carter *et al.* (1986), Yu and Houlsby (1991) and Collins *et al.* (1992) for cavity limit pressures, Randolph *et al.* (1994) suggested a satisfactory procedure for estimates end bearing capacity of piles in sand.

7. Apart from the axial pile capacity, examples are given in this chapter to show that cavity expansion theory can also be used to estimate later capacity of piles in clay, sand or rock. A closely related problem to pile design is to determine the

bearing capacity of sand with surcharge as shown in Figure 9.11. This is because most current design methods for determining the end bearing capacity of piles in sand are based on bearing capacity theory. A study was presented by Yeung and Carter (1989) in using spherical cavity expansion solutions to estimate the bearing capacity of a circular footing rested on both calcareous and silica sands. A comparative study suggests that the measured bearing capacities in calcareous and silica sands can be predicted by the theoretical spherical cavity expansion limit pressures to within 10–20%.

8. In the final part of this chapter, simple procedures are developed to apply cavity expansion solutions to estimate pull-out capacity of plate anchors buried in soils. It is assumed that the breakout of a plate anchor occurs if the boundary of the plastic zone due to the anchor pull-out action (calculated from cavity expansion theory) is sufficiently close to or on the ground surface. In other words, plate anchors break out when the plastic flow is not confined by the outer elastic zone and becomes free, Figure 9.14. Comparisons with available experimental data and more rigorous numerical results suggest that the simple cavity expansion procedures developed can be used to give satisfactory results for the pull-out capacity of plate anchors.

REFERENCES

Ali, J.I. (1968). *Pullout Resistance of Anchor Plates and Anchored Piles in Soft Bentonite Clay.* M.Sc. Thesis, Duke University, Durham, USA.

ASCE (1993). *Design of Pile Foundations.* Technical Engineering and Design Guides as Adapted from the US Army Corps of Engineers, No 1, ASCE Press.

Atkinson, J.H. and Bransby, P.L. (1978). *The Mechanics of Soils.* McGraw–Hill.

Baligh, M.M. (1976). Cavity expansion in sands with curved envelopes. *Journal of the Geotechnical Engineering Division*, ASCE, 102, 1131–1146.

Been, K. and Jefferies, M.G. (1985). A state parameter for sands. *Geotechnique*, 35, 99–112.

Bishop, R.F., Hill, R and Mott, N.F. (1945). The theory of indentation and hardness tests. *Proceedings of Physics Society,* 57, 147–159.

Bolton, M.D. (1986). The strength and dilatancy of sands. *Geotechnique*, 36(1), 65–78.

Bolton, M.D. (1987). Discussion on the strength and dilatancy of sands. *Geotechnique*, 37(2), 219–226.

Bond, A.J. and Jardine, R.J. (1991). Effect of installing displacement piles in a high OCR clay. *Geotechnique*, 41, 341–363.

Carter, J.P. and Kulhawy, F.H. (1992). Analysis of laterally loaded shafts in rock. *Journal of Geotechnical Engineering*, ASCE, 118(6), 839–855.

Carter, J.P., Randolph, M.F. and Wroth, C.P. (1979). Stress and pore pressure changes in clay during and after the expansion of a cylindrical cavity. *International Journal for Numerical and Analytical Methods in Geomechanics*, 3, 305–323.

Carter, J.P., Booker, J.R. and Yeung, S.R. (1986). Cavity expansion in cohesive-frictional soils. *Geotechnique*, 36, 349–358.

Chua, E.W. (1983). *Bearing Capacity of Shallow Founadstions in Calcareous Sand.* ME Thesis, University of Sydney, Australia.

Collins, I.F. and Yu, H.S. (1996). Undrained cavity expansions in critical state soils. *International Journal for Numerical and Analytical Methods in Geomechanics*, 20(7), 489–516.

Collins, I.F., Pender, M.J. and Yan Wang (1992). Cavity expansion in sands under drained loading condition. *International Journal for Numerical and Analytical Methods in Geomechanics*, 16, 3–23.

Cooke, R.W. and Price, G. (1978). Strains and displacements around friction piles. *Building Research Station* CP 28/78, October.

Coop, M.R. and Wroth, C.P. (1989). Field studies of an instrumented model pile in clay. *Geotechnique*, 39(4), 679–696.

Das, B. M. (1980). A procedure for estimation of ultimate capacity of foundations in clay. *Soils and Foundations,* 20(1), 77-82.

Das, B. M., Shin E. C., Dass R. N., and Omar M. T. (1994). Suction force below plate anchors in soft clay. *Marine Georesources and Geotechnology.,* 12, 71-81.

Davis, R.O., Scott, R.F. and Mullenger, G. (1984). Rapid expansion of a cylindrical cavity in a rate type soil. *International Journal for Numerical and Analytical Methods in Geomechanics*, 8, 125–140.

Fleming, W.G.K., Weltman, A.J., Randolph, M.F. and Elson, W.K. (1985). *Piling Engineering.* John Wiley and Sons.

Gibson, R.E. (1950). Correspondence. *Journal of Institution of Civil Engineers,* 34, 382–383.

Gibson, R.E. and Anderson, W.F. (1961). In situ measurement of soil properties with the pressuremeter. *Civil Engineering Public Works Review*, 56, 615–618.

Hill, R. (1950). *The Mathematical Theory of Plasticity.* Oxford University Press.

Kirby, R.C. and Esrig, M.I. (1979). Further development of a general effective stress method for prediction of axial capacity for driven piles in clay. *Recent Development in the Design and Construction of Piles*, ICE, London, 335–344.

Kodikara, J.K. and Moore, I.D. (1996). Axial response of tapered piles in cohesive frictional ground. *Journal of Geotechnical Engineering*, ASCE, 119(4), 675–693.

Kupferman, M. (1965). *The Vertical Holding Capacity of Marine Anchors in Clay subjected to Static and Cyclic Loading*. MSc Thesis, University of Massachusetts, Amherst, USA.

Ladanyi, B. and Johnston, G.H. (1974). Behaviour of circular footings and plate anchors embeded in permafrost. *Canadian Geotechnical Journal*, 11, 531–553.

Lo Presti, D. (1987). *Mechanical Behaviour of Ticino Sand From Resonant Column Tests*. PhD Thesis, Politecnico di Torino, Italy.

Mair, R.J. and Wood, D.M. (1987). *Pressuremeter Testing, Methods and Interpretation*. CIRIA report, Butterworths, London.

Merifield R. S., Sloan S. W. and Yu H. S. (1999). Stability of plate anchors in undrained clay. *Geotechnique* (submitted).

Merifield R. S., Lyamin, A.V., Sloan S. W. and Yu H. S. (2000). Three dimensional lower bound solutions for stability of plate anchors in clay. *Journal of Geotechnical and Geoenvironmental Engineering*, ASCE (submitted).

Muir Wood, D. (1990). *Soil Behaviour and Critical State Soil Mechanics*. Cambridge University Press.

Nystrom, G.A. (1984). Finite strain axial analysis of piles in clay. *Analysis and Design of Pile Foundations* (Editor: J.R. Meyer), ASCE, 1–20.

Randolph, M.F. and Houlsby, G.T. (1984). The limiting pressure on a circular pile loaded laterally in cohesive soil. *Geotechnique*, 34(4), 613–623.

Randolph, M.F., Carter, J.P. and Wroth, C.P. (1979). Driven piles in clay – the effects of installation and subsequent consolidation. *Geotechnique*, 29, 361–393.

Randolph, M.F., Dolwin, J. and Beck, R. (1994). Design of driven piles in sand. *Geotechnique*, 44(3), 427–448.

Richart, F.E., Hall, T.J. and Woods, R.D. (1970). *Vibrations of Soils and Foundatsions*. Englewood Cliffs: Prentice–Hall.

Seed, H.B. and Reese, L.C. (1955). The action of soft clay along friction piles. *Proceedings of ASCE*, Paper 842.

Schofield, A.N. and Wroth, C.P. (1968). *Critical State Soil Mechanics*. McGraw–Hill.

Teh, C.I. and Houlsby, G.T. (1991). An analytical study of the cone penetration test in clay. *Geotechnique*, 41(1), 17–34.

Vesic, A.S. (1972). Expansion of cavities in infinite soil mass. *Journal of the Soil Mechanics and Foundations Division,* ASCE, 98, 265–290.

Wroth, C.P., Carter, J.P. and Randolph, M.F. (1979). Stress changes around a pile driven into cohesive soil. *Recent Development in the Design and Construction of Piles*, ICE, London, 345–354.

Yasufuku, N. and Hyde, A.F.L. (1995). Pile end–bearing capacity in crushable sands. *Geotechnique*, 45(4), 663–676.

Yeung, S.K. and Carter, J.P. (1989). An assessment of the bearing capacity of calcareous and silica sands. *International Journal for Numerical and Analytical Methods in Geomechanics*, 13, 19–36.

Yu, H.S. (1990). *Cavity Expansion Theory and its Application to the Analysis of Pressuremeters*. DPhil Thesis, University of Oxford, England.

Yu, H.S. (1994). State parameter from self–boring pressuremeter tests in sand. *Journal of Geotechnical Engineering*, ASCE, 120(12), 2118–2135.

Yu, H.S. (1996). Interpretation of pressuremeter unloading tests in sands. *Geotechnique*, 46, 17–34.

Yu, H.S. and Houlsby, G.T. (1991). Finite cavity expansion in dilatant soil: loading analysis. *Geotechnique*, 41, 173–183.

Yu, H.S., Herrmann, L.R. and Boulanger, R.W. (2000). Analysis of steady cone penetration in clay. *Journal of Geotechnical and Geoenvironmental Engineering*, ASCE, 126(7).

CHAPTER 10

UNDERGROUND EXCAVATIONS AND TUNNELLING

10.1 INTRODUCTION

Underground excavations and tunnelling involve the removal of soil or rock masses from their initial locations. This action reduces and, in some cases (i.e. unlined tunnels), completely removes the initial stresses which existed in the area of tunnelling or excavations, Figure 10.1. Therefore, it may be reasonable to assume that the action of tunnelling and underground excavation can be modelled by the unloading of a cavity from the *in-situ* stress state.

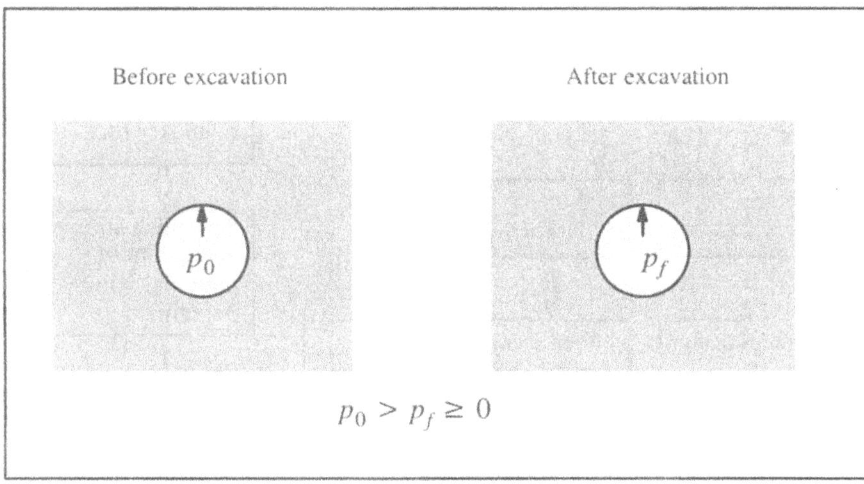

Figure 10.1: Action of tunnelling and underground excavation

The design and construction of tunnels and underground excavations have to satisfy two requirements: stability and serviceability. To ensure stability, it is often necessary to provide some support via linings to the internal boundary after a tunnel is excavated. To satisfy the serviceability requirement, the tunnelling-induced displacements must be small in order to avoid serious damage to the surrounding buildings and structures.

Cavity expansion theory has been used for decades for both the prediction of ground settlements due to tunnelling and the design of tunnel support systems to maintain stability (e.g. Ko *et al.*, 1980; Hoek and Brown, 1980; Brown *et al.*, 1983; Brady and Brown, 1993; Mair and Taylor, 1993; Yu and Rowe, 1999).

This chapter summarizes some of the main applications of cavity expansion/contraction solutions in the design and construction of tunnels and underground excavations in soil and rock. The emphasis is on the stability and ground settlements associated with tunnelling in soils and the design and construction of rock support systems in underground mining.

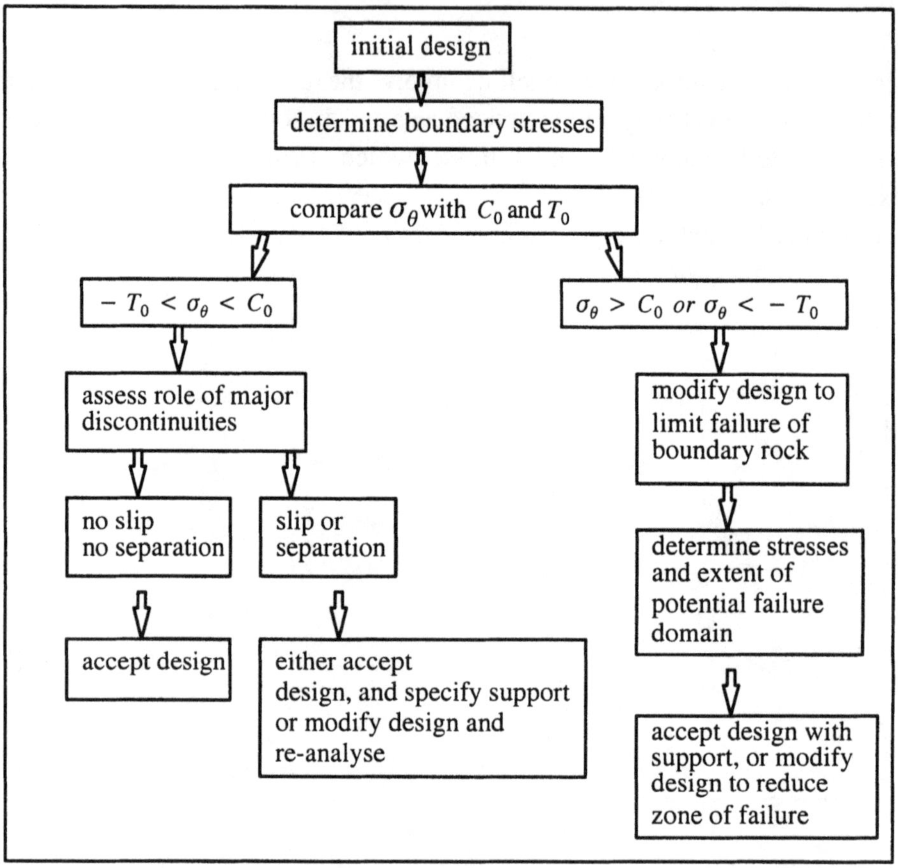

Figure 10.2: A design procedure for excavation in elastic rock (Brady and Brown, 1993)

10.2 EXCAVATION DESIGN IN MASSIVE ROCK

Excavation design in massive elastic rock is one of the simplest design problems in underground rock engineering. The basic procedure for excavation design has been discussed in detail by Brady and Brown (1993) and is illustrated simply in Figure 10.2 (where σ_θ is the tangential stress on the tunnel wall, and C_0 and T_0 denote the uniaxial compressive and tensile strengths of the rock). The design process begins with an initial configuration that will satisfy its duty requirements. A key step in the design process is the determination of the stress distribution around the excavation. Although the solution may be achieved by numerical methods, the simple closed-form solutions of cavity expansion theory have many advantages in the early stages of the design process.

To illustrate the usefulness of cavity expansion solutions in underground excavation engineering, this book focuses on the design of circular excavations in massive rock to allow the stress analysis to be carried out analytically.

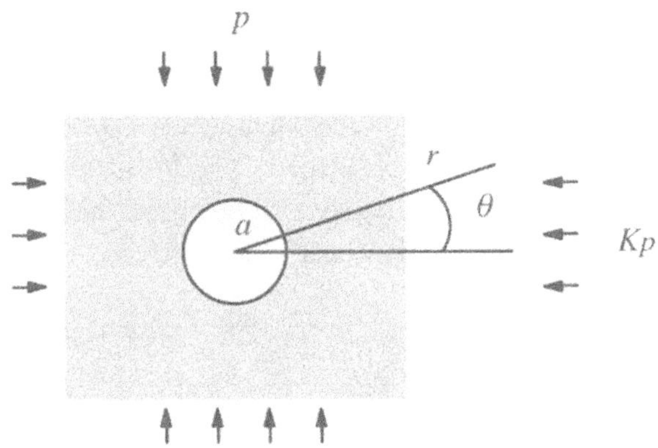

Figure 10.3: Problem geometry for circular excavation in rock

10.2.1 Elastic stress analysis

Assume that a circular excavation is made in a massive, elastic rock that has initial vertical and horizontal principal stresses of p and Kp, Figure 10.3. As presented in Chapter 2, if the rock behaves elastically, then the complete solutions for the stress distribution around the cavity are:

$$\sigma_r = \frac{p}{2}[(1+K)(1-\frac{a^2}{r^2})-(1-K)(1-4\frac{a^2}{r^2}+3\frac{a^4}{r^4})\cos 2\theta] \tag{10.1}$$

$$\sigma_\theta = \frac{p}{2}[(1+K)(1+\frac{a^2}{r^2})+(1-K)(1+3\frac{a^4}{r^4})\cos 2\theta] \tag{10.2}$$

$$\tau_{r\theta} = \frac{p}{2}[(1-K)(1+2\frac{a^2}{r^2}-3\frac{a^4}{r^4})\sin 2\theta] \tag{10.3}$$

Zone of influence of an excavation

The basic idea of a zone of influence is that it defines a domain of significant disturbance of the initial stress field by an excavation. It differentiates between the near field and far field of an opening. The extent of a near field for a circular opening may be illustrated by the simple case of the isotropic initial stress field.

If the initial stress state is isotropic (i.e. K=1), the stress expressions around the opening can be written as follows:

$$\frac{\sigma_r}{p} = 1-\left(\frac{a}{r}\right)^2 \tag{10.4}$$

$$\frac{\sigma_\theta}{p} = 1+\left(\frac{a}{r}\right)^2 \tag{10.5}$$

$$\tau_{r\theta} = 0 \tag{10.6}$$

These equations indicate that while the radial stress σ_r increases with radius, the tangential stress σ_θ decreases with radius. When the radius is greater than five times the radius of the opening, both the radial and tangential stresses become very close (within 5%) to the initial stress p. Consequently if the distance between the centers of two circular openings is greater than $6a$, for example, the mechanical interaction between the openings will be insignificant, Figure 10.4.

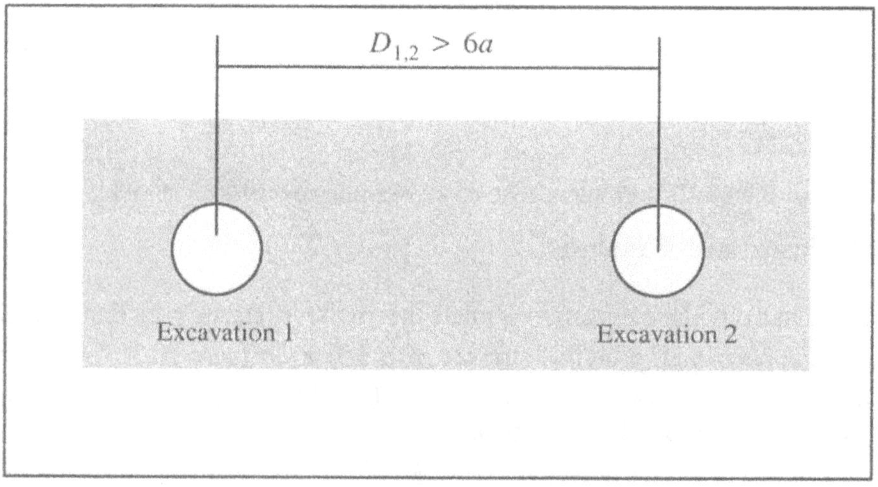

Figure 10.4: Interaction of two circular excavations

Effect of discontinuities on elastic stress distribution

When an underground excavation or tunnel is constructed in a rock mass with major discontinuities, it is questionable as to whether the elastic stress analysis described above will remain valid. This is because rock joints generally have much lower strengths than the intact rock itself. In particular, a rock discontinuity has zero or very low tensile strength. Two possible failure modes exist for a rock joint:

(a) Slip may occur if, in the simplest case, the normal and shear stresses acting on the joint satisfy the following criterion:

$$\tau \geq \sigma_n \tan\phi_j \tag{10.7}$$

where ϕ_j is the friction angle of the rock discontinuity.

(b) Separation may occur if the normal stress acting on the joint exceeds the tensile strength (which may often be assumed to be zero):

$$\sigma_n \leq 0 \tag{10.8}$$

Note that the notation of compression positive is used here.

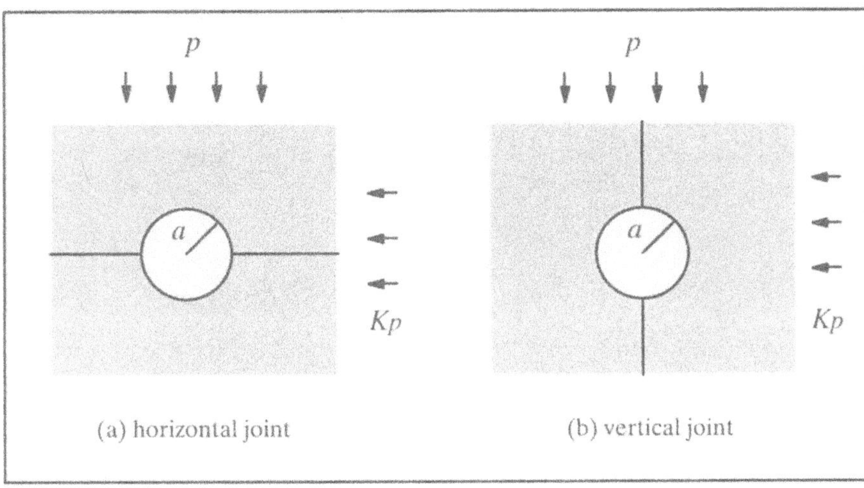

Figure 10.5: Excavation in rock with horizontal or vertical radial joints

Based on the above assumptions, Brady and Brown (1993) have used a number of simple cases to illustrate the fact that the effect of major rock joints on elastic stress distributions may be assessed easily. Some of these simple cases are repeated here, merely to show the usefulness of simple cavity expansion solutions.

Case 1 (Figure 10.5a). From the elastic stress solutions of (10.1)-(10.3), it can be shown that for a horizontal discontinuity ($\theta = 0^0$) the shear stress is zero. Thus there would be no possibility for shear slip to occur along the joint. In addition, it is also evident that the normal stress on the horizontal joint σ_θ will be compressive for all realistic values of K, hence eliminating the possibility of rock separation. It is therefore reasonable to conclude that the presence of a horizontal rock discontinuity would have no effect on the elastic stress field around the opening.

Case 2 (Figure 10.5b). From the elastic stress solutions of (10.1)-(10.3), it can be shown that for a vertical discontinuity ($\theta = 90^0$) the shear stress is zero. Thus there would be no possibility for the shear slip to occur along the vertical joint.

However, if K is less than 1/3, the normal stress to the vertical joint σ_θ will become tensile in the crown of the opening. Therefore the possibility of rock separation on the plane of weakness may arise. Based on this analysis, it may be concluded that for excavation in rock mass with a major vertical radial joint the elastic stress distribution will not be altered by slip if the *in-situ* principal stresses are vertical and horizontal. However, if the initial horizontal stress coefficient K is less than 1/3 (which is unlikely to be encountered in practice) the elastic stress field may be affected by the possibility of separation that may occur in the crown of the opening.

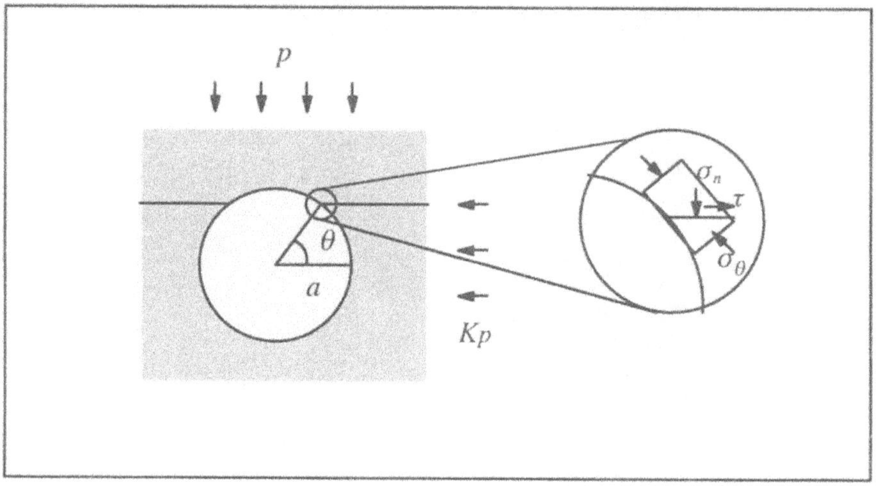

Figure 10.6: Excavation in rock with a horizontal, non-radial joint

Case 3 (Figure 10.6). From a simple stress transformation it can be shown that for a horizontal, non-radial discontinuity, the normal and shear stresses acting on the joint at the internal boundary are related to the tangential stress by

$$\sigma_n = \sigma_\theta \cos^2\theta \tag{10.9}$$

$$\tau = \sigma_\theta \sin\theta \cos\theta \tag{10.10}$$

Using the condition defined by (10.7), it can be shown that slip would occur for joints that satisfy the following condition:

$$\theta \geq \phi_j \tag{10.11}$$

As a result, special attention is needed in such cases to design a suitable support system to deal with potential slip failures in the crown areas.

Case 4 (Figure 10.7). To assess the effect of an inclined, radial joint on the elastic stress distribution around a circular opening, a simple case shown in Figure 10.7

is considered in detail. With the values of $\theta = 45^0$ and $K = 0.5$, the normal and shear stresses along the joint will be:

$$\sigma_n = \sigma_\theta = \frac{p}{2} \times 1.5 \times [1 + (\tfrac{a}{r})^2] \tag{10.12}$$

$$\tau = \tau_{r\theta} = \frac{p}{2} \times 0.5 \times [1 + 2(\tfrac{a}{r})^2 + 3(\tfrac{a}{r})^4] \tag{10.13}$$

The ratio of shear to normal stresses will be a function of the nondimensional radius r/a, Figure 10.7. For this special case, it can be shown that the ratio of τ/σ_n along the joint takes a maximum value of 0.375 (corresponding to a friction angle of 19.6 degrees) when the nondimensional radius is $r/a = 2.5$. In other words, if the angle of internal friction along the rock joint ϕ_j is greater than 19.6 degrees, slip can be precluded anywhere in the rock mass. Otherwise, slip will occur somewhere along the joint.

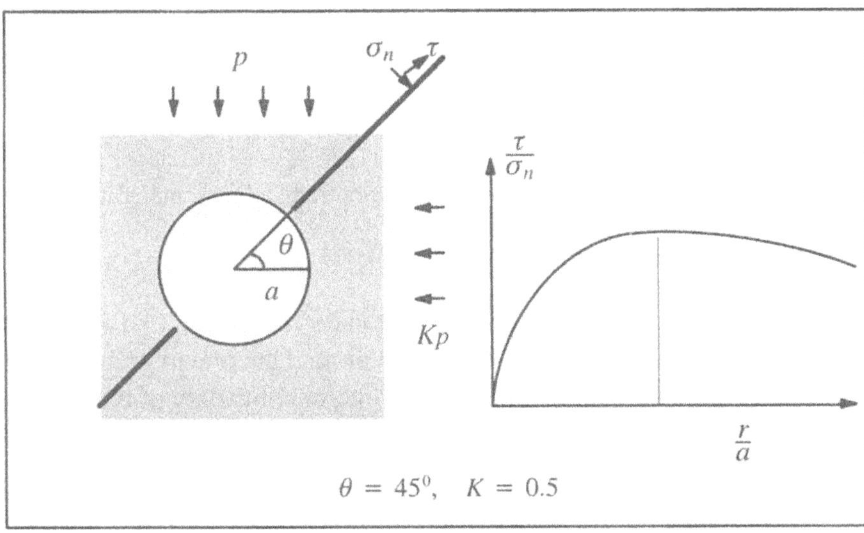

Figure 10.7: Excavation in rock with an inclined, radial joint

Case 5 (Figure 10.8). Brady and Brown (1993) used a simple case shown in Figure 10.8 to describe the procedure that can be utilised to assess the effect of a non-radial horizontal joint on the elastic stress field around the circular opening. In this case, the normal and shear stresses along the rock joint will be:

$$\sigma_n = p[1-(\tfrac{a}{r})^2 \cos 2\alpha] \tag{10.14}$$

$$\tau = p(\tfrac{a}{r})^2 \sin 2a \tag{10.15}$$

The above expressions can be used to indicate that the value of τ/σ_n varies with the location along the joint, Figure 10.8. In fact the peak ratio of τ/σ_n is 0.445, which corresponds to a friction angle of 24 degrees. This means that if the friction angle of the rock joint is higher than 24 degrees, the elastic stress field will not be altered by the presence of the joint as no slip will be possible in the rock mass.

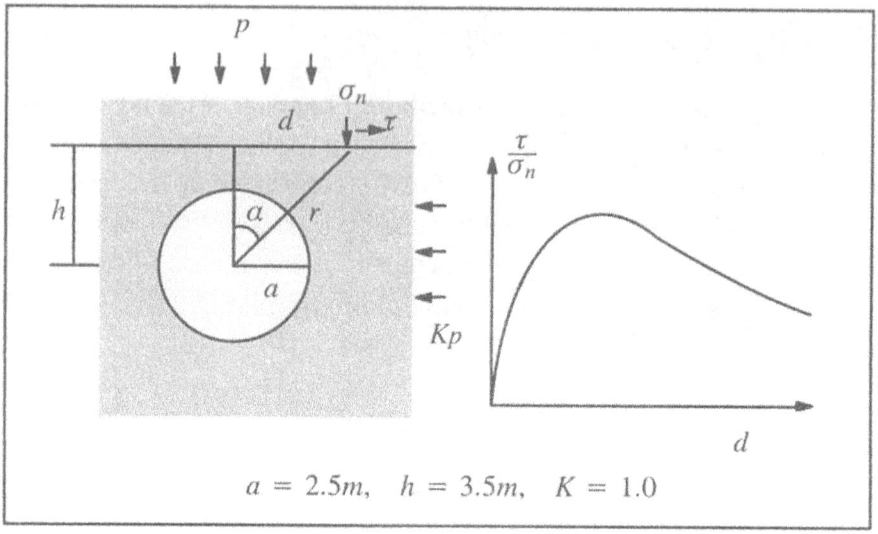

Figure 10.8: Excavation in rock with a non-radial, horizontal joint

10.2.2 Elastic-plastic (fracture) stress analysis

Many underground excavations are carried out under conditions that are sufficient to induce a failure (plastic or fractured) region around the opening. In such cases, it is necessary to provide some support on the internal boundary of the excavation to control the performance of fractured (or failed) rock.

To illustrate the function of rock support in an underground excavation, it is instructive to consider the simple problem shown in Figure 10.9. It is well known that when the internal pressure p is sufficiently lower than the *in-situ* stress, a failed zone denoted by radius c will form around the opening and, in this failed region, the strength of the rock may drop significantly.

The complete stress and displacement solutions for this elastic-plastic problem are presented in Chapter 5 (Section 5.3). To assess the effect of internal support on stresses and the size of the failure zone, some of the key results will be presented here for convenience. A cylindrical cavity is located in a infinitely large rock mass. Initially the rock is subject to a hydrostatic stress so that all the normal stress components are p_0. The stress and displacement fields around the cavity when the cavity pressure is gradually reduced to a lower value p are of primary interest. The analytical solutions presented here for brittle-plastic rock closely follow those given by Wilson (1980), Fritz (1984) and Reed (1986).

For most tunnel problems, it may be assumed that the axial stress σ_z is the intermediate stress between the radial and hoop stresses σ_r and σ_θ. For an unloading cavity, the hoop stress is the major principal stress.

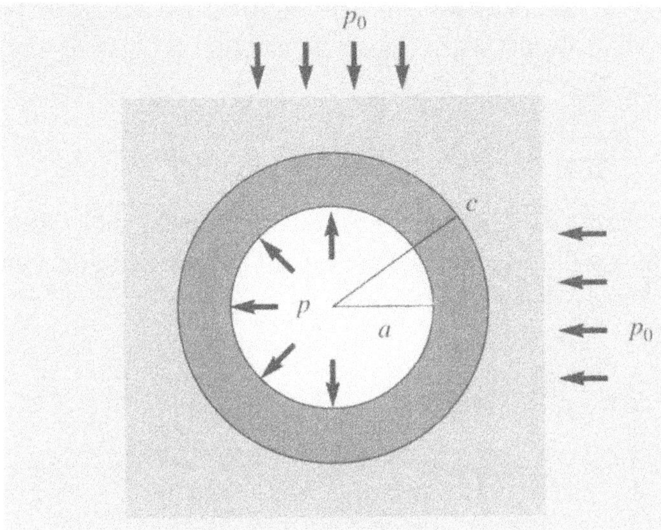

Figure 10.9: Circular opening with an internal support pressure

The following linear Mohr-Coulomb yield function may be used to indicate the initiation of yielding in the rock mass:

$$\sigma_\theta - \alpha\sigma_r = Y \tag{10.16}$$

where α and Y are related to the friction angle and cohesion ϕ and C by

$$\alpha = \frac{1 + \sin\phi}{1 - \sin\phi} \quad \text{and} \quad Y = \frac{2C\cos\phi}{1 - \sin\phi} \tag{10.17}$$

After yielding, the strength of rock may suddenly drop and, to allow for such a post-yield softening behaviour, it is assumed that the stresses after yielding satisfy the following residual failure function:

$$\sigma_\theta - \alpha'\sigma_r = Y' \tag{10.18}$$

where $\alpha' = (1 + \sin\phi')/(1 - \sin\phi')$ and $Y' = (2C'\cos\phi')/(1 - \sin\phi')$ are the residual strength parameters of the rock (Figure 10.10).

(a) Stress analysis

Elastic response and initial yielding

When the internal stress reduces, the rock initially behaves elastically and the stresses are given by the following expressions:

$$\sigma_r = p_0 - (p_0 - p)\left(\frac{a}{r}\right)^2 \tag{10.19}$$

$$\sigma_\theta = p_0 + (p_0 - p)\left(\frac{a}{r}\right)^2 \tag{10.20}$$

Yield will initiate at the cavity wall once the elastic stress fields satisfy the yield function (10.16) which occurs when the cavity pressure is equal to the following value:

$$p = p_{1y} = \frac{2p_0 - Y}{\alpha + 1} \tag{10.21}$$

If the internal pressure p is less than the value defined in equation (10.21), a plastic zone with radius c will form around the cavity. The rock in the zone outside the radius c will still behave elastically.

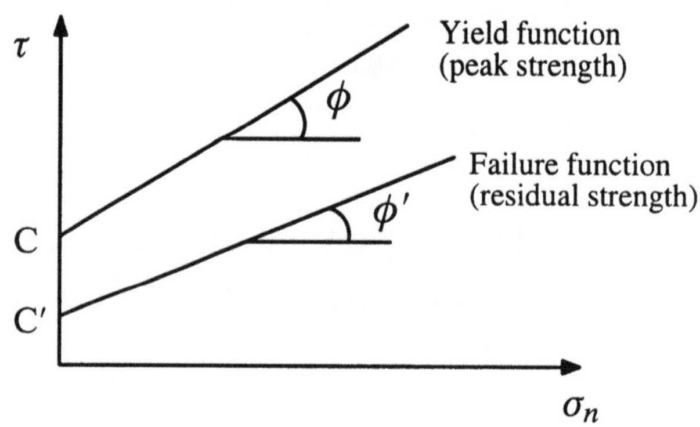

Figure 10.10: Rock strength parameters before and after yielding

Stresses in elastic zone $c \leq r \leq \infty$

After yielding, the stress fields in the outer elastic zone are easily shown to be

$$\sigma_r = p_0 - (p_0 - p_{1y})\left(\frac{c}{r}\right)^2 \tag{10.22}$$

$$\sigma_\theta = p_0 + (p_0 - p_{1y})\left(\frac{c}{r}\right)^2 \tag{10.23}$$

Stresses in plastic zone $a \leq r \leq c$

The stresses in the plastic region must satisfy the residual failure equation and the following equilibrium equation:

$$r\frac{d\sigma_r}{dr} = \sigma_\theta - \sigma_r \tag{10.24}$$

both of which can be combined to determine the stresses in the inner plastic zone. The derived stress expressions are:

$$\sigma_r = \frac{Y' + (a'-1)p}{a'-1}\left(\frac{r}{a}\right)^{a'-1} - \frac{Y'}{a'-1} \tag{10.25}$$

$$\sigma_\theta = a'\frac{Y' + (a'-1)p}{a'-1}\left(\frac{r}{a}\right)^{a'-1} - \frac{Y'}{a'-1} \tag{10.26}$$

The continuity of the radial stress at the elastic-plastic interface $r=c$ required by equilibrium can be used to determine the radius of the plastic region:

$$\frac{c}{a} = \left\{\frac{Y' + (a'-1)p_{1y}}{Y' + (a'-1)p}\right\}^{\frac{1}{a'-1}} \tag{10.27}$$

It is noted that the hoop (or tangential) stress is only continuous across the elastic-plastic boundary if the rock is perfectly plastic (i.e. $Y = Y'$).

To be informative, let us consider a simple calculation with the following rock strength properties: $Y = Y' = 0$, $\phi' = 30^0$, $\phi = 40^0$. The use of equation (10.27) gives the size of plastic (failed) zone as:

$$\frac{c}{a} = 0.6\sqrt{\frac{p_0}{p}} \tag{10.28}$$

The above expression indicates that the size of the plastic zone strongly depends on the relative magnitudes of the cavity internal pressure and the *in-situ* stress. Therefore by providing certain internal support (i.e. increasing the internal pressure p), the region of failed rock can be controlled and limited to a satisfactory level.

10.3 ROCK SUPPORT IN UNDERGROUND EXCAVATIONS

10.3.1 The principle of rock support and reinforcement

The basic principle used to design rock support in underground excavations is clearly illustrated by Brady and Brown (1993). For convenience, some of the basic ideas described in Brady and Brown (1993) will be described in this section.

For simplicity, let us consider a circular excavation shown in Figure 10.11 in which a heading is being advanced by conventional drill and blast methods. The pre-excavation state of stress is assumed to be hydrostatic and of magnitude p_0. Blocked steel sets are installed after each drill and blast cycle. Brady and Brown (1993) present a detailed discussion on the development of radial displacement and radial support 'pressure' at a point on the excavation periphery at section X–X as the heading progressively advances to and beyond X–X.

In step 1, the heading has not yet reached the section X–X and the rock mass on the periphery of the proposed profile is in equilibrium with an internal support pressure p, acting equal and opposite the pre-excavation pressure p_0.

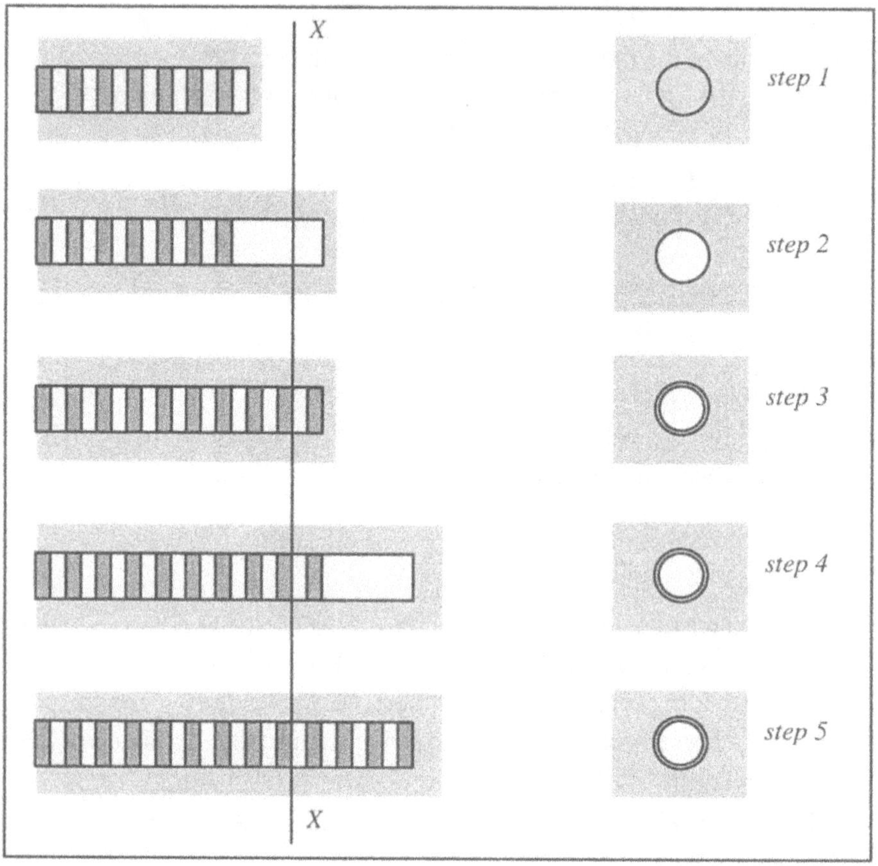

Figure 10.11: Basic ideas for rock support design

In step 2, the face has been advanced beyond the section X–X and the support pressure p, previously provided by the rock inside the excavation periphery, has been reduced to zero. However, the unsupported section between the face and the last steel set installed is constrained, to some extent, by the proximity of the face. The position at this stage is marked by point B in Figure 10.12.

In step 3, the heading has been mucked out and steel sets have been installed close to the face. At this stage, the steel sets carry little load because no deformation of the rock has occurred since their installation. The position at this stage is marked by points B for rock and D for steel support in Figure 10.12.

In step 4, the heading is advanced about one and a half tunnel diameters beyond the section X–X by a further cycle of drilling and blasting. The restraint offered by the proximity of the face is now negligible. At this stage, there will be a further radial displacement of the rock surface at X–X denoted by path BEF in Figure 10.12. This radial displacement in the rock then induces load in the steel sets which are assumed to show linear radial stress-displacement behaviour (path DEG). Equilibrium between the rock and the steel sets is reached at point E when the required support line intersects the support reaction line.

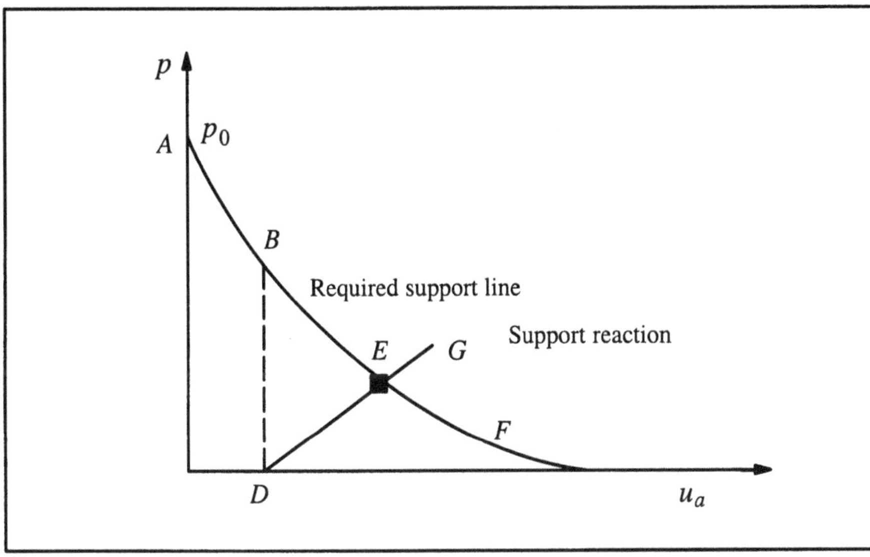

Figure 10.12: Rock-support interaction

From the above discussion, it is evident that we need to determine the required support lines (i.e. ground response curves – internal pressure and displacement correlation) for a rock mass in order to design a suitable support system. Hence the next section will be devoted to the use of cavity expansion theory for calculating the ground response curves with a realistic rock model.

10.3.2 Ground response curves

In addition to the linear Mohr-Coulomb yield criterion, solutions for ground response curves have also been derived by many researchers (e.g., Hobbs, 1966; Kennedy and Lindberg, 1978) using non-linear yield criteria. One of the most interesting developments is the cavity unloading solution derived by Brown *et al.* (1983) who used the empirical Hoek-Brown nonlinear criterion (Hoek and Brown, 1980) to describe the behaviour of rock masses. The cavity unloading solution of Brown *et al.* (1983) has been presented and improved in Chapter 5. In this section, only some of the key results that are relevant to ground response curves are given.

(a) Hoek-Brown yield and failure criteria

It is assumed that the yield of rock is governed by the Hoek-Brown criterion:

$$\sigma_1 = \sigma_3 + \sqrt{mY\sigma_3 + sY^2} \tag{10.29}$$

where σ_1 and σ_3 are the major and minor principal stresses; Y is the uniaxial compressive strength of the intact rock material; and m and s are constants depending on the nature of the rock mass and the extent to which it has been broken before being subjected to the principal stresses σ_1 and σ_3.

After yield, the strength parameters m and s drop to the residual values m' and s'. The unconfined compressive strength of the rock mass changes from its peak value $\sqrt{s}\,Y$ to its residual value $\sqrt{s'}\,Y$. The residual strength for the broken rock mass is therefore defined by

$$\sigma_1 = \sigma_3 + \sqrt{m'Y\sigma_3 + s'Y^2} \tag{10.30}$$

To ensure that a closed form solution can be obtained, it is necessary to further assume that after yield the strength of the rock drops suddenly to its residual value.

(b) Stress fields

The stresses in the outer elastic region can be shown to be:

$$\sigma_r = p_0 - (p_0 - p_{1y})(\tfrac{c}{r})^2 \tag{10.31}$$

$$\sigma_\theta = p_0 + (p_0 - p_{1y})(\tfrac{c}{r})^2 \tag{10.32}$$

where c is the radius of elastic-plastic boundary and p_{1y} is the radial stress at the elastic-plastic boundary. For tunnel unloading problems, the Hoek-Brown yield function takes the following form:

$$\sigma_\theta = \sigma_r + \sqrt{mY\sigma_r + sY^2} \tag{10.33}$$

The above yield criterion must be satisfied at the internal boundary of the elastic zone $r=c$. Putting equations (10.31) and (10.32) with $r=c$ into equation (10.33) leads to the following solution for the radial stress p_{1y} at the elastic-plastic boundary:

$$p_{1y} = p_0 - MY \tag{10.34}$$

where

$$M = \frac{1}{2}[(\frac{m}{4})^2 + m\frac{p_0}{Y} + s]^{1/2} - \frac{m}{8} \tag{10.35}$$

Within the inner plastic region (or broken zone), stresses must satisfy equilibrium and the following residual strength criterion:

$$\sigma_\theta = \sigma_r + \sqrt{m'Y\sigma_r + s'Y^2} \tag{10.36}$$

Both the equilibrium equation and the above failure criterion can be combined to give the following solution for the radial and hoop stresses in the plastic zone:

$$\sigma_r = p + A\ln(\frac{r}{a}) + B\ln^2(\frac{r}{a}) \tag{10.37}$$

$$\sigma_\theta = p + A + (A + 2B)\ln(\frac{r}{a}) + B\ln^2(\frac{r}{a}) \tag{10.38}$$

where p is the tunnel support pressure and A and B are defined as:

$$A = \sqrt{m'Yp + s'Y^2} \quad \text{and} \quad B = \frac{1}{4}m'Y \tag{10.39}$$

The radius of the elastic-plastic boundary c is solved by using the continuity of the radial stress at the elastic-plastic boundary as follows:

$$\frac{c}{a} = \exp\left\{N - \frac{2}{m'Y}\sqrt{m'Yp + s'Y^2}\right\} \tag{10.40}$$

in which

$$N = \frac{2}{m'Y}\sqrt{m'Yp_0 + s'Y^2 - m'Y^2M} \tag{10.41}$$

It should be noted that a plastic zone will form only if the tunnel support pressure reduces to the critical value $p \leq p_{1y}$.

(c) Elastic and plastic displacements

Displacement in the outer elastic zone

The displacements in the outer elastic zone can be shown to be:

$$u = \frac{1 + \nu}{E}(p_0 - p_{1y})\frac{c^2}{r} \tag{10.42}$$

In particular, for the elastic-plastic interface the displacement is:

$$u_c = u|_{r=c} = \frac{1 + \nu}{E}(p_0 - p_{1y})c \tag{10.43}$$

Displacement in the plastic zone

To determine the displacement field in the plastic zone, a plastic flow rule is needed. If a non-associated flow rule similar to that of the Mohr-Coulomb criterion can be used, this results in:

$$d\varepsilon_r^p + \beta d\varepsilon_\theta^p = 0 \tag{10.44}$$

where $\beta = (1 + \sin\psi)/(1 - \sin\psi)$ and ψ is the dilation angle of the rock material.

For small deformations using the plastic flow rule (10.44), the general solution for the radial displacement u (an inward radial displacement is taken as positive) around the tunnel may be shown to be:

$$u = n_1\frac{r}{1+\beta} + n_2\left[\frac{r\ln r}{1+\beta} - \frac{r}{(1+\beta)^2}\right] + D_3\left[\frac{r\ln^2 r}{1+\beta} - \frac{2r\ln r}{(1+\beta)^2} + \frac{2r}{(1+\beta)^3}\right] + C_0 r^{-\beta} \tag{10.45}$$

in which

$$D_1 = \frac{1+\nu}{E}\left\{(1+\beta)(1-2\nu)(p-p_0) + (\beta-\nu\beta-\nu)A\right\} \tag{10.46}$$

$$D_2 = \frac{1+\nu}{E}\left\{(1-\nu)(A+\beta A+2\beta B)-\nu(\beta A+A+2B)\right\} \tag{10.47}$$

$$D_3 = \frac{1+\nu}{E}(1+\beta)(1-2\nu)B \tag{10.48}$$

$$n_1 = D_1 - D_2\ln a + D_3\ln^2 a \tag{10.49}$$

$$n_2 = D_2 - 2D_3\ln a \tag{10.50}$$

$$C_0 = \frac{1+\nu}{E}(p_0-p_{1y})c^{1+\beta} - n_1\frac{c^{1+\beta}}{1+\beta} - n_2\left[\frac{\ln c}{1+\beta} - \frac{1}{(1+\beta)^2}\right]c^{1+\beta} - D_3\left[\frac{\ln^2 c}{1+\beta} - \frac{2\ln c}{(1+\beta)^2} + \frac{2}{(1+\beta)^3}\right]c^{1+\beta} \tag{10.51}$$

By combining equations (10.51) and (10.45), the displacement fields in the plastic zone are fully defined. In particular, the ground response curve of a tunnel is obtained as a special case by setting $r=a$ in equation (10.45).

10.4 TUNNELS IN COHESIVE SOILS

The design of shallow tunnels in soils needs to meet two requirements, serviceability and stability. The serviceability requirement is to ensure that the construction of a shallow tunnel will not cause excessive displacements in the soils that may damage the surrounding or adjacent buildings and utilities. Of particular importance are the surface settlements caused by tunnelling and their effects on the structures that are built directly above the tunnel, Figure 10.13.

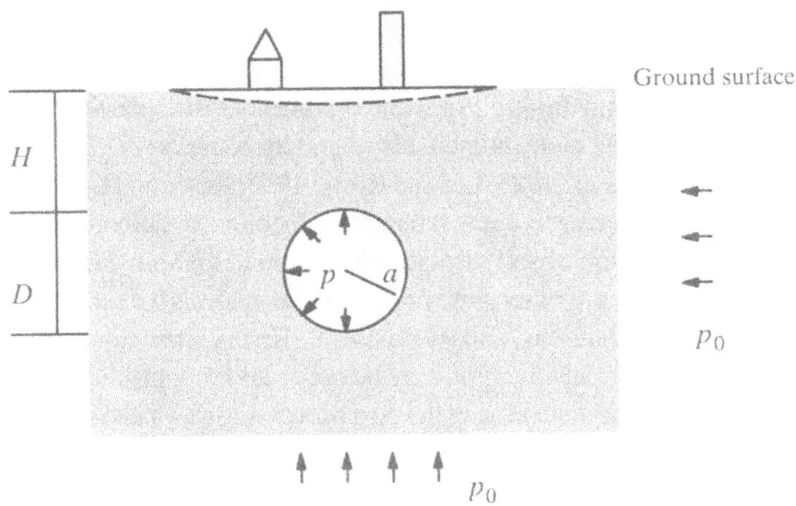

Figure 10.13: A shallow tunnel in soil

At the same time, the design of tunnels in soils must also ensure that the soils surrounding the tunnels do not collapse because the internal support pressure is not sufficiently large. Therefore, it is essential that stability calculations are made in the design stage so that adequate internal pressure is provided to maintain overall stability of the tunnel. The issues related to tunnel stability have been considered by Davis *et al.* (1980) and Sloan and Assadi (1993) among others.

Traditionally estimates of tunnelling-induced settlements have been based on an empirical relationship suggested by Peck (1969) who assumed that the transverse surface settlement profile follows a normal probability curve. To estimate the maximum surface settlement, Peck *et al.* (1972) further suggested that the volume of the surface settlement trough can be taken as 1% of the volume of soils or rock excavated. When applied with appropriate judgement based on similar past experience, this empirical procedure can be used to yield satisfactory designs. However, without past experience this empirical procedure may not be applicable to new situations which involve tunnels with different geometries, ground conditions and construction techniques. Furthermore, Peck's procedure does not provide information concerning the distribution of settlement other than at the surface.

Much research has since been conducted to develop a theoretically based procedure for predicting ground settlement due to tunnelling in soft soils. Typically this involves simulating construction of the tunnel by deducing the tractions that are acting around the surface of the tunnel (prior to excavation) and then removing these tractions. As discussed by Kulhawy (1974), this approach provides a reasonable approximation of the change in stress that occurs with the advance of the tunnelling machine. The theoretical research based on this assumption has involved both simple cavity unloading solutions (e.g., Pender, 1980; Lo *et al.*, 1980; Ogawa and Lo, 1987; Mair and Taylor, 1993) and nonlinear elastic-plastic 2-dimensional or 3-dimensional finite element methods (e.g., Ghaboussi *et al.*, 1978; Rowe and Kack, 1983; Clough *et al.*, 1985; Lee and Rowe, 1990; Rowe and Lee, 1992). Based on the results of this research, some satisfactory theoretical methods have been proposed for the prediction of soil displacements around tunnels. The method based on finite elements requires a separate finite element analysis for each tunnel geometry and set of soil conditions, and while useful, is very time consuming. While a rigorous 3D finite element analysis may be essential for some situations, the cost of data preparation, the analysis and the post processing may be too high to be justified in many cases. Nevertheless, the rigorous finite element analyses combined with field case records can be used to check the validity of other simpler methods such as cavity unloading solutions.

The comparative studies reported by Lo *et al.* (1980) and Ogawa and Lo (1987) indicate that after applying a correction factor, the simple plane strain cylindrical cavity unloading solution can be satisfactorily used to predict the measured soil displacements around tunnels. Research carried out by Mair and Taylor (1993) has also highlighted the usefulness of spherical cavity unloading solutions in modelling soil behaviour around an advancing tunnel heading. After comparing the cavity unloading solutions with some field and centrifuge measurements, Mair and Taylor concluded that the simple plasticity solution of the unloading of cavities can be successfully used to predict both displacements and pore pressure changes around tunnels in the clay they examined.

In the analysis of tunnels in soils, it is often important to investigate two separate cases. Firstly, the long term behaviour of tunnels in cohesive-frictional soils may be modelled as a fully drained problem. The drained analysis can be carried out in terms of effective stresses with the Mohr-Coulomb criterion (Rowe and Kack, 1983). The displacements obtained from the drained analyses include both the immediate and the time-dependent consolidation settlements. Previous cavity unloading work for modelling tunnels in Mohr-Coulomb materials has mainly concentrated on small strain, cylindrical cavity contraction models. As noted by Mair and Taylor (1993), however, for soil behaviour around an advancing tunnel heading the unloading of a spherical cavity is more realistic. At present, it is also unclear as to what effect the small strain assumption used in most previous cavity models will have on the predicted soil behaviour.

Secondly, tunnel construction in clay is usually sufficiently rapid that the clay behaviour around tunnels may be assumed to be undrained. This is relevant for the investigation of short term ground settlement due to tunnelling. In the past, the analysis of undrained tunnelling has often been carried out using a total stress approach with a linear elastic-perfectly plastic Tresca model. Despite its simplicity, the total stress approach suffers from the following drawbacks: (i) the effect of stress history (OCR) on the soil behaviour cannot be included; (ii) the variation of soil stiffness on stress level and void ratio is generally not taken into account; and (iii) the effect of soil strain-hardening/softening is not considered.

In the next two sections, it will be shown how simple cavity expansion solutions such as those developed in earlier chapters of this book can be used to assess the displacement behaviour and stability of tunnels.

10.4.1 Settlements due to tunnelling – total stress analysis

It is assumed that an unbound Tresca medium contains a single cylindrical or spherical cavity (tunnel). Initially the radius of the cavity (tunnel) is a_0 and a hydrostatic pressure p_0 acts throughout the soil which is assumed to be homogeneous. The pressure on the tunnel wall is then reduced sufficiently slowly for dynamic effects to be negligible. This section is concerned with the distribution of stress and displacement in the soil as the tunnel pressure reduces from its initial value. The problem defined here has been solved in Chapter 3 and some of the key results are repeated here for convenience in the application. Note that the symbol k is used to denote cylindrical (k=1) and spherical cavity models (k=2).

Elastic response

As the tunnel pressure p decreases from p_0, the deformation of the soil around the tunnel at first is purely elastic. The elastic solutions for stresses and displacement are:

$$\sigma_r = -p_0-(p-p_0)(\frac{a}{r})^{1+k} \tag{10.52}$$

$$\sigma_\theta = -p_0 + \frac{p-p_0}{k}(\frac{a}{r})^{1+k} \tag{10.53}$$

$$u = \frac{p-p_0}{2kG}(\frac{a}{r})^{1+k}r \tag{10.54}$$

Since the tensile positive notation is used for elastic-perfectly plastic solutions presented in this book, the yield equation takes the form:

$$\sigma_r - \sigma_\theta = Y = 2s_u \tag{10.55}$$

As the internal pressure decreases further, initial yielding occurs at the cavity wall when the tunnel pressure is equal to the following value:

$$p = p_{1y} = p_0 - \frac{kY}{1+k} \tag{10.56}$$

Elastic-plastic stress fields

After initial yielding at the cavity wall, a plastic zone within the region $a \leq r \leq c$ will form around the inner wall of the tunnel with the further decrease of the cavity pressure p.

The stresses in the elastic zone are shown to be:

$$\sigma_r = -p_0 + \frac{kY}{1+k}\left(\frac{c}{r}\right)^{(1+k)} \tag{10.57}$$

$$\sigma_\theta = -p_0 - \frac{Y}{1+k}\left(\frac{c}{r}\right)^{(1+k)} \tag{10.58}$$

On the other hand, the stresses in the plastic zone must satisfy equilibrium and the yield condition and can be shown to be as follows:

$$\sigma_r = -p_0 + \frac{kY}{1+k} + kY\ln\frac{c}{r} \tag{10.59}$$

$$\sigma_\theta = -p_0 - \frac{Y}{1+k} + kY\ln\frac{c}{r} \tag{10.60}$$

Applying equation (10.59) at the tunnel wall leads to a relationship between the tunnel pressure p and the plastic radius c:

$$\frac{p_0 - p}{Y} = k\ln\frac{c}{a} + \frac{k}{1+k} \tag{10.61}$$

Elastic-plastic displacements

The displacement in the elastic zone is shown to be:

$$u = \frac{p_{1y} - p_0}{2kG}\left(\frac{c}{r}\right)^{1+k} r = -\frac{Y}{2(1+k)G}\left(\frac{c}{r}\right)^{1+k} r \tag{10.62}$$

Therefore, at the elastic-plastic interface the displacement is:

$$u|_{r=c} = c - c_0 = -\frac{Yc}{2(1+k)G} \tag{10.63}$$

By using the elastic displacement solution at the plastic boundary (10.63) and the radius of the reverse plastic zone (10.61), the incompressibility condition in the plastic zone can be used to give the following displacement field in the plastic region:

$$(\frac{r_0}{r})^{1+k} = 1 + \exp[\frac{(1+k)(p_0-p)}{kY} - 1]$$

$$\times [(1 + \frac{Y}{2(1+k)G})^{1+k} - 1](\frac{a}{r})^{1+k} \qquad (10.64)$$

As a special case when $r = a$ and $r_0 = a_0$, equation (3.85) reduces to the following relationship between the tunnel pressure and the tunnel wall displacement:

$$(\frac{a_0}{a})^{1+k} = 1 + \exp[\frac{(1+k)(p_0-p)}{kY} - 1]$$

$$\times [(1 + \frac{Y}{2(1+k)G})^{1+k} - 1] \qquad (10.65)$$

For many tunnel problems, the strains are assumed to be small in the plastic zone, and for this case the large strain solutions (10.64) and (10.65) can be simplified to:

$$\frac{u}{a} = -\frac{Y}{2(1+k)G}(\frac{a}{r})^k \exp[\frac{(1+k)(p_0-p)}{kY} - 1] \qquad (10.66)$$

for displacement at any radius around the tunnel. Applying equation (10.66) at the tunnel wall leads to the following relationship between the tunnel wall movement and tunnel pressure:

$$\frac{u_a}{a} = -\frac{Y}{2(1+k)G}\exp[\frac{(1+k)(p_0-p)}{kY} - 1] \qquad (10.67)$$

Comparison with measured tunnel behaviour in clay

The small strain solutions (10.66) and (10.67) have been used by Mair and Taylor (1993) to predict the soil deformation around tunnels in clay. The cylindrical cavity expansion solution indicates that there is a linear correlation between the normalized soil displacement *u/a* with the value of *a/r*.

Figure 10.14 shows field measurements of sub-surface vertical ground movements (δ_v) above the centre-line of tunnels constructed in London clay for London Underground Ltd. The measurements were made on tunnels of 4.1m excavated diameter beneath Green Park at a depth of 29m and Regent's Park at depths of 20m and 24m. Also shown in Figure 10.14 are measurements of horizontal ground movements (δ_h) at the level of the tunnel axis perpendicular to the direction of the tunnel. Measured horizontal ground movements from Green Park and Regent's Park are shown, together with data from other tunnels as well as those from the Angel Station Reconstruction project.

The vertical and horizontal ground movement measurements from the different London clay sites shown in Figure 10.14 are reasonably consistent and are in general agreement with linear plots suggested by cavity expansion theory. This agreement supports the use of the simple unloading cylindrical cavity solution to predict the ground movements above and to the side of the tunnel. The plots of the vertical and horizontal movements are almost parallel but do not pass through the origin. This is because that the non-axisymmetric nature of the tunnel problems, as in reality the vertical movement above the tunnel at the radius r would be larger than the horizontal movement at the same radius to the side of the tunnel. It is also true that even at a large distance above the tunnel (i.e. a small value of *a/r*) measurable settlements are observed, while no measurable horizontal movements are observed to the side of the tunnel beyond a radius of 4*a*.

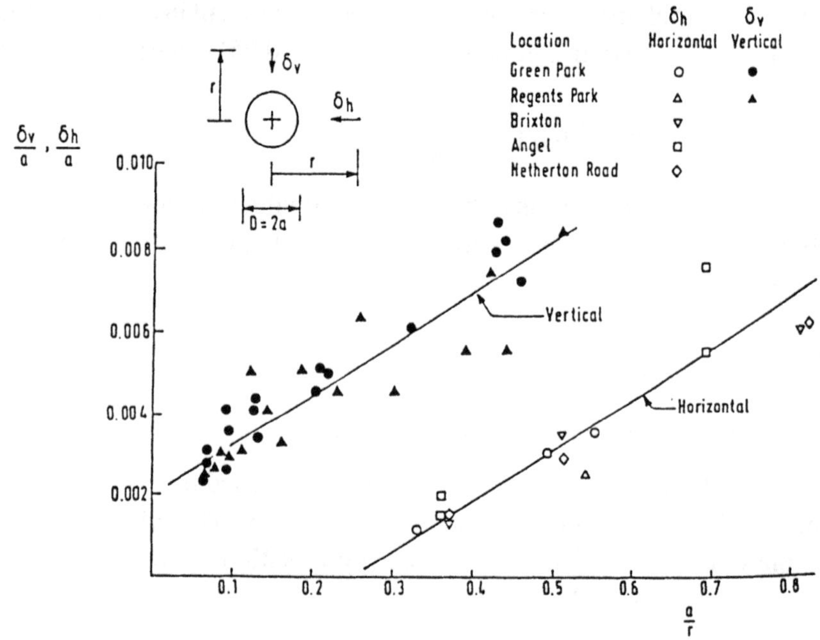

Figure 10.14: Vertical and horizontal subsurface movements in the vicinity of tunnels in London clay (Mair and Taylor, 1993)

As pointed out by Mair and Taylor (1993), the data shown in Figure 10.14 indicate that the idealization of axisymmetric conditions around a cylindrical cavity provides a valuable theoretical framework for prediction of ground movements around tunnels. The major advantage of this framework is its simplicity.

In addition to the estimation of soil movements around the tunnel, cavity expansion theory can also be used to predict the pore pressure changes around an excavated tunnel. Assumption of the linear elastic-perfectly plastic soil model gives the following expression for the pore pressure changes at a radius r within the plastic zone around an unloaded cylindrical cavity under axisymmetric conditions:

$$\frac{\Delta U}{s_u} = 1 - N + 2\ln(\frac{r}{a}) \tag{10.68}$$

where $N = (p_0 - p)/s_u$ is known as the stability number for tunnels.

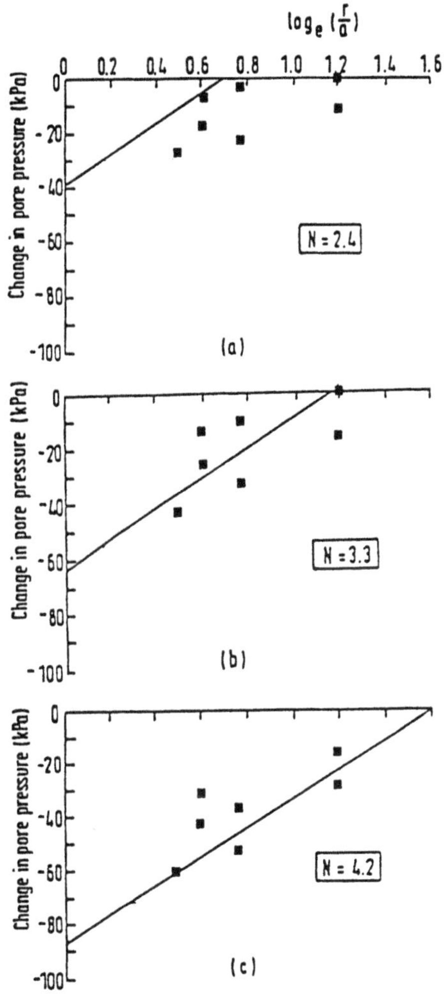

Figure 10.15: Changes in pore pressure around a centrifuge model tunnel in clay (Mair and Taylor, 1993)

The above equation is valid only in the plastically deforming zone. The radius of the plastic zone c is defined as:

$$\frac{c}{a} = \exp(\frac{N-1}{2}) \tag{10.69}$$

According to cavity expansion theory, the pore pressure changes in the outer elastic region will be zero.

Figure 10.15 shows measurements of pore pressure changes in a centrifuge test reported by Mair (1979) on a model tunnel in soft clay. The cover to diameter ratio was 3.1 and the internal support pressure was progressively reduced until collapse of the tunnel occurred. Measurements are shown as solid square points for three different stages of the model test, corresponding to stability numbers of $N = 2.4, 3.3$ and 4.2. It is noted that reasonably good agreement is obtained between equation (10.68) and measured pore pressures within the plastic zone.

10.4.2 Settlements due to tunnelling – effective stress analysis

Although most undrained soil mechanics problems can be more easily solved in terms of total stresses, this is no longer appropriate when using critical state models where the strength of the soil is a variable. This is because the soil strength is a function of the effective stresses rather than the total stresses. In particular, the total stress analysis does not take account of the influence of soil stress history on the soil behaviour. Thus, in this section we seek to find a solution in terms of effective stress. The effective stress cavity unloading solutions presented are those derived by Yu and Rowe (1999).

(a) Solution procedure

Kinematics of cavity unloading

The constant volume condition under undrained loading conditions gives the following relation between r, the current radius of a material element which was initially at r_0 and the current and initial radii of the cavity – a and a_0 respectively:

$$r_0^{\,k+1} - r^{k+1} = a_0^{k+1} - a^{\,k+1} \tag{10.70}$$

The radial speed of the element is hence related to the speed of contraction of the cavity by:

$$w = \dot{r} = (\tfrac{a}{r})^k \, \dot{a} \tag{10.71}$$

Following this the radial, circumferential, shear and volumetric strain rates can be expressed as follows:

$$\dot{\varepsilon}_r = -\frac{\partial w}{\partial r} = [\frac{ka^k}{r^{k+1}}] \, \dot{a} \tag{10.72}$$

$$\dot{\varepsilon}_\theta = -\frac{w}{r} = -[\frac{a^k}{r^{k+1}}] \, \dot{a} \tag{10.73}$$

$$\dot{\gamma} = \dot{\varepsilon}_r - \dot{\varepsilon}_\theta = [(k+1)\frac{a^k}{r^{k+1}}] \, \dot{a} \tag{10.74}$$

$$\dot{\delta} = \dot{\varepsilon}_r + k\dot{\varepsilon}_\theta = 0 \qquad (10.75)$$

Using equation (10.70) the shear strain rate can also be written in terms of the initial position of the particle r_0:

$$\dot{\gamma} = [\frac{(k+1)\, a^k}{(a^{k+1} + r_0^{k+1} - a_0^{k+1})}]\, \dot{a} \qquad (10.76)$$

Since r_0 is fixed for a given particle, equation (10.76) can be integrated to give the finite Lagrangean shear strain:

$$\gamma = \ln(\frac{a^{k+1} + r_0^{k+1} - a_0^{k+1}}{r_0^{k+1}}) = (k+1)\ln\frac{r}{r_0} \qquad (10.77)$$

associated with the particle originally at r_0. This relation can now be written back in terms of r, the current coordinate of the particle:

$$\gamma = -\ln[1 - \frac{(a^{k+1} - a_0^{k+1})}{r^{k+1}}] \qquad (10.78)$$

or inversely :

$$r^{k+1} = \frac{a_0^{k+1} - a^{k+1}}{\exp(-\gamma) - 1} \qquad (10.79)$$

by using (10.70) to eliminate r_0. This expression gives the distribution of shear strain with radius r at the instant when the current cavity radius is a.

From (10.77)-(10.79), it follows that the relations between radial and shear strain increments *(i)* for a given particle, and *(ii)* at a fixed instant of time are, respectively:

$$(k+1)\frac{dr}{r} = d\gamma \quad \text{and} \quad (k+1)\frac{dr}{r} = -\frac{d\gamma}{\exp(\gamma) - 1} \qquad (10.80)$$

Note that at the cavity wall the shear strain is:

$$\gamma_c = (k+1)\ \ln\frac{a}{a_0} \qquad (10.81)$$

which is infinite when the initial cavity radius is zero. These results are kinematic and apply in both the elastic and elastic/plastic phases of the cavity contraction.

Elastic unloading

The two effective stress invariants defined below are used for the analysis of cavity unloading problems:

$$q = \sigma'_r - \sigma'_\theta \quad \text{and} \quad p' = \frac{\sigma'_r + k\sigma'_\theta}{1+k} \qquad (10.82)$$

where σ'_r and σ'_θ are the effective radial and hoop stresses respectively. To fully take account of the scaling laws it is convenient to non-dimensionalise all stresses and moduli by some representative stress. Such nondimensional variables will be denoted by superposed bars. The usual convention in critical state soil mechanics is adopted and the equivalent consolidation pressure p'_e is used as this representative stress.

The elastic constitutive law is most conveniently expressed in rate form:

$$\dot{\delta}^e = \frac{\overset{o}{\bar{p}}{}'}{\bar{K}(\bar{p}', v)} \quad \text{and} \quad \dot{\gamma}^e = \frac{\overset{o}{\bar{q}}}{2\bar{G}(\bar{p}', v)} \tag{10.83}$$

where $\dot{\delta}^e$ and $\dot{\gamma}^e$ represent the elastic volumetric and shear strain rates respectively; and $\overset{o}{\bar{p}}{}'$ and $\overset{o}{\bar{q}}$ are the material rates of change of the non-dimensional, effective mean and shear stress invariants. The instantaneous bulk and shear moduli are both functions of the specific volume v and mean effective pressure p' in general, so that the elastic stress-strain relation obtained by integration will be nonlinear. The symbol $\overset{o}{()}$ denotes the material time derivative associated with a given solid material particle and it is related to the local time derivative $\dot{()}$, evaluated at fixed position r, by

$$\overset{o}{()} = \dot{()} + w\frac{\partial()}{\partial r} \tag{10.84}$$

where w is the radial speed of a solid material element.

In the initial, purely elastic phase of an undrained expansion the elastic volumetric strain rate $\dot{\delta}^e = 0$. From equation (10.83) the effective mean pressure remains constant and is equal to its initial value p'_0. The instantaneous elastic bulk and shear moduli hence also remain constant and equal to their initial values K_0 and G_0 respectively. The second part of equation (10.83) for the elastic shear strain rate can be integrated along a particle path, so that the shear stress invariant q is just twice the initial elastic shear modulus times the finite shear strain γ:

$$\gamma = \frac{\bar{q}}{2\bar{G}_0} < 0 \tag{10.85}$$

Hence, the radial and circumferential components of effective stress are given by

$$\bar{\sigma}_r' = \bar{p}'_0 + \frac{2\bar{G}_0\gamma k}{k+1} \quad \text{and} \quad \bar{\sigma}'_\theta = \bar{p}'_0 - \frac{2\bar{G}_0\gamma}{k+1} \tag{10.86}$$

These stresses can now be expressed in terms of the radial position coordinates by eliminating γ using either (10.77) or (10.78).

The effective stress distributions have been found without reference to the equilibrium equations and without the need to make any small strain assumptions. Since the effective mean pressure is constant the stress path of a material element in this elastic phase of the contraction is a vertical line in the $\bar{q} - \bar{p}'$ diagram, Figure 4.11.

The soil first goes plastic at the cavity wall, when the shear stress invariant reaches the yield value q_0, which will depend upon the particular yield criterion. The corresponding shear strain is:

$$\gamma_0 = (k + 1)\ln\frac{a_1}{a_0} = \frac{\bar{q}_0}{2\bar{G}_0} = \frac{q_0}{2G_0} \tag{10.87}$$

where a_1 is the cavity radius at the onset of yielding, γ_0 is the shear strain to yield and is a measure of the compliance of the material. In a perfectly plastic model $q_0 = -2s_u$ where s_u is the undrained shear strength.

These results are also valid in the outer elastic region during the elastic-plastic phase of the contraction. The radius of the elastic/plastic boundary c at the instant when the cavity has radius a is given by

$$\left(\frac{c}{a}\right)^{k+1} = \frac{1 - \left(\frac{a_0}{a}\right)^{k+1}}{1 - \exp(-\gamma_0)} \tag{10.88}$$

Using (10.78) or (10.79), this relation between the shear strain and radial coordinate can alternatively be written as:

$$\left(\frac{r}{c}\right)^{k+1} = \frac{\exp(-\gamma_0) - 1}{\exp(-\gamma) - 1} \tag{10.89}$$

Elastic-plastic unloading

In this section, the basic solution is developed in a general form appropriate to a wide class of materials for which the yield condition and flow rule can be written in the form:

$$\bar{q} = f(\bar{p}') \quad \text{and} \quad \frac{\dot{\delta}^p}{\dot{\gamma}^p} = g(\bar{p}') \tag{10.90}$$

In undrained deformation the total volumetric strain-rate is zero, so that $\dot{\delta}^e = -\dot{\delta}^p$. It follows from (10.83) and (10.90) that the total strain-rate is:

$$\dot{\gamma} = \dot{\gamma}^e + \dot{\gamma}^p = L(\bar{p}')\overset{\circ}{\bar{p}} \tag{10.91}$$

where

$$L(\bar{p}') = \frac{f'(\bar{p}')}{2\bar{G}(\bar{p}')} - \frac{1}{\bar{K}(\bar{p}')\, g(\bar{p}')} \tag{10.92}$$

Integrating (10.91) along a particle path starting at the elastic-plastic boundary gives a relationship between the finite shear strain and the effective mean pressure:

$$\gamma = \gamma_0 + I(\bar{p}') - I(\bar{p}'_0) \tag{10.93}$$

where

$$I(\bar{p}') = \int^{\bar{p}'} L(\bar{p}')\, d\bar{p}' \tag{10.94}$$

As a special case, (10.93) describes the relationship between the cavity pressure and the cavity shear strain after the cavity wall becomes plastic. As will be shown in a later section, the integral (10.94) can be evaluated analytically for the original Cam clay model and is readily evaluated numerically in other cases. The variation of $\bar{p}'$ with radius r can be obtained implicitly by eliminating γ between (10.93) and (10.77), (10.78) or (10.89).

Distribution of pore pressures

The distribution of pore pressure $U(r)$ can be calculated from the quasi-static radial equilibrium equation in terms of total stresses:

$$\frac{d\,\bar{\sigma}_r}{d\,r} + k\,\frac{\bar{\sigma}_r - \bar{\sigma}_\theta}{r} = 0 \tag{10.95}$$

Since $\bar{\sigma}_r = \bar{p} + (k/(k+1))\,\bar{q}$ and $\bar{p} = \bar{U} + \bar{p}'$, the non-dimensional pore pressure gradient is given by :

$$\frac{d\,\bar{U}}{d\,r} = -\frac{d\,\bar{p}'}{d\,r} - \frac{k}{k+1}\,\frac{d\,\bar{q}}{d\,r} - \frac{k\bar{q}}{r} \tag{10.96}$$

As the effective mean stress distribution is constant in the elastic region, the change in the pore pressure (i.e. excess pore pressure) in the elastic zone is given by

$$\Delta\,\bar{U} = -\frac{k}{k+1}\,\bar{q} - k\int \bar{q}\,\frac{d\,r}{r} \tag{10.97}$$

However $\bar{q}$ and (dr/r) can both be expressed in terms of γ from (10.85) and the second equation of (10.80), with (10.97) becoming:

$$\Delta\,\bar{U} = -\frac{2k\bar{G}_0}{k+1}\,[\,\gamma - \int_0^{\gamma} \frac{\gamma}{\exp(\gamma) - 1}\,d\gamma\,] \doteq -\frac{k\bar{G}_0\gamma^2}{2(k+1)} \tag{10.98}$$

The excess pore pressure at the elastic-plastic boundary is therefore:

$$\Delta \overline{U}_0 = -\frac{k\overline{G}_0\gamma_0^2}{2(k+1)} \tag{10.99}$$

Integrating (10.96) through the plastic region from the elastic/plastic boundary yields a relation between the excess pore pressure and the finite shear strain:

$$\Delta \overline{U} = \Delta \overline{U}_0 - (\overline{p}' - \overline{p}'_0) - \frac{k}{k+1}(\overline{q} - \overline{q}_0 - J) \tag{10.100}$$

where the integral J is most conveniently evaluated numerically by expressing both $\overline{q}$ and r in terms of $\overline{p}$ using (10.80), (10.90) and (10.91) :

$$J = \int_{\gamma_0}^{\gamma} \frac{\overline{q}}{(\exp(\gamma) - 1)}\, d\gamma = \int_{\overline{p}'_0}^{\overline{p}'} \frac{f(\overline{p}')\, L(\overline{p}')}{\exp(\gamma) - 1}\, d\overline{p}' \tag{10.101}$$

Again, as a special case, (10.100) represents the plastic relationship between the excess pore pressure and the finite shear strain for the cavity wall. Once the effective stress state has essentially reached the critical state, the value of $\overline{q}$ is effectively constant and (10.101) can then be integrated analytically to give:

$$J \doteq \overline{q}_{cs} \ln \frac{\exp(-\gamma) - 1}{\exp(-\gamma_0) - 1} \tag{10.102}$$

(b) Solutions for various soil models

The linear elastic-perfectly plastic Tresca model

Before discussing the solutions for various critical state models, it is useful to firstly derive the solution for a perfectly plastic Tresca model. This corresponds to the situation where the in situ soil under undrained loading conditions behaves purely elastically before reaching the critical state. Since the shear stress and effective mean pressure are now constant through the plastic annulus the J integral in (10.101) can be evaluated analytically, giving:

$$\Delta \overline{U} = \Delta \overline{U}_0 + \frac{k}{k+1}\overline{q}_{cs} \ln \frac{\exp(-\gamma) - 1}{\exp(-\gamma_0) - 1} \tag{10.103}$$

or, in terms of the radial coordinate:

$$\Delta \overline{U} = \Delta \overline{U}_0 + k\overline{q}_{cs} \ln \frac{R}{r} = \Delta \overline{U}_0 + \frac{k}{k+1}\overline{q}_{cs} \ln \frac{(\frac{a_0}{r})^{k+1} - (\frac{a}{r})^{k+1}}{\exp(-\gamma_0) - 1} \tag{10.104}$$

using (10.88) and (10.89). To a first order in the elastic limit strain γ_0 the excess pore pressure at the cavity wall is:

$$\Delta \,\overline{U}_c \doteq \frac{k}{k+1}\,\overline{q}_{cs}\,[\,\ln((\frac{a_0}{a})^{k+1} - 1) + \ln I_r\,] \tag{10.105}$$

Note that $I_r = G/s_u$ is the rigidity index and s_u is undrained shear strength. Using the above equation, the total radial stress solution at the cavity wall can be derived:

$$\overline{\sigma}_r|_c = \overline{p}_0 + \frac{k}{k+1}\,\overline{q}_{cs}\,[\,1 + \ln((\frac{a_0}{a})^{k+1} - 1) + \ln\frac{G}{s_u}\,] \tag{10.106}$$

With $q_{cs} = -2s_u$ the above equation can be reduced to the following cavity pressure-contraction relationship:

$$p = p_0 - \frac{2ks_u}{1+k}[1 + \ln\frac{G}{s_u}] - \frac{2ks_u}{1+k}\ln[(\frac{a_0}{a})^{k+1} - 1] \tag{10.107}$$

where p is the total cavity pressure and p_0 is the initial total stress in the soil.

For a complete unloading cavity which is relevant to unlined tunnels, the maximum displacement at the cavity wall is obtained by setting $p = 0$ in equation (10.107):

$$\frac{a_0}{a} = [1 + \exp(\frac{1+k}{2k} \times \frac{p_0}{s_u} - 1 - \ln\frac{G}{s_u})]^{\frac{1}{k+1}} \tag{10.108}$$

The original Cam clay model

The original Cam clay model, presented by Schofield and Wroth (1968), can be used to describe the stress-strain behaviour of clays. The yield surface of this model for cavity unloading problems can be shown to be :

$$\overline{q} = f(\overline{p}') = \frac{M}{\Lambda}\,\overline{p}'\ln\overline{p}' \tag{10.109}$$

where the stresses have been non-dimensionalised by the equivalent consolidation pressure at the same specific volume v:

$$p'_e = \exp(\frac{N - v}{\lambda}) \tag{10.110}$$

The constant $\Lambda = 1 - \varkappa/\lambda$, where $\varkappa$, λ are the slopes of the elastic swelling line and normal consolidation line, respectively, in $\ln p' - v$ space and N is the value of v on the normal consolidation line when $p' = 1$kPa. The final critical state constant M is the slope of the critical state line in $\overline{p}' - \overline{q}$ space. The original Cam clay yield surface defined in (10.109) is shown in Figure 10.16.

In this model the elastic moduli are given by

$$\overline{K} = \frac{v\overline{p}'}{\varkappa} \quad \text{and} \quad \overline{G} = \frac{(1+k)(1-2\mu)}{2(1+(k-1)\mu)}\overline{K} \tag{10.111}$$

where μ denotes Poisson's ratio. Some authors assume μ to be constant whilst others keep G fixed and use (10.111) to calculate Poisson's ratio (Gens and Potts, 1988). For comparison, the next section demonstrate that a constant shear modulus can also be incorporated in the present cavity unloading solution procedure.

The overconsolidation ratio (OCR) in terms of the mean effective stress is:

$$n_p = (\bar{p}'_0)^{-\frac{1}{\Lambda}} \tag{10.112}$$

At the critical state $\bar{q}/\bar{p}' = M$, $n_p = e = 2.718$ and $\bar{p}' = e^{-\Lambda}$, whilst the undrained stress path in $\bar{p}' - \bar{q}$ space has a maximum value when $\bar{p}' = 1/e = 0.368$ and $\bar{q}/\bar{p}' = M/\Lambda$ as shown in Figure 10.16.

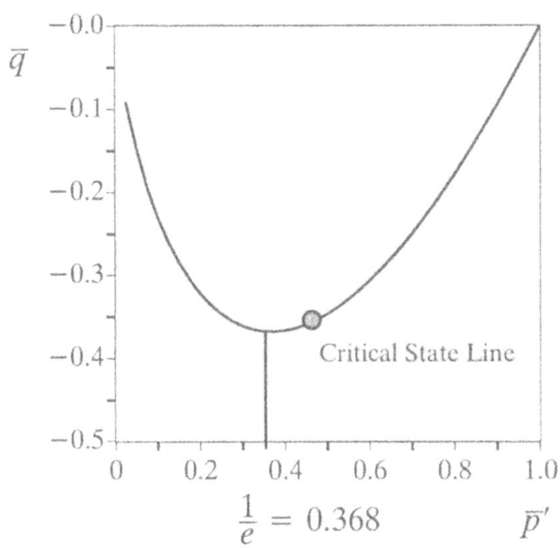

Figure 10.16: The original Cam clay yield surface for cavity unloading

The ratio of the plastic volumetric and shear strain rates calculated from the normal flow rule is:

$$\frac{\dot{\delta}^p}{\dot{\gamma}^p} = g(\bar{p}') = -\frac{kM}{(k+1)\Lambda}(\Lambda + \ln \bar{p}') \tag{10.113}$$

The function $L(\bar{p}')$ needed to calculate the effective pressure distributions in (10.92) is hence:

$$L(\bar{p}') = \frac{A(1 + \ln \bar{p}')}{\bar{p}'} + \frac{B}{\bar{p}'(\Lambda + \ln \bar{p}')} \tag{10.114}$$

and upon integration the function needed in (10.93) to calculate the shear strain is:

$$I(\bar{p}') = A(\ln\bar{p}' + \frac{1}{2}(\ln\bar{p}')^2) + B\ln|(\Lambda + \ln\bar{p}')| \quad (10.115)$$

where $A = M\kappa/(2\Lambda\alpha v)$ and $B = (k+1)\Lambda\kappa/(kMv)$ are constants. Note that the value of the integral I and hence the shear strain is inversely proportional to the specific volume v.

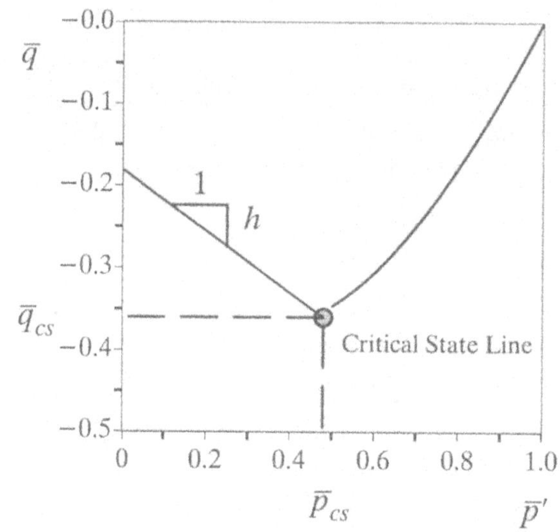

Figure 10.17: The original Cam clay – Hvorslev yield surface for cavity unloading

The original Cam clay – Hvorslev model

It is well established that the original Cam clay yield surface overpredicts the soil strength significantly for heavily overconsolidated (OC) clays. In this case, the Hvorslev surface has often been used as the yield function. The Hvorslev yield surface is a straight line in $\bar{p}' - \bar{q}$ space (Atkinson and Bransby, 1978):

$$\bar{q} = -h\bar{p}' - (M - h)\exp(-\Lambda) \quad (10.116)$$

where h is the slope of the Hvorslev yield surface, Figure 10.17.

It is also noticed that the use of (10.111) for heavily overconsolidated clays would result in unrealistically low values of the elastic moduli. To overcome this limitation, Randolph *et al.* (1979) proposed a more realistic hypothesis that is to select G as half of the maximum value of the elastic bulk modulus, K_{max} that was ever reached during the stress history of the soil. In their proposal, the bulk modulus is still assumed to be pressure dependent, and so the resulting model is conservative for elastic behaviour (Zytynski *et al.*, 1978). Equation (10.111) is then replaced by the following expression for heavily OC clays:

$$\overline{K} = \frac{v\overline{p}'}{\varkappa} \quad \text{and} \quad \overline{G} = \frac{v + \lambda(\Lambda - 1)\ \ln n_p}{2\varkappa}\ (n_p)^{1-\Lambda} \tag{10.117}$$

Using (10.116) as the yield function, (10.109) as the plastic potential and (10.117) for the elastic moduli, we obtain:

$$L(\overline{p}') = -\frac{h}{2\overline{G}} + \frac{(1+k)\varkappa}{kv(M-h)}\ \frac{1}{(\overline{p}' - \exp(-\Lambda))} \tag{10.118}$$

and

$$I(\overline{p}') = -\frac{h\overline{p}'}{2\overline{G}} + \frac{(1+k)\varkappa}{kv(M-h)}\ \ln(\overline{p}' - \exp(-\Lambda)) \tag{10.119}$$

where the constant $\overline{G}$ is given by equation (10.117).

The modified Cam clay model

To improve the performance of the original Cam clay model for normally consolidated clays, Roscoe and Burland (1968) proposed the modified Cam clay model. The modified Cam clay yield surface for cavity unloading problems is :

$$\overline{q} = f(\overline{p}') = -M\overline{p}'\sqrt{\ (\overline{p}'^{-\frac{1}{\Lambda}} - 1\ )} \tag{10.120}$$

At the critical state $\overline{q}/\overline{p}' = M$, $n_p = 2$ and $\overline{p}' = 2^{-\Lambda}$, as shown in Figure 10.18.

The ratio of the plastic volumetric and shear strain rates calculated from the normal flow rule is:

$$\frac{\dot{\delta}^p}{\dot{\gamma}^p} = g(\overline{p}') = \frac{k}{(k+1)}\left(\frac{M^2 - \eta^2}{2\eta}\right) \tag{10.121}$$

where η can be expressed as a function of mean effective stress using (10.120). The function $L(\overline{p}')$ needed to calculate the effective pressure distributions in (10.92) can be obtained using (10.120) and (10.121). Unlike the original Cam clay model, the function needed in (10.94) to calculate the shear strain for the modified Cam clay model cannot be obtained in closed form and instead a simple numerical integration must be used.

(c) Comparison with a centrifuge tunnel test in clay

To demonstrate the relevance of the present cavity unloading solutions for modelling soil behaviour around tunnels, it is necessary to compare the predicted behaviour with some experimental data. A comparison between the theoretical solutions and experimental results can provide valuable insights into the applicability of cavity unloading solutions to tunnel modelling.

In this section, we will consider a comparison between predicted and observed crown and mid-surface settlements for a centrifuge tunnel test reported by Mair (1979). The tunnel test chosen for the comparison is Test 2DP, which has a cover to diameter ratio of 1.67. The tunnel test was carried out quickly so that soil behaviour can be assumed to be undrained. The measured crown and mid-surface settlements for Test 2DP are presented in Figure 7.8 of Mair (1979). According to Mair, the critical state soil properties of the clay used for Test 2DP may be assumed to be: $\Gamma = 3.92$ $\lambda = 0.3$, $\kappa = 0.05$, $M = 0.8$. Poisson's ratio is assumed to be 0.3. Although the clay around the tunnel in Test 2DP is lightly overconsolidated, the exact value of overconsolidation ratio in term of mean effective stress, n_p, is not known. This is further complicated by the fact that the overconsolidation ratio n_p defined in terms of mean effective stress is invariably greater than the one-dimensional overconsolidation ratio OCR defined in terms of vertical effective stress (Muir Wood, 1990). It is also not clear what the specific volume of the clay was before the tunnel test was carried out. As a result, some assumptions must be made as to what values should be used for the initial specific volume and the overconsolidation ratio n_p. From the value of undrained shear strength $s_u = 26kPa$ and $s_u = 0.5M\exp((\Gamma - v)/\lambda)$, the initial specific volume may be estimated as $v = 2.67$. After the specific volume is assumed, the overconsolidation ratio n_p is then estimated in such a way that the predicted initial tunnel support pressure is equal to that used in the centrifuge tunnel test.

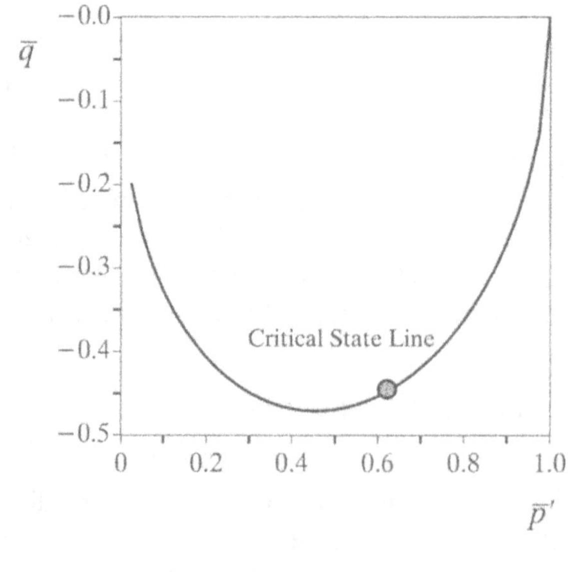

Figure 10.18: The modified Cam clay yield surface for cavity unloading

Figure 10.19 and Figure 10.20 show comparisons between the predicted and observed crown and mid-surface settlements when the tunnel support pressure is reduced. The tunnel support pressure and settlements are normalized by the undrained shear strength and the initial tunnel radius respectively. Although the predicted settlements shown in Figure 10.19 and Figure 10.20 are obtained from the original Cam clay soil model, similar results can be obtained from other critical state models. The differences in prediction with different critical state models are found to be insignificant for this particular case as the clay used in Test 2DP is only lightly overconsolidated.

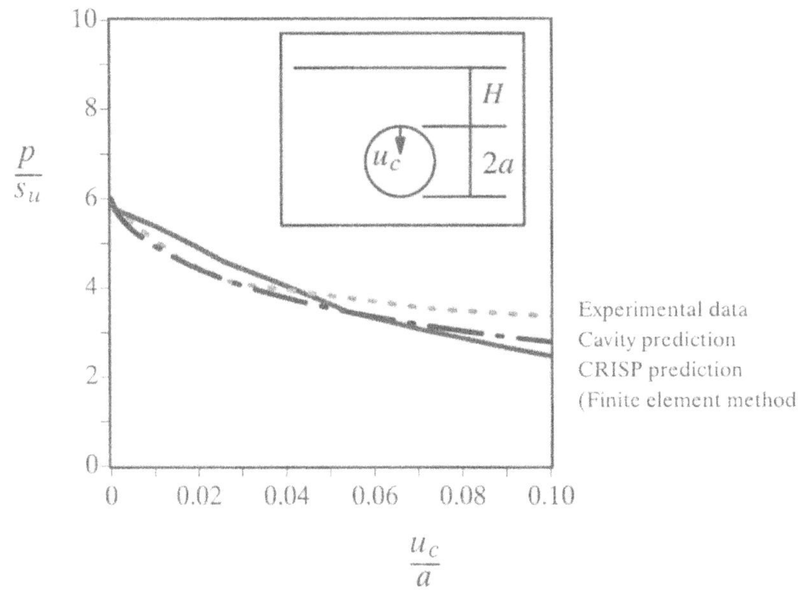

Figure 10.19: Comparison of predicted and measured crown displacement as tunnel support pressure is reduced

From the comparison it can be concluded that the present cavity unloading solutions can be used to accurately predict crown settlements around the tunnel. In contrast, the cavity unloading solutions tend to under-predict the observed mid-surface settlement. This is due to the fact that the model test 2DP is a shallow tunnel test with a very low cover to diameter ratio ($H/2a = 1.67$). For very shallow tunnels, solutions for contracting cavities in an infinite soil medium are expected to be less accurate for modelling the behaviour of soils far away from the cavity wall. For relatively deep tunnels, however, the present cavity unloading solutions should give more accurate predictions for surface settlements around the tunnel.

Effect of ground surface on cavity unloading solutions

It has been shown that while the solution of cavity unloading in an infinite medium can be used to accurately predict the observed tunnel wall movement, it tends to underestimate the surface displacement considerably for shallow tunnels. This discrepancy is, to a large extent, due to the effect of the free ground surface that is not accounted for in cavity unloading in an infinite soil mass. To take into account the effect of free ground surface, the analytical solution of Sagaseta (1987) and Verruijt and Booker (1996) for displacements due to the unloading of a cavity in a half space can be used.

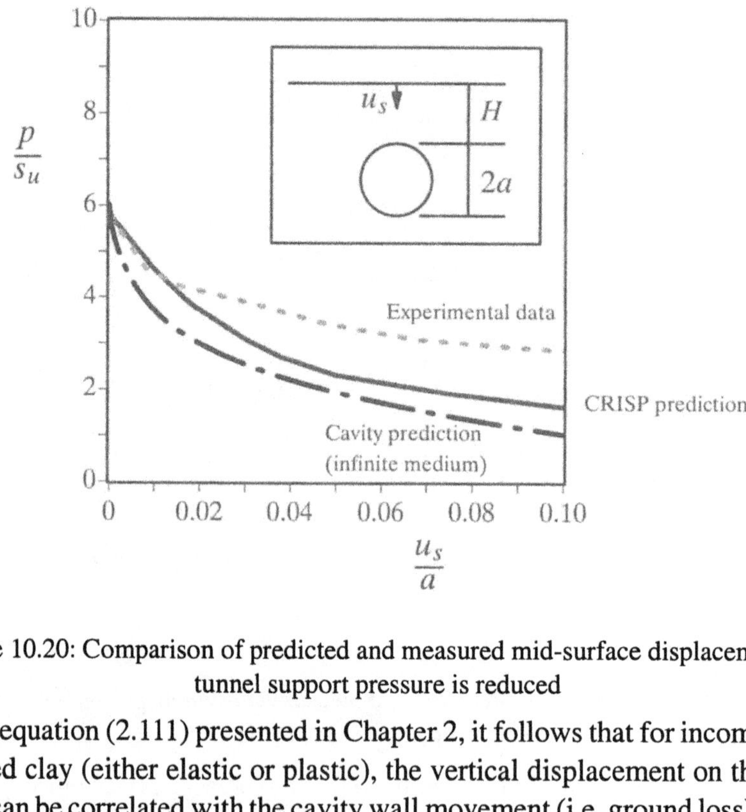

Figure 10.20: Comparison of predicted and measured mid-surface displacement as tunnel support pressure is reduced

From equation (2.111) presented in Chapter 2, it follows that for incompressible undrained clay (either elastic or plastic), the vertical displacement on the ground surface can be correlated with the cavity wall movement (i.e. ground loss) in a simple manner:

$$\frac{u_z|_{z=0}}{u_0} = \frac{2\frac{h}{a}}{(\frac{x}{a})^2 + (\frac{h}{a})^2} \tag{10.122}$$

where $h = H + a$ and x is the distance from the cavity centre-line (see Figure 2.5).

Applying equation (10.122) to the centrifuge test 2DP reported by Mair (1979), we obtain the result that the mid-surface settlement is related to the cavity wall movement as follows:

$$u_s = 2\frac{a}{H+a}u_c = 0.46u_c \tag{10.123}$$

Using the cavity unloading solution in an infinite medium for the cavity wall movement u_c, equation (10.123) may be used to estimate mid-surface settlement. Figure 10.21 gives a comparison of measured mid-surface displacement and that predicted by equation (10.123) as the tunnel support pressure is reduced. By comparing Figure 10.20 and Figure 10.21, it is clear that while still underestimating the measured settlement, the unloading solution of a cavity in a half-space gives a better estimate of surface settlement than the solution of cavity unloading in an infinite medium.

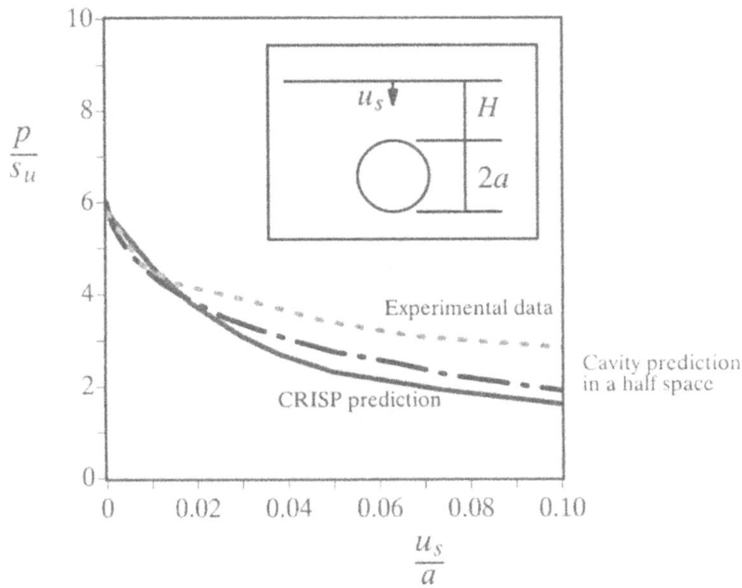

Figure 10.21: Predicted and measured mid-surface displacements as tunnel support pressure is reduced

In addition to the cavity expansion predictions, the crown and mid-surface settlements predicted by Mair (1979) using a two-dimensional finite element model are also plotted in Figure 10.19, Figure 10.20 and Figure 10.21. It is interesting to note that although the present cavity unloading solutions are very simple, they give very similar predictions to those made by Cambridge University's finite element program CRISP, which is much more complicated. It is therefore logical to suggest that the simple analytical cavity unloading solutions presented in this section can have a very useful role to play in the analysis and the design of tunnels in soils.

10.4.3 Stability of tunnels

In this section, we present procedures that may be used to apply cavity unloading solutions to conduct stability calculations for tunnels in cohesive soils. In doing so, we will basically follow Caquot and Kerisel (1966) by assuming that collapse of a tunnel will occur when the plastic zone reaches the ground surface. In other words, tunnels will collapse when the plastic flow ceases to be confined and becomes free, Figure 10.22.

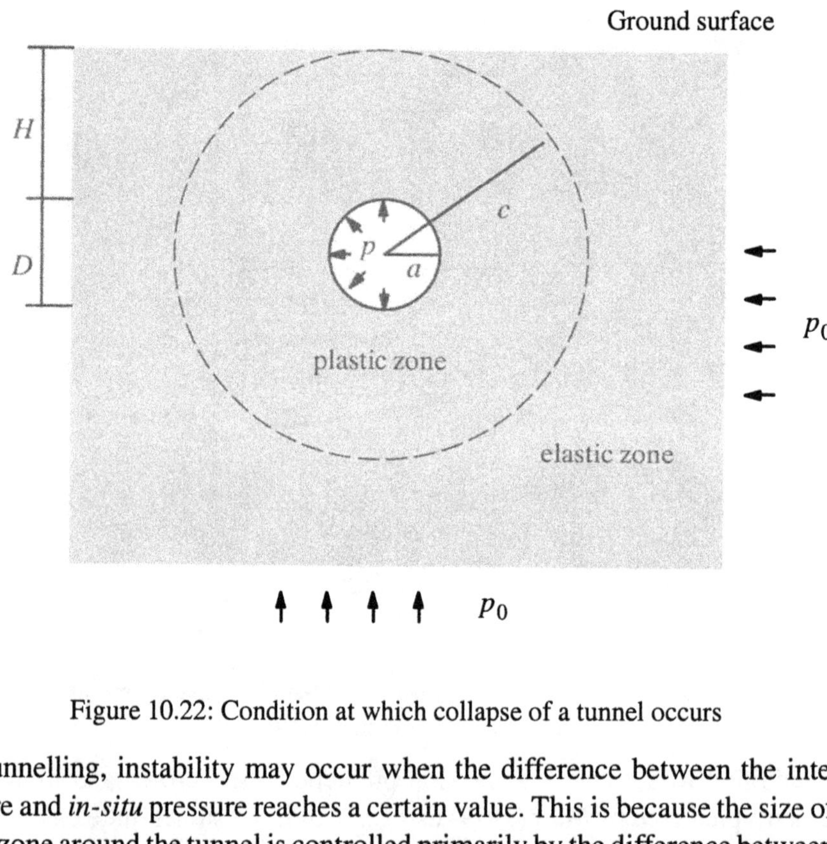

Figure 10.22: Condition at which collapse of a tunnel occurs

In tunnelling, instability may occur when the difference between the internal pressure and *in-situ* pressure reaches a certain value. This is because the size of the plastic zone around the tunnel is controlled primarily by the difference between the internal tunnel pressure and the pre-tunnelling *in-situ* pressure. With reference to Figure 10.13 or Figure 10.22, the cavity unloading solution in cohesive soils defined by equation (10.61) gives the relationship between internal pressure p and plastic radius c:

$$\frac{p_0 - p}{s_u} = 2k \ln \frac{c}{a} + \frac{2k}{1+k} \tag{10.124}$$

where p and p_0 are current tunnel support and pre-tunnelling pressures respectively, and s_u is the undrained shear strength of clay. For plane strain modelling of tunnels $k=1$, equation (10.124) gives the following expression for the 'stability number' at the onset of tunnel failure (i.e. $c = H + D/2$ and $a = D/2$):

$$N = \frac{p_0 - p}{s_u} = 2\ln(2\frac{H}{D} + 1) + 1 \tag{10.125}$$

The pre-tunnelling pressure p_0 may be estimated as follows:

$$p_0 = q + \gamma(H + \frac{D}{2}) \tag{10.126}$$

with q as the surcharge acting on the ground surface and γ being the unit weight of clay.

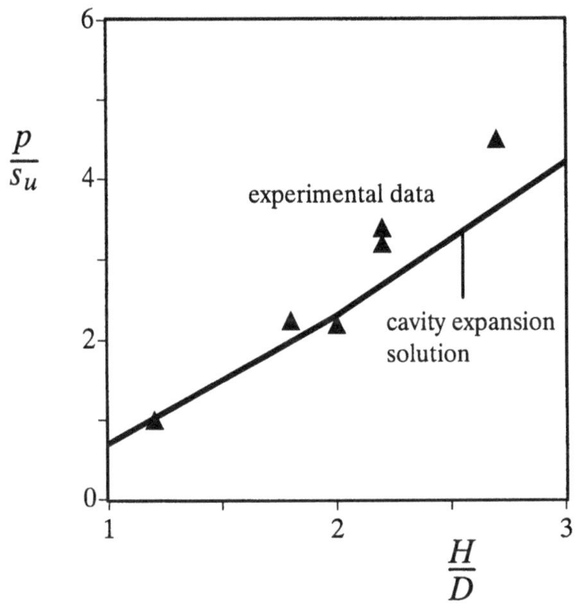

Figure 10.23: Comparison of predicted and measured stability results for tunnels in soil with $\gamma D/s_u = 2.6$

It is interesting to note that by combining equilibrium and the yield function for a soil wedge directly above the tunnel, a similar solution for the stability number has been obtained by Bolton (1991) for tunnels in a clay soil:

$$N = \frac{p_0 - p}{s_u} = 2\ln(2\frac{H}{D} + 1) \tag{10.127}$$

In deriving the solution (10.127), the effect of the free surface has been taken into account. By comparing equations (10.125) and (10.127), it is clear that both of these approaches give very similar results.

To assess the suitability of the cavity expansion solution (10.125) in predicting the stability of tunnels in practice, it is necessary to compare its predictions with observed behaviour. For this purpose, the centrifuge results reported by Mair (1979) provide a valuable source of data. The stability results of Mair (1979) are shown in Figure 10.23 for the case of zero surcharge on the ground surface. The undrained shear strength was found to be uniform with depth. It is clear from this figure that the stability results predicted by the cavity expansion solution (10.125) agree very well with the measured results. Using the results of Sloan and Assadi (1993), it is also found that the cavity expansion results are generally very similar to the rigorous upper and lower bound stability solutions. It is therefore believed that the stability number defined by (10.125) derived from the simple cavity expansion solutions can provide a very useful tool in assessing the stability of tunnels in clay.

10.5 TUNNELS IN COHESIVE-FRICTIONAL SOILS

10.5.1 Settlements due to tunnelling

Once again, it is assumed that an unbound Mohr-Coulomb medium contains a single cylindrical or spherical cavity (tunnel). Initially the radius of the tunnel is a_0 and a hydrostatic pressure p_0 acts throughout the soil which is assumed to be homogeneous. This section is concerned with the distribution of stress and displacement in the soil around the tunnel as the tunnel pressure p reduces from its initial value. The complete solutions for this problem have been detailed in Chapter 3 and only key equations relevant to tunnel deformations are presented here.

Elastic response

As the pressure p decreases from p_0, the deformation of the soil is at first purely elastic. The elastic solutions for stresses and displacement are as follows:

$$\sigma_r = -p_0-(p-p_0)(\frac{a}{r})^{1+k} \tag{10.128}$$

$$\sigma_\theta = -p_0 + \frac{p-p_0}{k}(\frac{a}{r})^{1+k} \tag{10.129}$$

$$u = \frac{p-p_0}{2kG}(\frac{a}{r})^{1+k}r \tag{10.130}$$

For the unloading of cavities, the Mohr-Coulomb yield equation takes the form:

$$\alpha\sigma_r-\sigma_\theta = Y \tag{10.131}$$

As the internal pressure decreases further, initial yielding occurs at the cavity wall when the tunnel pressure is reduced to:

$$p = p_{1y} = \frac{1 + k}{1 + ak}p_0 - \frac{kY}{1 + ak} \tag{10.132}$$

Elastic-plastic stress fields

After initial yielding at the tunnel wall, a plastic zone within the region $a \le r \le c$ will form around the inner wall of the tunnel with a further decrease in the tunnel pressure p.

The stresses in the elastic zone are shown to be as follows:

$$\sigma_r = -p_0 - \frac{k[(1 - a)p_0 - Y]}{1 + ka}\left(\frac{c}{r}\right)^{(1+k)} \tag{10.133}$$

$$\sigma_\theta = -p_0 + \frac{[(1 - a)p_0 - Y]}{1 + ka}\left(\frac{c}{r}\right)^{(1+k)} \tag{10.134}$$

On the other hand, the stresses in the plastic zone must satisfy equilibrium and the yield condition, resulting in:

$$\sigma_r = \frac{Y}{a - 1} + Ar^{k(a-1)} \tag{10.135}$$

$$\sigma_\theta = \frac{Y}{a - 1} + Aar^{k(a-1)} \tag{10.136}$$

where

$$A = -\frac{(1 + k)[Y + (a-1)p_0]}{(a-1)(1 + ka)}c^{(1-a)k} \tag{10.137}$$

Applying $\sigma_r = -p$ at the tunnel wall leads to the following relationship between the tunnel pressure p and the plastic radius c:

$$\frac{c}{a} = \left\{\frac{(1 + ka)[Y + (a-1)p]}{(1 + k)[Y + (a-1)p_0]}\right\}^{\frac{1}{k(1-a)}} \tag{10.138}$$

Elastic-plastic displacements

The displacement in the elastic zone is shown to be:

$$u = \frac{(1-a)p_0 - Y}{2G(1 + ka)}\left(\frac{c}{r}\right)^{1+k}r \tag{10.139}$$

Therefore, at the elastic-plastic interface the displacement is:

$$u|_{r=c} = c - c_0 = -\frac{[(1-a)p_0 + Y]c}{2(1 + ka)G} \tag{10.140}$$

If the elastic contribution in the plastically deforming region is ignored, the solution can be considerably simplified. By adopting logarithmic strains, the plastic flow rule can be integrated to give the following large strain solution linking the tunnel pressure and tunnel wall movement:

$$\frac{1-(\frac{a_0}{a})^{1+k\beta}}{1-(\frac{c_0}{c})^{1+k\beta}} = \left\{\frac{(1+k\alpha)[Y+(\alpha-1)p]}{(1+k)[Y+(\alpha-1)p_0]}\right\}^{\frac{1+k\beta}{k(1-\alpha)}} \tag{10.141}$$

where c_0/c can be determined from equation (10.140).

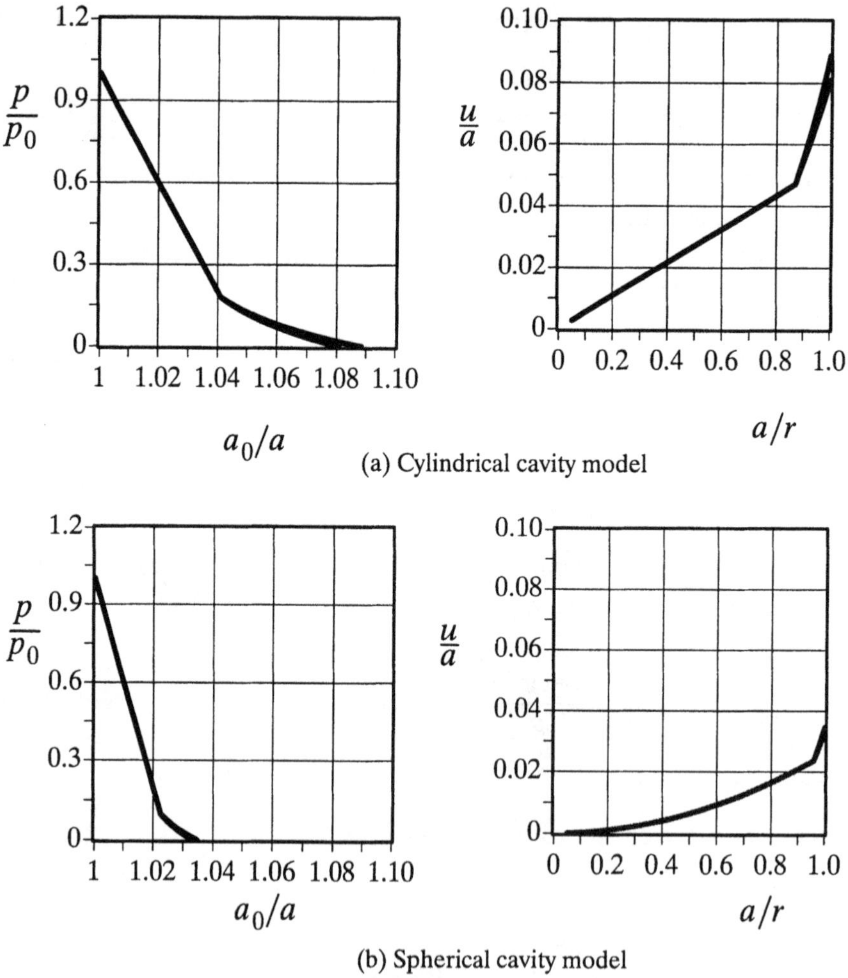

Figure 10.24: Predicted soil behaviour around tunnels in cohesive-frictional soils using both small and large strain solutions ($\phi = 40^0, \psi = 40^0, Y/p_0 = 1.0, G/p_0 = 10$)

With the small strain assumption, the displacement field in the plastic region can be shown to be:

$$u = r - r_0 = \frac{(1-\alpha)p_0 - Y}{2G(1+\alpha k)}\left(\frac{c}{r}\right)^{1+k\beta} r \tag{10.142}$$

In particular, the displacement at the tunnel wall is:

$$\frac{u_a}{a} = \frac{[(1-\alpha)p_0 - Y]}{2G(1+\alpha k)}\left\{\frac{(1+k\alpha)[Y+(\alpha-1)p]}{(1+k)[Y+(\alpha-1)p_0]}\right\}^{\frac{1+k\beta}{k(1-\alpha)}} \tag{10.143}$$

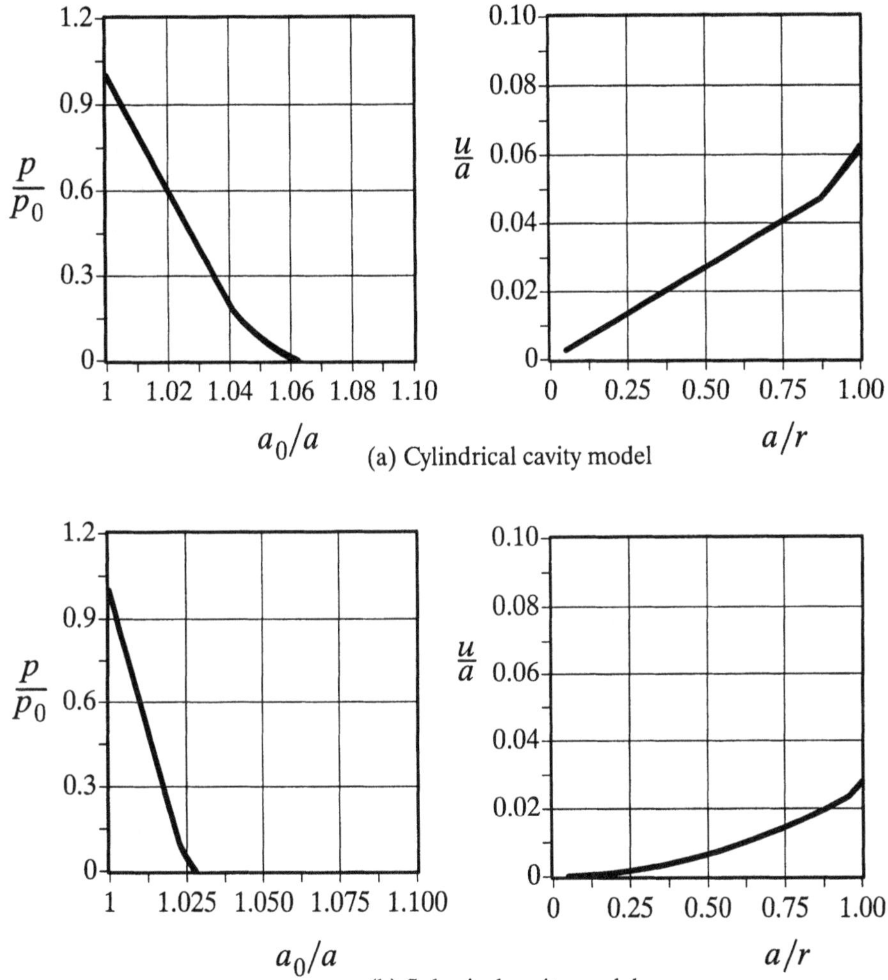

Figure 10.25: Predicted soil behaviour around tunnels in cohesive-frictional soils using both small and large strain solutions ($\phi = 40^0, \psi = 20^0, Y/p_0 = 1.0, G/p_0 = 10$)

Soil behaviour around tunnels in fully drained cohesive-frictional soils

Selected results of soil behaviour around tunnels in a Mohr-Coulomb soil with a friction angle of $\phi' = 40^0$, a cohesion parameter of $Y/p_0 = 1.0$ and a Poisson's ratio of 0.3 are presented in this subsection. The effect of soil dilatancy on the predicted soil behaviour around tunnels is studied using two dilation angles of 20 and 40 degrees. While the high values chosen for the angles of internal friction and dilation may not be generally encountered in practice, the use of the high values for example calculations is to illustrate the maximum possible effects of soil strength and dilatancy on soil behaviour around tunnels.

The cavity contraction curves (i.e. the ground response curves) and soil displacement distribution at the instant of complete unloading (i.e. the cavity pressure is reduced to zero and this is the case for unlined tunnels) for the case of $G/p_0 = 10$ are plotted in Figure 10.24 and Figure 10.25. For comparison, solutions for both cylindrical and spherical cavities are presented together. Based on the selected results shown in these two figures and those detailed by Yu and Rowe (1999), the following observations can be made:

(a) The small difference between solutions from small strain and large strain formulations shown in the results suggests that for tunnelling problems the small strain solutions would be adequate for practical purposes.

(b) The soil displacements induced by tunnelling from the plane strain cylindrical cavity theory are about twice as large as those from the spherical cavity theory.

(c) In the elastic region, the cylindrical cavity theory predicts that the normalized displacement u/a varies linearly with a/r. It is interesting to note that in the plastic region, the variation may also be assumed to be linear. However, the slope of the displacement variation in the plastic zone is much larger than that in the elastic zone.

(d) The soil displacements induced by tunnelling increase with increasing dilation angle. Also the size of plastic zone increases when the value of the dilation angle is increased.

(e) The ground response curves and soil displacements are very sensitive to the value of soil stiffness. In fact, by increasing the stiffness index from 10 to 50, the soil displacements due to tunnelling are reduced by about five times.

10.5.2 Stability of tunnels

This section presents a simple procedure in which existing cavity expansion solutions are used to conduct stability calculations for tunnels in cohesive-frictional soils. For dilatant cohesive-frictional soils it may be reasonable to assume that collapse of tunnels will occur as long as the unloading plastic zone extends sufficiently close to the free surface. This is due to the fact that dilation of sandy soils tends to cause a very large deformation in the soil around the plastic zone and this would accelerate the collapse of tunnels.

As a result, the plastic radius at tunnel collapse may be defined by $c = xH + D/2$, where x indicates the closeness of the plastic boundary to the ground surface at tunnel collapse, and is in the range of 0 to 1. The cavity unloading solution in cohesive-frictional soils defined by equation (10.138) gives the relationship between internal pressure p and plastic radius c:

$$\frac{(1 + k\alpha)[Y + (\alpha-1)p]}{(1 + k)[Y + (\alpha-1)p_0]} = (\frac{c}{a})^{k(1-\alpha)} \qquad (10.144)$$

where p and p_0 are internal and pre-tunnelling pressures respectively. α and Y are functions of cohesion C and friction angle ϕ:

$$Y = \frac{2C\cos\phi}{1 - \sin\phi} \quad \text{and} \quad \alpha = \frac{1 + \sin\phi}{1 - \sin\phi} \qquad (10.145)$$

For plane strain modelling of tunnels $k=1$, equation (10.144) should give the following expression for stability number at the onset of tunnel failure:

$$N = \frac{Y + (\alpha-1)p}{Y + (\alpha-1)p_0} = \frac{2}{1 + \alpha}[2x\frac{H}{D} + 1]^{(1-\alpha)} \qquad (10.146)$$

As in purely cohesive soils, the pre-tunnelling pressure p_0 may be estimated as follows:

$$p_0 = q + \gamma(H + \frac{D}{2}) \qquad (10.147)$$

with q as the surcharge acting on the ground surface and γ being the unit weight of the cohesive-frictional soil. For the case of zero surcharge, equation (10.146) can also be written in the following form:

$$\frac{p}{C} = \frac{2}{1 + \alpha}(2x\frac{H}{D} + 1)^{1-\alpha} \times [\cot\phi + \frac{\gamma D}{C}(\frac{H}{D} + 0.5)] - \cot\phi \qquad (10.148)$$

To illustrate the relevance of the above cavity expansion approach, it is compared with the upper and lower bound solutions shown in Figure 10.26 for the case of $\phi = 15^0$ and $\gamma D/C = 3.0$. The figure shows that the cavity expansion results obtained using equation (10.148) with x=0.375 agree well with the upper and lower bound solutions. It must be pointed out that the cavity expansion prediction is quite sensitive to the value of x. It is also expected that a reasonable value of x for use will tend to decrease with the dilation angle of the soil.

For tunnels in a purely frictional soil, the cohesion may be assumed to be zero (Y=0). If it is further assumed that the pre-tunnelling pressure is mainly due to the self-weight of soils ($q = 0$ and $p_0 = \gamma(H + \frac{D}{2})$) and that the stability number can be simplified as follows:

$$N = \frac{p}{\gamma H} = \frac{2}{1+\alpha}(1 + \frac{D}{2H})[2x\frac{H}{D} + 1]^{(1-\alpha)} \tag{10.149}$$

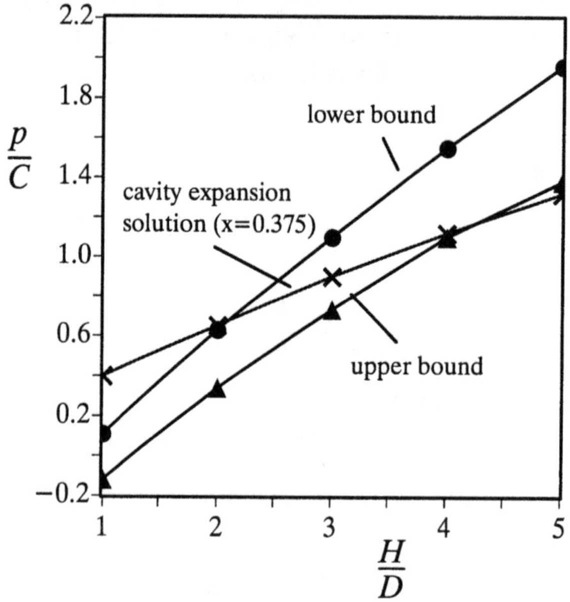

Figure 10.26: Comparison of stability results for tunnel in cohesive-frictional soil with $\gamma D/C = 3$

10.6 SUMMARY

1. Underground excavations and tunnelling involve the removal of soil or rock masses from their initial locations. This action reduces and, in some cases (i.e. unlined tunnels), completely removes the initial stresses which exist in the area of tunnelling or excavations. Accordingly the action of tunnelling and underground excavations may be accurately modelled by the unloading of a cavity from the *in-situ* stress state. For decades, cavity expansion theory has been used to predict ground settlements due to tunnelling and to design tunnel support systems to maintain stability.

2. Excavation design in elastic rock is a relatively simple problem in underground rock engineering. As discussed by Brady and Brown (1993), the basic procedure

for excavation design can be illustrated simply in Figure 10.2 (where σ_θ is the tangential stress on the tunnel wall, and C_0 and T_0 denote the uniaxial compressive and tensile strengths of the rock). A key step in the design process is to dertermine the stress distribution around the excavation. Although the stress solution may be achieved by numerical methods, the simple closed-form solutions of cavity expansion theory have many advantages in the early stages of the design process. In particular, elastic cavity expansion solutions can be used to determine a zone of influence of an excavation. When an underground excavation is constructed in a rock mass with major discontinuities, it is questionable if the elastic stress analysis will be still valid. This is because rock joints generally have much lower strengths than the intact rock itself. It is, however, shown that the effect of discontinuities on elastic stress distribution can be easily assessed.

3. Many underground excavations are carried out under conditions that are sufficient to induce a failure (plastic or fractured) region around the opening. In such cases, it is necessary to provide some support on the internal boundary of the excavation to control the performance of fractured (or failed) rock. To illustrate the function of rock support in an underground excavation, it is useful to consider the simple problem shown in Figure 10.9. When the internal pressure p is sufficiently lower than the *in-situ* stress, a failed zone denoted by radius c forms around the opening and, in this failed region, the strength of the rock drops significantly. According to the brittle/plastic cavity expansion solutions developed in Chapter 5, the relationship between the internal pressure and the size of the failed region is defined by equation (10.27). Hence, by providing certain internal support (i.e. increasing the internal pressure p), the region of failed rock can be controlled and limited to a satisfactory level.

4. The basic principle used to design rock support in underground excavations is clearly illustrated by Brady and Brown (1993). In order to design a suitable support system, it is essential to determine the required support lines for a rock mass (i.e. ground response curves – internal pressure and displacement correlation). Cavity expansion solutions developed in Chapter 5 can be used to calculate the ground response curves with a realistic rock model. For example, using the Hoek-Brown criterion, the ground response curve is given by equation (10.45).

5. Cavity expansion theory can also be used as a simple tool in the design of shallow tunnels in soils in order to meet the serviceability and stability requirements. The serviceability requirement ensures that the construction of a shallow tunnel does not cause excessive displacements in the soil which may damage the surrounding or adjacent buildings and utilities. Of particular importance are the surface settlements caused by tunnelling and their effects on the structures that are built direct-

ly above the tunnel. For example, Mair and Taylor (1993) demonstrated the relevance of total stress cavity unloading solutions, defined by (10.66) and (10.67), in the prediction of soil movements around tunnels in undrained clay. Yu and Rowe (1999), however, noted that the chief advantage of an effective stress analysis is it can take into account the effect of stress history (i.e. overconsolidation ratios) on the soil behaviour around tunnels. Using the results of a centrifuge tunnel test carried out by Mair (1979), it is shown the cavity unloading solutions of Yu and Rowe (1999) give very similar predictions to those made by Cambridge University's finite element program CRISP.

6. Relatively few studies have been carried out using cavity expansion theory to predict the stability of tunnels in soils. In this chapter, we present procedures that may be used to apply cavity unloading solutions to conduct stability calculations for tunnels in soils. In doing so, we basically follow Caquot and Kerisel (1966) in assuming that collapse of a tunnel will occur when the plastic zone reaches the ground surface. In other words, tunnels will collapse when the plastic flow ceases to be confined and becomes free, Figure 10.22. The suitability of this assumption for predicting the stability of tunnels in cohesive soils is confirmed by good agreement with the centrifuge results reported by Mair (1979). For dilatant cohesive-frictional soils, however, it is more realistic to assume that collapse of tunnels occurs if the unloading plastic zone spreads sufficiently close to the ground surface. This is because dilation of sandy soils tends to cause a very large deformation in the soil around the plastic zone and therefore accelerates the collapse of tunnels.

REFERENCES

Atkinson, J.H. and Bransby, P.L. (1978). *The Mechanics of Soils*. McGraw-Hill.

Brady, B.H.G. and Brown, E.T. (1993). *Rock Mechanics for Underground Mining*. 2nd edition, Chapman & Hall, London.

Bolton, M.D. (1991). *A Guide to Soil Mechanics*. M. D. and K. Bolton, Cambridge.

Brown, E.T., Bray, J.W., Ladanyi, B. and Hoek, E. (1983). Ground response curves for rock tunnels. *Journal of Geotechnical Engineering*, ASCE, 109(1), 15–39.

Caquot, A. and Kerisel, J. (1966). *Traite de Mecanique des Sols*. Gauthier–Villars.

Clough, G.W., Shirasuna, T. and Finno, R.J. (1985). Finite element analysis of advanced shield tunnelling in soils. *Proceedings of the 5th International Conference on Numerical Methods in Geomechanics*, Nagoya, 1167–1174.

Collins, I.F. and Yu, H.S. (1996). Undrained cavity expansions in critical state soils. *International Journal for Numerical and Analytical Methods in Geomechanics*, 20, 485–516.

Davis, E.H. (1968). Theories of plasticity and the failure of soil masses. *Soil Mechanics: Selected Topics* (Editor: I.K. Lee), Butterworths, Sydney.

Davis, E.H., Gunn, M.J. Mair, R.J. and Seneviratne, H.N. (1980). The stability of shallow tunnels and underground openings in cohesive material. *Geotechnique,* 30(4), 397–416.

Fritz, P. (1984). An analytical solution for axisymmetric tunnel problems in elasto-visco-plastic media. *International Journal for Numerical and Analytical Methods in Geomechanics,* 8, 325–342.

Gens, A. and Potts, D.M. (1988). Critical state models in computational geomechanics. *Engineering Computations,* 5, 178–197.

Ghaboussi, J., Ranken, R.E. and Karshenas, M. (1978). Analysis of subsidence over soft ground tunnels. *ASCE International Conference on Evaluation and Prediction of Subsidence*, Pensacola Beach, 182–196.

Hobbs, D.W. (1966). A study of the behaviour of broken rock under triaxial compression and its application to mine roadways. *International Journal for Rock Mechanics and Mining Science & Geomechanics Abstracts,* 3, 11–43.

Hoek, E. and Brown, E.T. (1980). *Underground Excavations in Rock.* The Institution of Mining and Metallurgy, London, England.

Kennedy, T.C. and Lindberg, H.E. (1978). Tunnel closure for nonlinear Mohr-Coulomb functions. *Journal of the Engineering Mechanics Division*, ASCE, 104(EM6), 1313–1326.

Kulhawy, F.H. (1974). Finite element modelling criteria for underground openings in rock. *International Journal of Rock Mechanics and Mining Sciences*, 11, 465–472.

Lee, K.M. and Rowe, R.K. (1990). Finite element modelling of the three dimensional ground deformations due to tunnelling in soft cohesive soils: I – method of analysis. *Computers and Geotechnics*, 10, 87–109.

Lo, K.Y., Ng, M.C. and Rowe, R.K. (1984). Predicting settlement due to tunnelling in clays. *Tunnelling in Soil and Rock, ASCE Geotech III Conference, Atlanta*, 48–76.

Mair, R.J. (1979). *Centrifuge Modelling of Tunnel Construction in Soft Clay.* PhD Thesis, University of Cambridge, England.

Mair, R.J. and Taylor, R.N. (1993). Prediction of clay behaviour around tunnels using plasticity solutions. *Predictive Soil Mechanics* (Editors: G.T. Houlsby and A.N. Schofield), Thomas Telford, London, 449–463.

Muir Wood, D. (1990). *Soil Behaviour and Critical State Soil Mechanics*. Cambridge University Press.

Ogawa, T. and Lo, K.Y. (1987). Effects of dilatancy and yield criteria on displacements around tunnels. *Canadian Geotechnical Journal*, 24, 100–113.

Peck, R.B., Hendron, A.J. and Moheraz, B. (1972). State of the art of soft ground tunnelling. *Proceedings of the 1st Rapid Excavation Tunnelling Conference*, Chicago, AIME, Vol 1, 259–286.

Peck, R.B. (1969). Deep excavations and tunnelling in soft ground. *Proceedings of the 7th International Conference on Soil Mechanics and Foundation Engineering*, Mexico City, 225–290.

Pender, M. (1980). Elastic solutions for a deep circular tunnel. *Geotechnique,* 30(2), 216–222.

Reed, M.B. (1986). Stresses and displacement around a cylindrical cavity in soft rock. *IMA Journal of Applied Mathematics,* 36, 223–245.

Reed, M.B. (1988). The influence of out-of-plane stress on a plane strain problem in rock mechanics. *International Journal for Numerical and Analytical Methods in Geomechanics,* 12, 173–181.

Roscoe, K.H. and Burland, J.B. (1968). On the generalized stress-strain behaviour of wet clay. In: *Engineering Plasticity,* Cambridge University Press, 535–609.

Rowe, R.K. and Kack, G.J. (1983). A theoretical examination of the settlements induced by tunnelling: four case histories. *Canadian Geotechnical Journal*, 20 (2), 299–314.

Rowe, R.K. and Lee, K.M. (1992). An evaluation of simplified techniques for estimating three dimensional undrained ground movements due to tunnelling in soft soils. *Canadian Geotechnical Journal*, 29, 39–52.

Rowe, R.K. (1986). The prediction of deformations caused by soft ground tunnelling – Recent trends. *Canadian Tunnelling*, D. Eisenstein (ed.), 91–108.

Sagaseta, C. (1987). Analysis of undrained soil deformation due to ground loss. *Geotechnique,* 37(3), 301–320.

Schofield, A.N. and Wroth, C.P. (1968). *Critical State Soil Mechanics*. McGraw–Hill.

Sloan, S.W. and Assadi, A. (1993). The stability of shallow tunnels in soft ground. *Predictive Soil Mechanics* (Editors: G.T. Houlsby and A.N. Schofield), Thomas Telford, London, 644–663.

Verruijt, A. and Booker, J.R. (1996). Surface settlements due to deformation of a tunnel in an elastic half plane. *Geotechnique,* 46(4), 753–756.

Wilson, A.H. (1980). A method of estimating the closure and strength of lining required in drivages surrounded by a yield zone. *International Journal for Rock Mechanics and Mining Sciences,* 17, 349–355.

Yu, H.S. and Houlsby, G.T. (1991). Finite cavity expansion in dilatant soil: loading analysis. *Geotechnique*, 41, 173–183.

Yu, H.S. and Houlsby, G.T. (1995). A large strain analytical solution for cavity contraction in dilatant soils. *International Journal for Numerical and Analytical Methods in Geomechanics,* 19, 793–811.

Yu, H.S. and Rowe, R.K. (1999). Plasticity solutions for soil behaviour around contracting cavities and tunnels. *International Journal for Numerical and Analytical Methods in Geomechanics,* 23, 1245–1279.

Zytynski, M., Randolph, M.F., Nova, R. and Wroth, C.P. (1978). On modelling the unloading–reloading behaviour of soils. *International Journal for Numerical and Analytical Methods in Geomechanics,* 2, 87–93.

CHAPTER 11

WELLBORE INSTABILITY

11.1 INTRODUCTION

Wellbore instability during drilling is a major problem in the application of rock mechanics in petroleum engineering. Opinions may vary as to the value of lost equipment and time arising directly from wellbore instability, but no one would dispute estimates of over $500 millions per year worldwide (Dusseault, 1994). Better analysis and prediction methods for wellbore instability will definitely help reduce the costs associated with the wellbore breakout and instability problems.

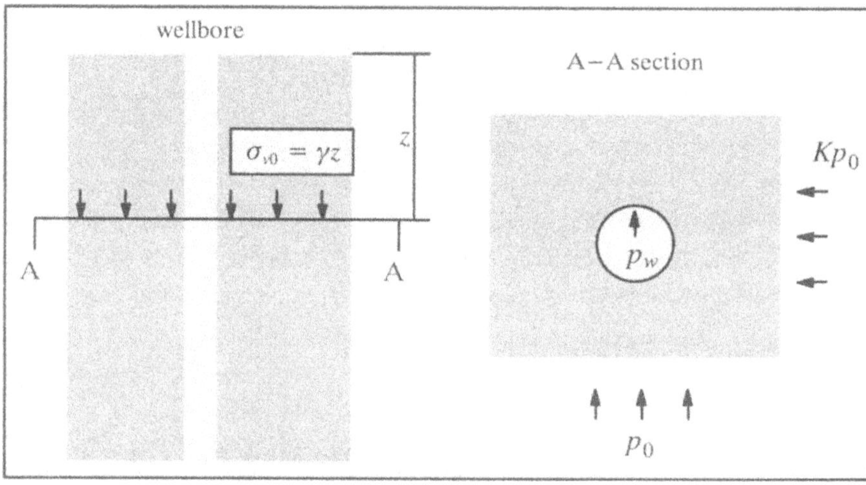

Figure 11.1: A vertical wellbore in rock

Stress boundary conditions around a vertical wellbore are shown in Figure 11.1. For simplicity, the horizontal stresses may be assumed to be the same in all directions (i.e. K=1). As noted by Bradley (1979) and Santarelli *et al.* (1986), stress-induced wellbore instabilities are of three general types:

(a) hole size reduction due to ductile yield of the rock;

(b) hole enlargement due to brittle rock fracture or rupture; and

(c) unintenional hydraulic fracturing induced by excessive mud pressure.

It is common practice to adjust the internal wellbore pressure (i.e. mud pressure p_w, Figure 11.1) to avoid borehole instabilities due to rock fracturing and rupture.

Cavity expansion solutions obtained with elastic, poroelastic and plastic models have been used widely to study the problem of wellbore instabilities in rock (e.g. Woodland, 1990; Santarelli *et al.*, 1986; Wu and Hudson, 1991; Detournay and Cheng, 1988; Charlez and Heugas, 1991).

This chapter will review some of the main applications of cavity expansion solutions in the analysis and prediction of stress-induced wellbore instabilities in rock. While all three major types of wellbore instabilities mentioned earlier can be considered, attention will be focused mainly on the instability associated with hole enlargement due to brittle rock fracture or rupture around the wellbore. Borehole size reduction due to ductile yield of the rock will also be considered towards end of the chapter when plastic analysis is used to predict wellbore instability.

11.2 ELASTIC ANALYSIS OF WELLBORE INSTABILITY

In this section, wellbore instability is studied using elastic analysis of the stresses around the wellbore concerned. The problem is shown in Figure 11.2. The basic steps which must be followed in the elastic analysis are: (1) determine the elastic stress field around the wellbore, (2) choose a suitable failure criterion for the rock concerned, (3) compare the elastic stresses and the selected rock failure criterion. If the failure criterion is satisfied anywhere in the rock, then the wellbore is regarded as unstable.

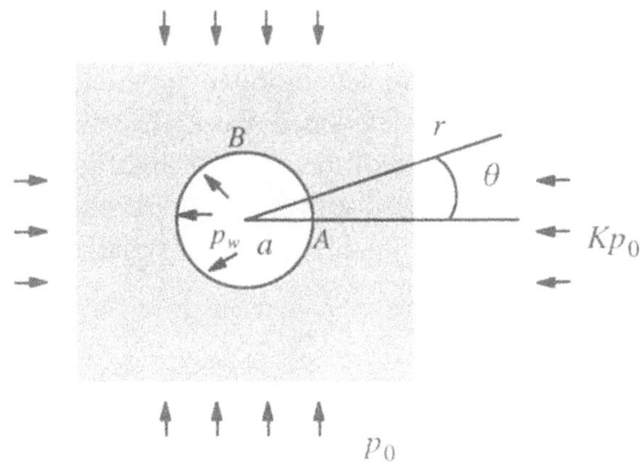

Figure 11.2: Problem geometry for circular wellbore with non-hydrostatic *in-situ* stresses

In elastic analysis, various assumptions can be made of the form of variation of stiffness parameters (e.g. Young's modulus) on stress and strain level. It is also possible to consider the solution in cross-anisotropic rock.

11.2.1 Stress analysis using constant stiffness elasticity

Stress distribution

As shown in Chapter 2, if the rock behaves in a linear elastic fashion, a complete solution for the stress distribution around a wellbore is:

$$\sigma_r = \frac{p_0}{2}[(1 + K)(1 - \frac{a^2}{r^2}) - (1 - K)(1 - 4\frac{a^2}{r^2} + 3\frac{a^4}{r^4})\cos 2\theta] + \frac{p_w a^2}{r^2} \tag{11.1}$$

$$\sigma_\theta = \frac{p_0}{2}[(1 + K)(1 + \frac{a^2}{r^2}) + (1 - K)(1 + 3\frac{a^4}{r^4})\cos 2\theta] - \frac{p_w a^2}{r^2} \tag{11.2}$$

$$\tau_{r\theta} = \frac{p_0}{2}[(1 - K)(1 + 2\frac{a^2}{r^2} - 3\frac{a^4}{r^4})\sin 2\theta] \tag{11.3}$$

For linear elastic analysis the above solutions can be used in conjunction with a rock fracture or failure criterion to assess the stability of the wellbore for a given mud pressure p_w.

Mohr-Coulomb failure criterion and wellbore instability prediction

As mentioned earlier, there are three major types of wellbore instabilities that may be encountered in practice. Different types of instabilities require different criteria. While elastic analysis may be acceptable for studying the second and third types of wellbore instabilities concerning hole enlargement due to brittle rock fracture or rupture, it may not be suitable for predicting wellbore instability concerning hole closure (inward displacements).

Brittle rock fracture or rupture is traditionally predicted by the Mohr-Coulomb criterion, although the newly-developed Hoek-Brown criterion may be better suited for predicting the imitation of shear failure in the jointed rock mass. To illustrate the basic procedure described above, some case studies are presented here to investigate the possibility of rock failure occurring on the surface of the wellbore.

On the surface of the wellbore, $r = a$, the stresses are:

$$\sigma_r = p_w \tag{11.4}$$

$$\sigma_\theta = p_0[1 + K + 2(1 - K)\cos 2\theta] - p_w \tag{11.5}$$

$$\tau_{r\theta} = 0 \tag{11.6}$$

It is likely and will therefore be assumed here that for the wellbore stability problem, the radial and tangential stresses are the minor and major principal stresses respectively. Therefore the Mohr-Coulomb criterion can be written as follows:

$$\sigma_\theta - \alpha\sigma_r = Y \tag{11.7}$$

where

$$a = \frac{1 + \sin\phi}{1 - \sin\phi} \quad \text{and} \quad Y = \frac{2C\cos\phi}{1 - \sin\phi} \tag{11.8}$$

and ϕ and C are the friction angle and cohesion respectively.

Applying the stresses on the surface of the wellbore to the Mohr-Coulomb criterion (11.7) leads to the following condition for the initiation of rock fracture failure in the rock mass:

$$p_w < \frac{[(1 + K) + 2(1-K)\cos 2\theta]p_0 - Y}{1 + a} \tag{11.9}$$

For the hydrostatic *in-situ* stress state, K=1, the above condition reduces to:

$$P_w < \frac{2p_0 - Y}{1 + a} \tag{11.10}$$

For the case of a non-hydrostatic *in-situ* stress state, we can consider the following two points:

Point A: At point A, $\theta = 0^0$ and $r = a$, the condition for wellbore instability to occur is:

$$P_w < \frac{(3-K)p_0 - Y}{1 + a} \tag{11.11}$$

Point B: At point B, $\theta = 90^0$ and $r = a$, the condition for wellbore instability to occur is:

$$P_w < \frac{(3K-1)p_0 - Y}{1 + a} \tag{11.12}$$

Hoek-Brown failure criterion and wellbore instability prediction

If the Hoek-Brown failure criterion is selected to predict the initiation of rock fracture and failure, the condition for wellbore instability can be obtained in the same way as for the Mohr-Coulomb criterion.

For the wellbore stability problem, the Hoek-Brown failure criterion can be expressed as follows:

$$\sigma_\theta = \sigma_r + \sqrt{mq_c\sigma_r + sq_c^2} \tag{11.13}$$

where q_c is the uniaxial compressive strength of the intact rock material and m and s are constants depending on the nature of the rock mass and the extent to which it had been broken.

By applying the Hoek-Brown failure criterion (11.13) to the stress field on the surface of wellbore, the following condition is obtained. This condition must be satisfied in order for the wellbore instability to occur:

$$p_w < \frac{d}{2}p_0 - [\frac{1}{8}\sqrt{m^2 + 8dmp_0/q_c - 16s} - \frac{m}{8}]q_c \tag{11.14}$$

where d is defined as

$$d = 1 + K + 2(1-K)\cos 2\theta \tag{11.15}$$

For the case of a hydrostatic stress state K=1, d=2 and the above wellbore instability condition reduces to:

$$p_w < p_0 - [\frac{1}{8}\sqrt{m^2 + 16mp_0/q_c - 16s} - \frac{m}{8}]q_c \tag{11.16}$$

11.2.2 Analysis using pressure-dependent elasticity

The solutions obtained in the previous section are valid for linear elastic models with constant stiffness. The prediction of wellbore instability by assuming linear elasticity with constant elastic parameters may be in error. It has been established that the pre-yield and pre-peak stress strain properties of some rocks are nonlinear and that the elastic properties are pressure dependent (Kulhawy, 1975).

To investigate the effect of pressure dependent elastic parameters on the prediction of wellbore instability, Santarelli *et al.* (1986) presented a numerical study on the analysis of borehole stresses using a nonlinear elastic model with a simple pressure dependent Young's modulus. Based on the experimental data obtained for dry Carboniferous sandstone, Santarelli *et al.* assumed that Young's modulus depends on the minor principal stress σ_3 in the following way:

$$E = E_0(1 + 0.043\sigma_3^{0.78}) \tag{11.17}$$

where $E_0 = 17.49$GPa, σ_3 is in MPa and E is in GPa.

It is noted that with the nonlinear elasticity defined by (11.17), it is no longer possible to obtain a closed form solution for the elastic stress distribution around a wellbore. In fact, it is necessary to use a numerical method such as the finite element methods or finite difference methods. In Santarelli *et al.* (1986), a finite difference technique was used to calculate elastic stress distribution in a thick cylinder subject to internal and external pressures, Figure 11.3.

As shown in Figure 11.4, the main conclusion of the analysis by Santarelli *et al.* (1986) is that close to the cavity wall, the tangential stress derived from a nonlinear elastic model is considerably smaller than that from a linear constant stiffness model. The linear elastic model predicts that the maximum tangential stress occurs at the borehole wall, and that the tangential stress is concentrated near the wall when the internal pressure is reduced. On the other hand, the nonlinear elasticity results of Santarelli *et al.* (1986) suggest that the maximum tangential stress does not necessarily develop at the wall, and that the borehole will influence the stresses in a large volume of rock. The fact that the maximum tangential stress can develop away from the borehole wall could explain the observed initiation of fracture within the rock surrounding borehole and the production of discrete, largely intact rock chips (see, for example, Bradley, 1979).

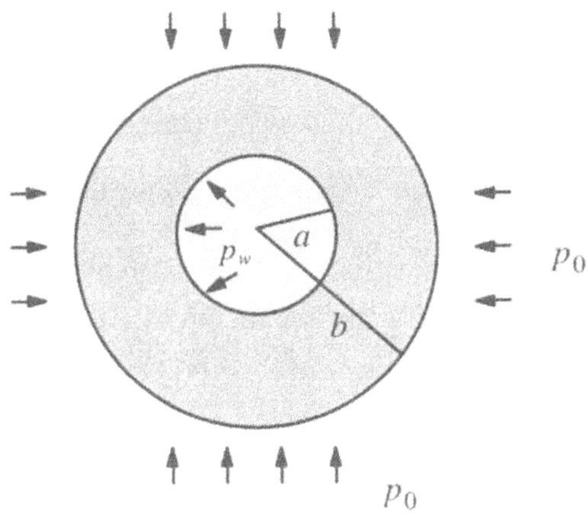

Figure 11.3: Problem of a thick cylinder subject to internal and external pressures

In addition to the pressure dependent model defined by (11.17), McLean (1987) and Nawrocki and Dusseault (1994) have also used the following pressure-dependent modulus to study its effect on wellbore stability.

$$E = E_0 + h\sigma_3 \tag{11.18}$$

where E_0 is Young's modulus at zero strain and h is a material constant. The results presented in Nawrocki and Dusseault (1994) are in agreement with the conclusion of Santarelli *et al.* (1986) (i.e. Figure 11.4).

In summary, all these nonlinear elastic studies suggest that if the Mohr-Coulomb failure criterion is applied, the rock around the borehole would appear to be a lot stronger than that predicted by the linear elastic theory.

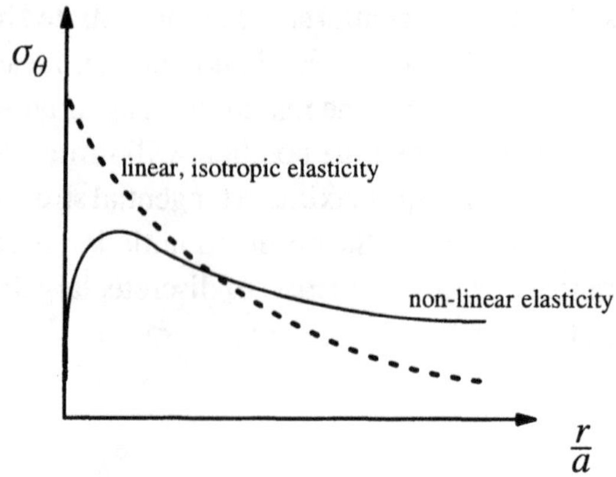

Figure 11.4: Results of tangential stress with linear and non-linear elastic models

11.2.3 Effect of stress-induced anisotropy on wellbore instability

Wu and Hudson (1991) studied the effect of stress-induced anisotropy on wellbore stability. The elastic solutions for the expansion of a thick-walled cylinder with a cross-anisotropic model derived by Lekhnitskii (1963) and Wu *et al.* (1991) were presented in Chapter 2.

The solution presented in this section for a cylinder with both internal and external pressures follows the procedure used by Wu *et al.* (1991). This is a special case of the general solutions derived by Lekhnitskii (1963) for a variety of loading conditions.

This analytical solution can be used to deal with the case in which any axis (r, z, θ) may be treated as the axis of symmetry. In other words, the following three possibilities exist:

(a) the r (radial) direction is the axis of symmetry and the (z, θ) plane is the isotropic plane;

(b) the z (axial) direction is the axis of symmetry and the (r, θ) plane is the isotropic plane;

(c) the θ (tangential) direction is the axis of symmetry and the (r, z) plane is the isotropic plane.

It must be noted that while the first two cases can find applications to geotechnical engineering, the third possibility may not be very relevant in geomechanics.

Stress-strain relations

The general stress-strain relationship for a cross-anisotropic material can be written as follows (Lekhnitskii, 1963; Van Cauwelaert, 1977):

$$\varepsilon_r = a_{11}\sigma_r + a_{12}\sigma_\theta + a_{13}\sigma_z \tag{11.19}$$

$$\varepsilon_\theta = a_{12}\sigma_r + a_{22}\sigma_\theta + a_{23}\sigma_z \tag{11.20}$$

$$\varepsilon_z = a_{13}\sigma_r + a_{23}\sigma_\theta + a_{33}\sigma_z \tag{11.21}$$

where the coefficients a_{ij} can be expressed as simple functions of Young's moduli and Poisson's ratios.

For plane strain in the z direction:

$$\sigma_z = -\frac{1}{a_{33}}(a_{13}\sigma_r + a_{23}\sigma_\theta) \tag{11.22}$$

Substituting equation (11.22) into equations (11.19) and (11.20) leads to:

$$\varepsilon_r = \beta_{11}\sigma_r + \beta_{12}\sigma_\theta \tag{11.23}$$

$$\varepsilon_\theta = \beta_{12}\sigma_r + \beta_{22}\sigma_\theta \tag{11.24}$$

where the coefficients β_{ij} are:

$$\beta_{ij} = a_{ij} - \frac{a_{i3}a_{j3}}{a_{33}}, \quad (i,j = 1,2) \tag{11.25}$$

Solution procedure

For the cylindrical cavity, the strains are expressed as functions of the radial displacement u

$$\varepsilon_r = -\frac{du}{dr} \quad , \quad \varepsilon_\theta = -\frac{u}{r} \tag{11.26}$$

which can be used to eliminate the displacement u to give the following compatibility condition:

$$\varepsilon_r = \frac{d}{dr}(r\varepsilon_\theta)$$

The equilibrium equation is

$$r\frac{d\sigma_r}{dr} + (\sigma_r - \sigma_\theta) = 0$$

If the radial stress is treated as the fundamental variable, we can obtain the following differential equation by combining the stress-strain relations, compatibility condition and the equilibrium equation:

$$\beta_{22}r^2\frac{d^2\sigma_r}{dr^2} + 3\beta_{22}r\frac{d\sigma_r}{dr} - (\beta_{11} - \beta_{22})\sigma_r = 0 \tag{11.27}$$

The above equation can be solved to give the following general solution for the radial stress:

$$\sigma_r = Ar^{n-1} + \frac{B}{r^{n+1}} \tag{11.28}$$

in which n is defined by:

$$n = \sqrt{\frac{\beta_{11}}{\beta_{22}}} \tag{11.29}$$

Using the boundary conditions:

$$\sigma_r|_{r=a} = p_w$$

$$\sigma_r|_{r=b} = p_0$$

the integration constants A and B can be derived. The final solutions for the stresses are:

$$\sigma_r = \frac{p_0 - p_w(\frac{a}{b})^{n+1}}{1-(\frac{a}{b})^{2n}}(\frac{r}{b})^{n-1} + \frac{p_w - p_0(\frac{a}{b})^{n-1}}{1-(\frac{a}{b})^{2n}}(\frac{a}{r})^{n+1} \tag{11.30}$$

$$\sigma_\theta = n\frac{p_0 - p_w(\frac{a}{b})^{n+1}}{1-(\frac{a}{b})^{2n}}(\frac{r}{b})^{n-1} - n\frac{p_w - p_0(\frac{a}{b})^{n-1}}{1-(\frac{a}{b})^{2n}}(\frac{a}{r})^{n+1} \tag{11.31}$$

It can be shown that for an isotropic material $n = 1$, the above solutions reduce to the previous cylinder expansion solutions for isotropic materials.

The main result presented by Wu and Hudson (1991) is summarised in Figure 11.5. It suggests that the effect of stress-induced anisotropy on stiffness has a very significant effect on the elastic stress distribution around the borehole. In particular, the tangential stress obtained by assuming cylindrical anisotropy is always lower than that obtained by assuming isotropy provided that Young's modulus in the tangential direction is greater than that in the radial direction. It should be stressed that this is normally the case for a borehole problem, as the tangential stress is generally the maximum compressive stress which would generate a maximum modulus in that direction.

Further studies of the analytical stress solutions defined by (11.30) and (11.31) indicate that like linear isotropic elasticity, the linear anisotropic model would predict that the Mohr-Coulomb failure criterion is first satisfied at the wall of borehole. It is also true that the rock surrounding the borehole would appear to be stronger than that predicted by the isotropic linear elastic theory. This is because the difference between tangential and radial stresses close to the borehole wall obtained from the linear anisotropic model is smaller than that given by the linear elastic model.

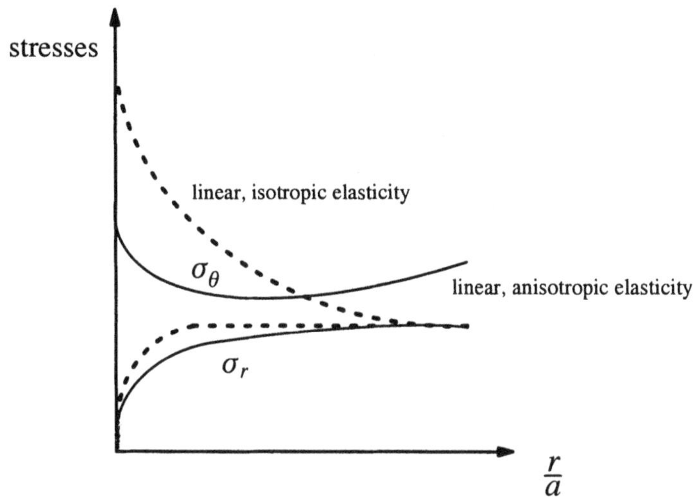

Figure 11.5: Comparison of radial and tangential stresses with isotropic and anisotropic elastic models

11.3 POROELASTIC ANALYSIS OF WELLBORE INSTABILITY

11.3.1 Semi-analytical solutions

The elastic solutions presented in the previous sections assume that the rock behaves as a single phase material and hence the solutions obtained are relevant only to the short term (or undrained) condition or the long term (or fully drained) condition. If, however, a borehole is created in a saturated rock the resulting displacements and stress changes will be time dependent. This time dependence is due to the two phase nature of a saturated elastic material. Volume change takes place only as fluid is expelled from the voids between the solid particles. Movement of fluid through the soil or rock cannot occur instantaneously and so any deformations which involve a change in volume require a finite time to occur.

Assuming a two phase model, Carter and Booker (1982) presented a semi-analytical solution for the displacements and stress changes around a long circular opening in a saturated elastic medium. For simplicity, it was assumed that the compressibility of water and particles was small and may therefore be ignored. In this section, we will summarise the basic solution procedure used by Carter and Booker (1982) for deriving the time-dependent stress and displacements around a created hole. The next section will illustrate how Carter and Booker's solutions can be used for the prediction of wellbore instability in elastic, saturated rock material.

Problem description

The problem of the creation of a borehole is shown in Figure 11.6. The initial stress field before the borehole boring is defined by a non-uniform horizontal stress state. It is also assumed that the initial porewater pressure is hydrostatic with a magnitude of U_0. We are interested in the stress and pore pressure changes as well as the displacements due to the creation of the borehole.

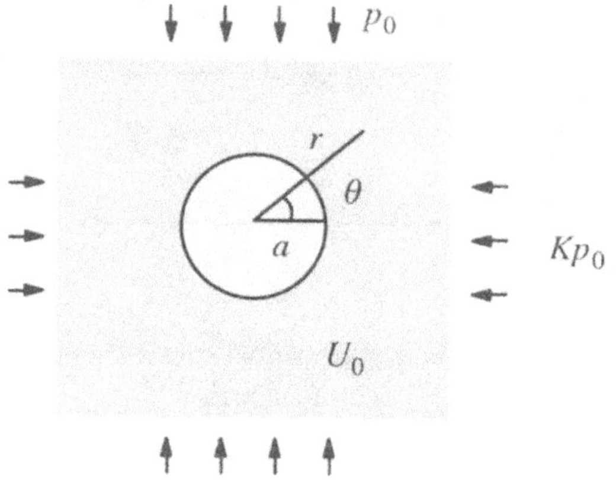

Figure 11.6: Problem of the creation of a borehole in a two-phase elastic, saturated medium with a non-hydrostatic stress field

Before the borehole is bored, the total stresses acting on a circular boundary of the borehole are given by:

$$\sigma_r = \sigma_m + \sigma_d \cos 2\theta \tag{11.32}$$

$$\sigma_\theta = \sigma_m - \sigma_d \cos 2\theta \tag{11.33}$$

$$\tau_{r\theta} = -\sigma_d \sin 2\theta \tag{11.34}$$

where

$$\sigma_m = \frac{1}{2}(K+1)p_0 \tag{11.35}$$

$$\sigma_d = \frac{1}{2}(K-1)p_0 \tag{11.36}$$

After boring, the normal total stress σ_r and the shear stress $\tau_{r\theta}$ are both removed from the borehole boundary. Hence the removal of the material within the boundary of the borehole is equivalent to the application of the following stress increments on the borehole boundary:

$$\Delta\sigma_r|_{r=a} = -\sigma_m - \sigma_d \cos 2\theta \tag{11.37}$$

$$\Delta\tau_{r\theta}|_{r=a} = \sigma_d \sin 2\theta \tag{11.38}$$

where the symbol Δ indicates stress increments.

Carter and Booker (1982) also considered two extreme hydraulic boundary conditions at the borehole wall. The first case is for a permeable borehole surface, which is defined by the following boundary condition on pore pressure:

$$U|_{r=a} = 0 \tag{11.39}$$

The other extreme case that was considered is an impermeable borehole, which would give the following boundary condition:

$$\frac{\partial U}{\partial r}|_{r=a} = 0 \tag{11.40}$$

Equations (11.37)-(11.40) combined define the complete boundary conditions at the borehole surface for modelling the removal of material to form a circular borehole.

To simplify the analysis, it is better to solve equations (11.37)-(11.40) as three (for permeable borehole) or two (for impermeable borehole) separate simple problems. The final solution can then be obtained using the principle of superposition.

For a permeable borehole, the creation of the borehole is divided into the following three simple loading modes:

Loading Ia: The removal of the volumetric component of the total radial stress on the borehole surface:

$$\Delta\sigma_r|_{r=a} = -\sigma_m \tag{11.41}$$

$$\Delta\tau_{r\theta}|_{r=a} = 0 \tag{11.42}$$

$$U|_{r=a} = 0 \tag{11.43}$$

Loading IIa: The removal of the initial porewater pressure on the borehole surface:

$$\Delta\sigma_r|_{r=a} = 0 \tag{11.44}$$

$$\Delta\tau_{r\theta}|_{r=a} = 0 \tag{11.45}$$

$$U|_{r=a} = -U_0 \tag{11.46}$$

Loading IIIa: The removal of the devetoric components of total radial stress and shear stress on the borehole surface:

$$\Delta\sigma_r|_{r=a} = -\sigma_d \cos 2\theta \tag{11.47}$$

$$\Delta\tau_{r\theta}|_{r=a} = \sigma_d \sin 2\theta \tag{11.48}$$

$$U|_{r=a} = 0 \tag{11.49}$$

On the other hand, for an impermeable borehole, the creation of the borehole may be divided into the following two simple loading modes:

Loading Ib: The removal of the volumetric component of the total radial stress on the borehole surface:

$$\Delta\sigma_r|_{r=a} = -\sigma_m \tag{11.50}$$

$$\Delta\tau_{r\theta}|_{r=a} = 0 \tag{11.51}$$

$$\frac{\partial U}{\partial r}|_{r=a} = 0 \tag{11.52}$$

Loading IIIb: The removal of the devetoric components of total radial stress and shear stress on the borehole surface:

$$\Delta\sigma_r|_{r=a} = -\sigma_d \cos 2\theta \tag{11.53}$$

$$\Delta\tau_{r\theta}|_{r=a} = \sigma_d \sin 2\theta \tag{11.54}$$

$$\frac{\partial U}{\partial r}|_{r=a} = 0 \tag{11.55}$$

It can be shown that loading Ia of a permeable borehole and loading Ib of an impermeable borehole are identical. In other words, both are independent of the hydraulic boundary conditions.

General solutions

Carter and Booker (1982) presented a general solution for the basic problem outlined above by assuming that both pore fluid and particles are incompressible. This is particularly applicable to soil mechanics problems. For application to wellbore instability, Detournay and Cheng (1988) extended the Carter and Booker solution to include the compressibility of the pore fluid and particles. However in the solution of Detournay and Cheng (1988), only the permeable borehole was considered.

As the general solution procedure is quite tedious, it is not repeated here. However, interested readers can refer to the papers by Carter and Booker (1982) and Detournay and Cheng (1988) for further details.

11.3.2 Application to wellbore instability prediction

A brief summary of the analytical solution derived by Detournay and Cheng (1988) and its application to wellbore instability prediction is presented in this section. Before presenting the solutions, it is necessary to define all the materials parameters required for two-phase elastic, saturated materials.

Material constants

As discussed by Detournay and Cheng (1988, 1993), a set of five bulk material constants are needed to provide a full description of an isotropic rock-fluid system. These include: two elastic constants G and ν (shear modulus and drained Poisson's ratio); two poroelastic coefficients B (ratio of the induced pore pressure to the variation of confining pressure under undrained conditions) and ν_u (undrained Poisson's ratio); and the parameter $\varkappa$ which, for the case of incompressible fluid and solids, is related to the permeability coefficient k by $\varkappa = k/\gamma_w$.

For convenience, some other common parameters that will be used in the solution are given:

$$c = \frac{2\varkappa B^2 G(1-\nu)(1+\nu_u)^2}{9(1-\nu_u)(\nu_u-\nu)} \tag{11.56}$$

$$\eta = \frac{3(\nu_u-\nu)}{2B(1-\nu)(1+\nu_u)} \tag{11.57}$$

$$S = \frac{B(1+\nu_u)}{3(1-\nu_u)} \tag{11.58}$$

Loading Ia

For loading Ia, the solution corresponds to the classical Lame solution in elasticity, which is given below:

$$\frac{2Gu_r^{(1)}}{\sigma_m a} = \frac{1}{\varrho} \tag{11.59}$$

$$\frac{\sigma_r^{(1)}}{\sigma_m} = -\frac{1}{\varrho^2} \tag{11.60}$$

$$\frac{\sigma_\theta^{(1)}}{\sigma_m} = \frac{1}{\varrho^2} \tag{11.61}$$

where $\varrho = \frac{r}{a}$ is a dimensionless radial coordinate. As the displacement field is characterised by zero volumetric strain, there is no mechanism for pore pressure generation and its subsequent dissipation. The stress and displacement fields are therefore time independent.

Loading IIa

In the case of loading IIa, the pressure field is governed by a homogeneous consolidation equation:

$$\frac{\partial^2 U}{\partial r^2} + \frac{1}{r}\frac{\partial U}{\partial r} = \frac{1}{c}\frac{\partial U}{\partial t} \tag{11.62}$$

where the consolidation coefficient c is defined by equation (11.56).

The pore pressure can be solved by taking the Laplace transform of equation (11.62). The Laplace transform of pore pressure is:

$$\frac{s\overline{U}^{(2)}}{U_0} = -\frac{K_0(\xi)}{K_0(\beta)} \tag{11.63}$$

where K_0 is the modified Bessel function of the second kind of order zero, $\xi = r\sqrt{s/c}$, and $\beta = a\sqrt{s/c}$.

The Laplace transforms of the displacement and stress fields are:

$$\frac{2Gs\overline{u}_r^{(2)}}{aU_0} = 2\eta\left[\frac{K_1(\xi)}{\beta K_0(\beta)} - \frac{a}{r}\frac{K_1(\beta)}{\beta K_0(\beta)}\right] \tag{11.64}$$

$$\frac{s\overline{\sigma}_r^{(2)}}{U_0} = -2\eta\left[\frac{a}{r}\frac{K_1(\xi)}{\beta K_0(\beta)} - \frac{a^2}{r^2}\frac{K_1(\beta)}{\beta K_0(\beta)}\right] \tag{11.65}$$

$$\frac{s\overline{\sigma}_\theta^{(2)}}{U_0} = 2\eta\left[\frac{a}{r}\frac{K_1(\xi)}{\beta K_0(\beta)} - \frac{a^2}{r^2}\frac{K_1(\beta)}{\beta K_0(\beta)} + \frac{K_0(\xi)}{K_0(\beta)}\right] \tag{11.66}$$

Loading IIIa

As discussed by Carter and Booker (1982) and Detournay and Cheng (1988, 1993), the solution for loading IIIa is much more involved. Using symmetry considerations, it may be shown that the dependence of the stresses on the polar angle θ is of the following form:

$$[\overline{\sigma}_r^{(3)}, \overline{\sigma}_\theta^{(3)}, \overline{U}^{(3)}] = [\overline{S}_r, \overline{S}_\theta, \overline{P}]\cos 2\theta \tag{11.67}$$

$$\overline{\tau}_{r\theta}^{(3)} = \overline{S}_{r\theta}\sin 2\theta \tag{11.68}$$

where $\overline{S}_r$, $\overline{S}_\theta$, $\overline{S}_{r\theta}$, and $\overline{P}$ are purely functions of r and s, as defined below:

$$\frac{s\overline{P}}{\sigma_d} = -\frac{C_1}{2\eta}K_2(\xi) + S\frac{C_2}{\varrho^2} \tag{11.69}$$

$$\frac{s\overline{S}_r}{\sigma_d} = C_1\left[\frac{1}{\xi}K_1(\xi) + \frac{6}{\xi^2}K_2(\xi)\right] - \frac{1}{1-\nu_u}\frac{C_2}{\varrho^2} - \frac{3C_3}{\varrho^4} \tag{11.70}$$

$$\frac{s\bar{S}_\theta}{\sigma_d} = -C_1[\frac{1}{\xi}K_1(\xi) + (1 + \frac{6}{\xi^2})K_2(\xi)] + \frac{3C_3}{\varrho^4} \tag{11.71}$$

$$\frac{s\bar{S}_{r\theta}}{\sigma_d} = 2C_1[\frac{1}{\xi}K_1(\xi) + \frac{6}{\xi^2}K_2(\xi)] - \frac{1}{2(1-\nu_u)}\frac{C_2}{\varrho^2} - \frac{3C_3}{\varrho^4} \tag{11.72}$$

in which

$$C_1 = -\frac{4\beta(\nu_u-\nu)}{D_2-D_1} \tag{11.73}$$

$$C_2 = \frac{4(1-\nu_u)D_2}{D_2-D_1} \tag{11.74}$$

$$C_3 = -\frac{\beta(D_2 + D_1) + 8(\nu_u-\nu)K_2(\beta)}{\beta(D_2-D_1)} \tag{11.75}$$

and

$$D_1 = 2(\nu_u-\nu)K_1(\beta) \tag{11.76}$$

$$D_2 = \beta(1-\nu)K_2(\beta) \tag{11.77}$$

It is noted that for the special case of incompressible fluid and solids (i.e., B=1, $\nu_u = 0.5$), the above solution reduces to that of Carter and Booker (1982).

Results in time domain

The results in time domain can be obtained by applying an approximate Laplace inversion to the analytical solutions defined by equations (11.59)-(11.77). If the method of Stehfest is used for the inversion, the results in time domain are given by the formula:

$$f(t) = \frac{\ln 2}{t}\sum_{n=1}^{N} C_n\bar{f}(n\frac{\ln 2}{t}) \tag{11.78}$$

where the coefficient C_n is defined by

$$C_n = (-1)^{n+N/2}\sum_{k=(n+1)/2}^{\min(n,N/2)} \frac{k^{N/2}(2k)!}{(N/2-k)!k!(k-1)!(n-k)!(2k-n)!} \tag{11.79}$$

in which the number of terms N in the series is even and is typically in the range of 10–20.

Application to the prediction of wellbore instability

A major effect of the creation of a borehole is to make the tangential stress more compressive due to the removal of the normal stress along the borehole surface, with the potential of causing compressive failure at the wall.

It is worth noting that for constant boundary conditions, loading IIIa is the only loading component to introduce a time-dependant variation of the stress concentration around the borehole. In fact, the borehole response to loading Ia is purely elastic and the maximum stress concentration caused by loading IIa is reached instantaneously. As pointed out by Detournay and Cheng (1988, 1993), the maximum compressive stress takes place at points connected a diameter orthogonal to the direction of the most compressive *in-situ* stress. For example, if K is less than 1, initially the stress concentration will occur at $\theta = 0, \pi$ and the tangential stress is given by:

$$\sigma_\theta(a, 0^+) = 2\sigma_m + 2\eta U_0 - 4\frac{1-\nu_u}{1-\nu}\sigma_d \tag{11.80}$$

where the initial stress parameters σ_m, σ_d are defined by equations (11.35) and (11.36). As time goes by, the tangential stress becomes more compressive and eventually reaches the following large time asymptotic value:

$$\sigma_\theta(a, \infty) = 2\sigma_m + 2\eta U_0 - 4\sigma_d \tag{11.81}$$

This time-dependent increase of the stress concentration may account for delayed borehole failure.

A simple comparison of stress fields obtained by Detournay and Cheng (1988) for selected points with the effective stress based Mohr-Coulomb criterion indicates that failure can initiate at some distance away from the borehole surface, and not right at the borehole wall as predicted by linear, isotropic elastic analysis. According to the results presented by Detournay and Cheng (1988), failure occurs away from the wall at a distance ranging from 5% to 10% of the borehole radius.

11.4 PLASTIC ANALYSIS OF WELLBORE INSTABILITY

11.4.1 Stability criteria

Approximately 80% of the wellbore instability problems are encountered when drilling soft shales or clay layers overlaying the payzone (Charlez and Heugas, 1991). For this type of soft rock, the stability criterion based on brittle failure or initial yielding is either difficult to apply or too conservative to use in practice. It is therefore important to search for other stability criteria which may be more suitable for predicting wellbore instability in soft rock.

Charlez (1997) suggested two criteria that may be used for wellbore instability. Both of these criteria require the analysis of wellbore instability as an elastic-plastic problem. In particular, Charlez (1997) used critical state models to investigate the wellbore instability problem. It is logical considering that in recent years much work has been carried out into the applicability of the critical state concept to the modelling of soft rock.

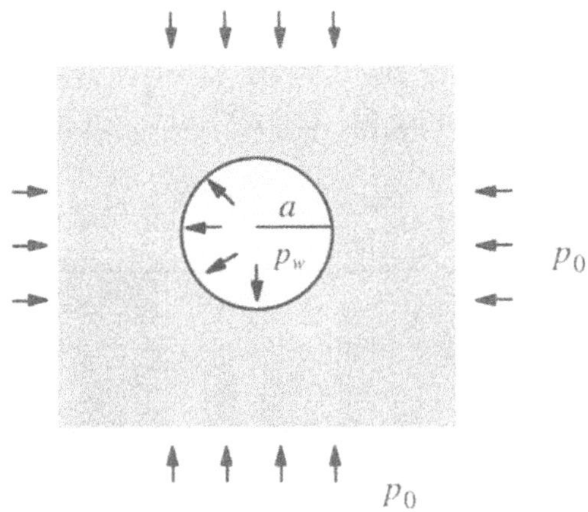

Figure 11.7: Problem of borehole stability in critical state materials

The first stability criterion suggested by Charlez (1997) states that in order to maintain stability no points in the rock surrounding the borehole are at the critical state. The basic idea of the second stability criterion is to limit the inward borehole displacement due to borehole creation. The values that have been suggested by Charlez and Heugas (1991) and Charlez (1997) for the maximum inward borehole displacement are somewhere between 2%–5% of the borehole radius.

11.4.2 Stability analysis using critical state models

In this section, a borehole stability study using the analytical solutions developed by Yu and Rowe (1999) for cavity contraction in critical state materials is presented. To compare the two different stability criteria, the results of a simple parametric study using material constants that are close to those of London clay are considered. The critical state properties are $\Gamma = 2.759$, $\lambda = 0.161$, $\varkappa = 0.062$, $\phi'_{cs} = 22.75^o$. In obtaining the following results, Possion's ratio is assumed to be 0.3 and the initial specific volume is set 2.0.

In these analytical solutions, it is assumed that the initial stress state is isotropic with a value of p_0, Figure 11.7. The solutions are obtained for cavity pressure contraction curves when the pressure on the cavity wall p_w reduces slowly from its initial value p_0. For simplicity, the solutions of Yu and Rowe (1999) have been derived assuming an undrained condition.

Figure 11.8 presents the results of a parametric study using the original Cam clay model. In the figure, the ratios of internal cavity pressure (i.e. mud pressure) to the initial pressure are shown for the cases when cavity strains are –2% and –5% (negative means inward displacement) as well as when the cavity wall just reaches the critical state. Note that the cavity strain is defined by $\varepsilon_c = (a - a_0)/a_0$, where a and a_0 are current and initial cavity radii respectively. To study the effect of past stress history on the the prediction, various values of the overconsolidation ratio (OCR) are considered.

OCR(n_p)	$\varepsilon_c = -2\%$	$\varepsilon_c = -5\%$	Cavity wall reaches critical state
1	0.82	0.67	0.47 (–12.8%)
1.5	0.63	0.41	0.1 (–15.2%)
2	0.55	0.27	–
3	0.49	0.1	–
5	0.48	–	–

Figure 11.8: Values of p_w/p_0 for the instants when the cavity strains are 2% and 5% as well as when the cavity wall just reaches the critical state – original Cam clay model

Figure 11.9 summparises the results of a parametric study using the modified Cam clay model. As for the original Cam clay model, the ratios of internal cavity pressure to the initial pressure are obtained for the cases when cavity strains are –2% and –5%, as well as when the cavity wall just reaches the critical state.

Both of these figures indicate that for heavily overconsolidated soft rock, little internal pressure (i.e. mud pressure) is needed in order for the borehole to be stable. This is true for both stability criteria mentioned before, i.e. the magnitude of the cavity strain is less than 5% or the cavity wall does not reach the critical state. On the other hand, however, for normally or lightly overconsolidated soft rock, it is obvious that considerable mud pressure must be provided in order to make the borehole stable.

$OCR(n_p)$	$\varepsilon_c = -2\%$	$\varepsilon_c = -5\%$	Cavity wall reaches critical state
1	0.66	0.44	0.15 (–15.5%)
1.5	0.55	0.26	–
3	0.48	–	–
5	0.47	–	–

Figure 11.9: Values of p_w/p_0 for the instants when the cavity strains are 2% and 5% as well as when the cavity wall just reaches the critical state – modified Cam clay model

For the material properties used, it is true that a very large cavity strain must be mobilised before the cavity wall reaches the critical state, and therefore this stability criterion is not controlling. Of course, which stability criterion (cavity strain or cavity wall reaching the critical state) will control the design will be dependent on the actual material properties. To illustrate this point, results are now presented for a new set of critical state properties for soft rock used by Charlez and Heugas (1991): $\Gamma = 2.759$, $\lambda = 0.077$, $\varkappa = 0.0071$, $M = 1.3$.

$OCR(n_p)$	$\varepsilon_c = -2\%$	$\varepsilon_c = -5\%$	Cavity wall reaches critical state
1	0.06	–	0.13 (–1.4%)

Figure 11.10: Values of p_w/p_0 for the instants when the cavity strains are 2% and 5% as well as when the cavity wall just reaches the critical state – modified Cam clay model

It was found that for this new set of critical state properties the cavity strain required for the cavity wall to reach the critical state was very small (1.4%). Therefore unlike the results presented in Figure 11.8 and Figure 11.9, the result in Figure 11.10 suggests that the stability criterion based on the cavity wall reaching the critical state is critical.

11.5 SUMMARY

1. Wellbore instability during drilling is a major problem in the application of rock mechanics in petroleum engineering. It is well recognised that the value of lost equipment and time arising directly from wellbore instability is significant with estimates of losses over $500 million per year worldwide (Dusseault, 1994). The

stress-induced wellbore instabilities consist of three general types: (a) hole size reduction due to ductile yield of the rock; (b) hole enlargement due to brittle rock fracture or rupture; and (c) unintentional hydraulic fracturing induced by excessive mud pressure. It is common practice to adjust the internal wellbore pressure (i.e. mud pressure) to avoid borehole instabilities caused by rock fracturing and rupture.

2. Cavity expansion solutions obtained with elastic, poroelastic and plastic models are used to study the problem of wellbore instabilities in soft rock. For example, the basic steps in the elastic analysis of borehole instability are: (a) determine the elastic stress field in rock around the wellbore; (b) choose a suitable failure criterion for the rock concerned; (c) compare the elastic stresses and the selected rock failure criterion. If the failure criterion is satisfied anywhere in the rock the wellbore is regarded as unstable.
3. When an elastic analysis is used to predict borehole instability, care must be taken in selecting realistic failure criteria and elasticity models as they both have a significant effect on the final results. In particular, studies show that the inclusion of pressure dependent stiffness and stress-induced anisotropy in the elastic models can considerably affect the predicted stress fields around the borehole.
4. Time-dependent borehole behaviour can be predicted using cavity expansion theory in conjunction with poroelastic models. In particular, poroelastic analysis predicts a time-dependent increase of stress concentration around the borehole which account for delayed borehole failure.
5. Approximately 80% of the wellbore instability problems occur when drilling soft shales or clay layers overlaying the payzone. For this soft rock, the stability criterion based on brittle failure or initial yielding may be either difficult to apply or too conservative to use in practice. Charlez (1997) suggested two alternative criteria that may be used for predicting wellbore instability in soft rock. The first stability criterion states that in order to maintain stability no points in the rock surrounding the borehole are at the critical state. The second criterion is to limit the inward borehole displacement due to borehole creation, say by between 2%–5% of the borehole radius. Both of these criteria require the analysis of wellbore instability as an elastic-plastic problem. To illustrate the procedure of plastic analysis, the cavity unloading solution of Yu and Rowe (1999) was used to investigate the wellbore instability.

REFERENCES

Bradley, W.B. (1979). Failure of inclined boreholes. *Journal of Energy Resource and Technology, Transactions of ASME,* 101, 232–239.

Carter, J.P. and Booker, J.R. (1982). Elastic consolidation around a deep circular tunnel. *International Journal of Solids and Structures,* 18(12), 1059–1074.

Charlez, Ph. (1991). *Rock Mechanics. Vol 1: Theoretical Fundamentals.* Editions Technip.

Charlez, Ph. (1997). *Rock Mechanics. Vol 2: Petroleum Applications.* Editions Technip.

Charlez, Ph. and Heugas, O. (1991). Evaluation of optimal mud weight in soft shale levels. *Rock Mechanics as a Multidisciplinary Science* (Editor: Roegiers), Balkema, 1005–1014.

Detournay, E. and Cheng, A.H.D. (1988). Poroelastic response of a borehole in a non-hydrostatic stress field. *International Journal for Rock Mechanics and Mining Sciences,* 25(3), 171–182.

Detournay, E. and Cheng, A.H.D. (1993). Fundamentals of poroelasticity. In: *Comprehensive Rock Engineering* (Editor: J.A. Hudson), Pergamon Press, Oxford, Vol. 2, 113–171.

Dusseault, M.B. (1994). Analysis of borehole stability. *Computer Methods and Advances in Geomechanics,* (Editors: Siriwardane and Zaman), Balkema, 125–137.

Kulhawy, F.H. (1975). Stress deformation properties of rock and rock discontinuities. *Engineering Geology,* 9, 325–350..

Lekhnitskii, S.G. (1963). *Theory of Elasticity of an Anisotropic Elastic Body.* Holden-Day, Inc.

McLean, M.R. (1987). *Wellbore Stability Analysis.* PhD Thesis, University of London, England.

Nawrocki, P.A. and Dusseault, M.B. (1994). Effect of material nonlinearity on deformations around openings in geomaterials. *Computer Methods and Advances in Geomechanics,* (Editors: Siriwardane and Zaman), Balkema, 2119–2124.

Santarelli, F.J., Brown, E.T. and Maury, V. (1986). Analysis of borehole stresses using pressure-dependent, linear elasticity. *International Journal for Rock Mechanics and Mining Sciences,* 23(6), 445–449.

Woodland, D.C. (1990). Borehole instability in the Western Canadian overthrust belt. *SPE Drilling Engineering,* March, 27–33.

Wu, B. and Hudson, J.A. (1991). Stress–induced anisotropy in rock and its influence on wellbore stability. *Rock Mechanics as a Multidisciplinary Science,* (Editor: Roegiers), Balkema, 941–950.

Wu, B.L., King, M.S. and Hudson, J.A. (1991). Stress-induced ultrasonic wave velocity anisotropy in a sandstone. *International Journal for Rock Mechanics and Mining Scinces,* 28(1), 101–107.

Yu, H.S. and Rowe, R.K. (1999). Plasticity solutions for soil behaviour around contracting cavities and tunnels. *International Journal for Numerical and Analytical Methods in Geomechanics,* 23, 1245–1279.

INDEX